DESIGN

and

COMMUNICATION

DESIGN

and

COMMUNICATION

A Yarwood

J White

HODDER AND STOUGHTON
LONDON SYDNEY AUCKLAND TORONTO

Other books by A. Yarwood

Design and Craft by A. Yarwood and S. Dunn
Design and Technology by A. Yarwood and A. H. Orme
Design and Woodwork by A. Yarwood
Tools and Processes by A. Yarwood
Teach Yourself Woodwork by A. Yarwood
Teach Yourself Graphical Communication by A. Yarwood

ISBN 0 340 40710 7

First published 1989

Typeset in Monotype Lasercomp Ehrhardt and Optima by
Latimer Trend & Company Ltd, Plymouth
Printed in Hong Kong
for Hodder and Stoughton Educational
a division of Hodder and Stoughton Ltd, Mill Road
Dunton Green, Sevenoaks, Kent by Colorcraft Ltd

Preface

This book has been written as a text for pupils and students who are preparing for examinations in the GCSE subject CDT: Design and Communication. It is the fifth in a series of books covering those GCSE syllabuses which have been designed to meet the requirements of the CDT National Criteria. The other four books are texts for CDT: Design and Realisation and CDT: Technology.

There is inevitably some overlap between the contents of this book and the contents of the others in the series. This is because the CDT National Criteria insist that technical graphics is to be examined as the design medium for Craft, Design and Technology. However, this overlap has been kept to a minimum, although at the same time the text aims at complete cover of the CDT: Design and Communication syllabuses as set by the five examining groups.

Prior to the introduction of the GCSE, technical drawing was examined mainly as a craft and engineering drawing language. Within recent years and as an interim stage, technical drawing syllabuses were enlarged to become graphical communication. Graphical communication syllabuses confirmed that the skills of technical drawing could be applied in a much wider context than just craft and engineering drawing. However, the idea of examining technical graphics as the means by which design ideas and solutions in craft, technology and graphics are communicated, has now been adopted and so has overtaken graphical communication. At last, or so it seems to us, technical graphics has come home to its rightful place – as the design 'tool' for craft, technology and graphics.

As in other books in the series, the design process – situation; design brief; investigation; ideas for solution; chosen solution; model; realisation; evaluation – has been followed. In fact, this design routine is repeated over and over again throughout all five books.

All syllabuses prepared to conform with the requirements of the CDT National Criteria are based on the assumption that the examinations are for candidates who have completed a five-year CDT course. This book has been designed with this in mind. It has been written specifically for pupils and students studying during their final two years prior to entry for an examination in CDT: Design and Communication. However, it is assumed that readers will already have an elementary knowledge of materials, processes, graphics and design resulting from their participation in CDT courses during the earlier part of their five secondary school years.

Another requirement of all syllabuses prepared to meet the CDT National Criteria is that project and/or coursework forms an important part of examinations. This requirement is such as to demand that 30% to 40% of total marks for the examinations be awarded for project and/or coursework. The contents of this book clearly reflect this requirement. A number of complete or partly complete graphic projects are included, together with numerous suggestions for project topics and methods of designing and producing projects.

A. Yarwood
J. White

Acknowledgements

The authors wish to acknowledge with grateful thanks, the permissions granted by the following firms to reproduce copyright photographs in the pages of this book:

Apple Computers Ltd. – for the photograph of an Apple IIGS computer on page 165.

Commotion Technology Supplies – for the colour photograph of a model being made from 'Corriflute' sheet appearing on page 120.

Economatics (Education) Ltd. – for supplying a photograph which appears on page 78 and for allowing photographs to be taken in their offices at Handsworth, Sheffield.

LEGO® (UK) Ltd. (Education Division) – for the photograph of an interface appearing on page 159.

Morris and Ingram (London) Ltd – for supplying photographs of their Badger 200 airbrush appearing on page 107 and for giving freely of their advice concerning the use of airbrushes.

The undermentioned examining boards have kindly allowed questions to be reproduced from specimen examination papers set by them:
The Welsh Joint Education Committee (WJEC)

CONTENTS

SECTION 1 A graphics design process

This book deals with the graphics associated with the process of designing. It is important to realise that such graphics are themselves the result of a design process. It does not matter whether the graphics which are being considered are single sheets such as one of the working drawings for the making of, say, a part of a car seat adjuster, or whether they are the finished full colour artwork from which thousands of copies for an advertising campaign are to be printed – the sheets of graphics must first be designed. The first pages of this book therefore deal with the processes involved in the designing of graphics.

In all societies, visual images play an important part in the communication of ideas between people and in fact they always have done so. You have only to consider your own dependence upon visual images to realise this. Newspapers, magazines, books, television all rely very much on graphics for the conveying of ideas to large audiences. Advertising depends largely upon graphics for its effectiveness. Without drawings many industries engaged in, say, engineering or building could not exist. So let us first consider a graphic design process.

A graphics design process

All graphic design is intended to communicate ideas in a visual form. If the designing of your graphics is poor then you will not be communicating those ideas as effectively as you would if the graphics had been well designed. Because of this, the planning of the design of any piece of graphical work should proceed as carefully and as thoroughly as possible. A simplified flow diagram on page 4 indicates a method, a routine or a process which can be followed to assist in the production of well designed graphics.

This flow diagram illustrates designing as a continuous process. When a graphics design has been realised (i.e. when the graphics have been completed) there is always the possibility that changes to the design will be required as a result of evaluation.

1. The design may not meet the requirements of the design brief.
2. The introduction of new materials may force amendments to the design.
3. Changes in recent fashion may lead to the conclusion that the design is not modern in appearance.
4. If the graphics are designs for artefacts that must be made and the resulting artefact does not meet the requirements of the design brief, or fails as a result of evaluation testing, then the graphics for the design will need re-designing, even if only partially.

Hence the circular aspect of this flow diagram. A summary and revision of the graphic design process is given in the form of a flow chart on page 8.

Situation

What is the situation which has caused the wish to communicate graphically? It may be as simple as the need to draw orthographic projections of an engineering component in order that its functioning can be explained to others. It may be the need to draw isometric views of a corner of a garden shed in order to understand how that corner can be effectively constructed. The situation may be that a number of freehand sketches are required to enable a selection to be made from which a good shape for some particular purpose can be chosen. Think about the situation for which you are designing your graphics. Also, when you have completed an item of graphics, think back to determine whether your work has satisfied the situation which it has been designed to satisfy.

Design brief

A design brief arises from the design situation. Design briefs should be written down and should clearly define what it is that is to be designed. From the situations given above, design briefs such as the following could arise:

- Draw a full size, three-view, Third Angle orthographic projection of an assembled bicycle bottom bracket, complete with its spindle.
- Make isometric drawings, scale 1:2, of constructions suitable for the top right hand corner of a garden shed.
- Produce a number of freehand drawings of identically shaped interlocking pieces which can be formed into a pattern.

Investigation

Before graphics of any kind can be produced, a variety of details need to be investigated. Some of the answers to these details must be found before the graphics can be started. Other details need to be investigated when the process of producing the work is under way.

Purpose

What is the purpose of the work you are about to undertake? Look back at your written design brief. Is the work for your own personal use? Is it to be seen and used by others? Will it need to be on display? Is a small or large drawing(s) needed?

Method

Consider the most effective techniques which can be employed. Remember, above all, that you are communicating ideas graphically and that some graphic techniques are not suited to some audiences, e.g. orthographic projection is a suitable technique only for those who can understand such projections; for a general audience pictorial techniques such as isometric drawing or cabinet drawing may be more appropriate. Some pictorial drawings communicate ideas better if in planometric, say, rather than in perspective.

Having determined the techniques to be employed, now consider the methods by which those techniques are to be applied. For example, freehand or instrument aided; strictly geometrical methods or estimated methods; using drawing aids or relying upon good freehand lines where necessary.

Planning

How many drawings are required to communicate the message you wish to convey? How large or how small will the drawings need to be?

Layout

Good, well planned layout is essential. Such details as size of paper or drawing material, borders, titles, printing and the positioning of blocks of print, the balance and proportions of any colour or shading. Once again freehand sketches made on scrap paper could assist in the determination of a good layout.

Display

Is your work to be displayed? If you are making up a project as part of an examination, then you must assume that all the parts of your project are 'on display'. Consider mounting your work on mounting boards. Consider covering drawings with transparent plastic sheet or tracing paper. Some work which would not normally be coloured may require added colour if the work is to be displayed. When making up a project from a number of drawings remember that a well designed folder cover might go a long way to convincing others that the folder contents are really worth looking at.

Materials

Which is the best paper to use? Is card or cardboard required? Will materials such as tracing paper, grid papers be necessary? If a model is to be made or some form of collage to be part of the work, will glues, pastes or other adhesives be necessary? Will adhesive tapes prove to be of value?

Shape and form

You will have to design the shape and form of parts of your graphics. Freehand sketches on scrap paper can be a useful aid in determining the shapes and forms to be adopted in your work.

Proportions

Consider all proportions very carefully, not only the proportions of the whole drawing you are engaged on, but also its parts. The good appearance of a piece of graphics is often judged mainly on its proportions.

Media

Before commencing work on the graphics, consider the various media you can choose from and select that which you consider to be most suitable – pencil (which grades?), pens (technical, ball, penstik), crayons, colours, colour wash, transfer materials such as letter transfers.

Colour

Is colouring necessary? Will colour add to the effectiveness of the communication which is being attempted? Is colour best added with paints, inks, crayons or by pasting coloured paper on parts of the drawings?

Shading

Consider whether shading is necessary and, if so, which shading methods would be most suited to the graphics being undertaken. Should the shading be in pencil, with colour, with dry transfer toning?

Economics

Economics must be considered. For example, should an expensive art board be used or will cartridge paper be suitable? Art board is much more expensive than cartridge paper and the added expense may be worthwhile, but must be considered. Economy in time must also be considered. If, in an examination say, extra marks can be gained by applying shading, this should only be considered if the time spent is worth the marks gained. Of particular importance is economy of space and line. Over-decoration, too much colour, unnecessary shapes, may spoil a well proportioned, nicely laid out graphical design.

Solutions

As the investigation has been proceeding you will have been making a series of freehand or instrument assisted sketches and diagrams, with plentiful notes describing what the graphics are intended to show. You may find it advisable to make a series of freehand sketches to check whether the investigation you have carried out will produce a well designed piece of graphics. These freehand sketches will check on points such as whether the most suitable paper or drawing sheet has been chosen; which are the most suitable media for this particular design; whether colour is needed; which colour; whether shading should be applied; whether the most suitable technique is being employed to communicate the ideas in your mind; sketch shapes, layouts; proportions.

The production of ideas for solutions is possibly the most important part of the graphic design process. It is here that ideas are tried out in sketch and note form. The more numerous and varied the ideas, the better will the final design be. Of course, it may well be that your chosen solution will be already in your 'mind's eye' and the ideas at the solutions stage may be experiments with a single, original idea.

Model

Is a model necessary? This could well be a form which requires a surface development, or a mock-up of the graphics you have designed. If your graphics are associated with the designing of an artefact, it will be virtually essential to make a model.

Chosen solution

When you have finally decided on all the details covered by the investigation, you should be able to choose from the results and from the solution sketches your final solution for the design of the graphics you are producing.

Realisation

Now produce the design. Make the necessary drawings and complete them. In this part of the book the term 'realisation' is used to refer to the graphics which are being designed. Later in the book the term will be applied to the making of artefacts which have been designed.

Evaluation

Always look critically at any graphics you have produced. Is your work good or is it bad? Does it communicate clearly what you wish to show? Unless you are candid with your work at this stage, you may not make good progress. If you have to go through the work again because the results are not as good as you desire, then it is advisable to make the attempt at improving your work.

A questionnaire such as the following may assist in the evaluation of your graphic design.

(a) Does it work, i.e. does it communicate effectively?
(b) Does it meet the demands of the situation and the design brief?
(c) Could other graphical methods have been employed to give a better effect?
(d) Could colour have been used more effectively?
(e) Were the best materials used – papers, boards, media?
(f) Have the layouts been well planned?
(g) Are the proportions of the graphics good?
(h) Is the design economically viable?
(i) Would a model have proved useful in evaluating the design?

Christmas card project

This project consists of nine sheets of drawings (given on pages 5, 6 and 7), together with a reproduction of the finished artwork (shown on page 108). The finished artwork appears to meet the requirements of the given design brief. The nine sheets of drawings, together with the resulting completed piece of artwork are a fairly typical example of a sequence of drawings resulting in the production of a graphical design. The sequence within this project follows the route given by the flow diagram on this page.

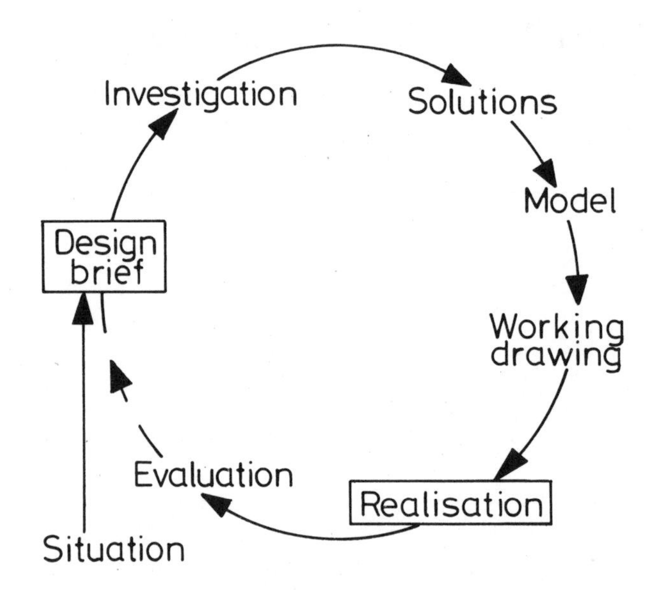

A flow diagram showing an outline process for graphic design

Situation and design brief
These are given below.

Investigation
Sheets 1 to 8 include details of an investigation into the problems arising from attempts at finding an answer to the given design brief. Details such as method, layout, display, materials, media, shape and form, proportions, colours, economics are analysed and suggestions made for solving problems arising from the investigation.

Solutions
As the investigation proceeds, so a number of ideas for solutions suggest themselves. These are analysed and discussed in a series of sketches and notes linked to the details arising from the investigation.

Chosen solution
A possible solution which appears to satisfy the design brief is eventually suggested on sheet 6 and this chosen idea is developed in sheets 7, 8 and 9 into a final design.

Realisation
After making a line drawing of the chosen solution (Sheet 9) the final item of graphics is produced mainly with the aid of air-brush techniques. The finished design is shown on page 108.

Evaluation
Because this design was accepted by the firm which commissioned the work and was subsequently printed and published by them and sold in large numbers, the design could be said to have been successful.

Situation

A firm which publishes and sells greeting cards requires a new Christmas card to add to their existing stock of designs. I have been commissioned to produce a new design, together with the artwork from which the design can be printed. The design of the card should be based on the theme of Christmas trees.

Design brief

Design a Christmas card based on the theme of Christmas trees, but possibly to include other details showing a Christmas 'spirit'.

1. Study Christmas Trees.

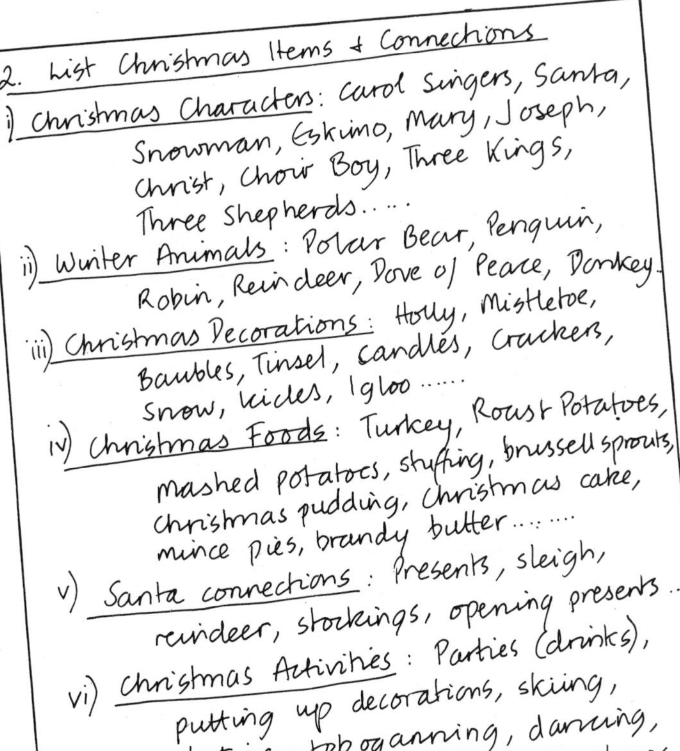
2. List Christmas Items + Connections
i) Christmas Characters: Carol Singers, Santa, Snowman, Eskimo, Mary, Joseph, Christ, Choir Boy, Three Kings, Three Shepherds.....
ii) Winter Animals: Polar Bear, Penguin, Robin, Reindeer, Dove of Peace, Donkey.
iii) Christmas Decorations: Holly, Mistletoe, Baubles, Tinsel, Candles, Crackers, Snow, Icicles, Igloo......
iv) Christmas Foods: Turkey, Roast Potatoes, mashed potatoes, stuffing, brussell sprouts, christmas pudding, christmas cake, mince pies, brandy butter.......
v) Santa connections: Presents, sleigh, reindeer, stockings, opening presents...
vi) Christmas Activities: Parties (drinks), putting up decorations, skiing, skating, tobogganing, dancing, discos, pulling crackers, christmas shopping, snowballing......

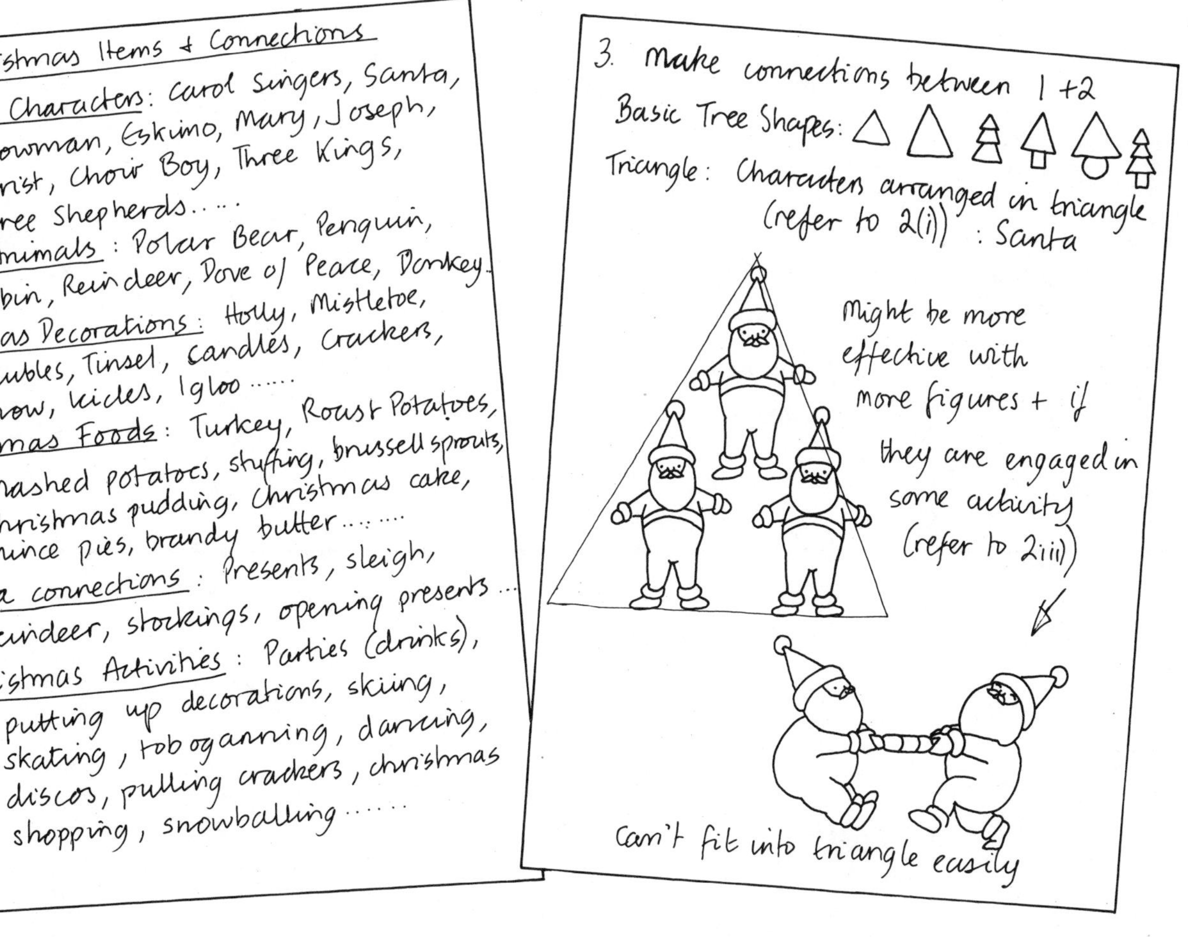
3. Make connections between 1 +2
Basic Tree Shapes:
Triangle: Characters arranged in triangle (refer to 2(i)) : Santa
Might be more effective with more figures + if they are engaged in some activity (refer to 2iii)
Can't fit into triangle easily

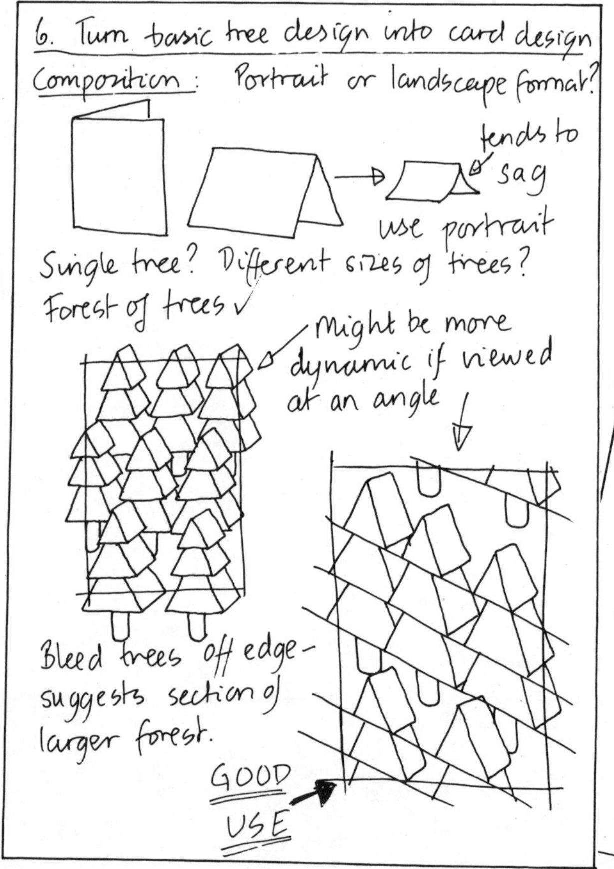

7. Colours

Red and Green are very important colours for symbolising Christmas. What scope in this design for using red and green?

Components of cake:

icing: white - good - like snow
Marzipan: yellow
Decorations icing rossettes: white
cherries: red
Cake itself: raisins brown/black
cherries: red

Could add holly as decorations, has both red & green.

Background: no suitable red/green object. Use snow i.e. white.

To introduce large area of red to imply Christmas: ADD BORDER

8. Red Border

Materials:

Material has to be paper as design is to be reproduced by scanner + therefore has to to go round drum of scanner

Media: Inks/Gouache, Watercolour. Technique: airbrushed areas of flat colours in gouache + inks - will need masking film. Will need a hammer paper for good surface for airbrushing. Detail to be painted in by brush with gouache

Costs: Liable to be 7-10% of anticipated payment received for design. Estimated time taken to complete design (judged from experience of previous jobs) realistic for meeting deadline. Therefore job can be completed on time with adequate profit : PROCEED

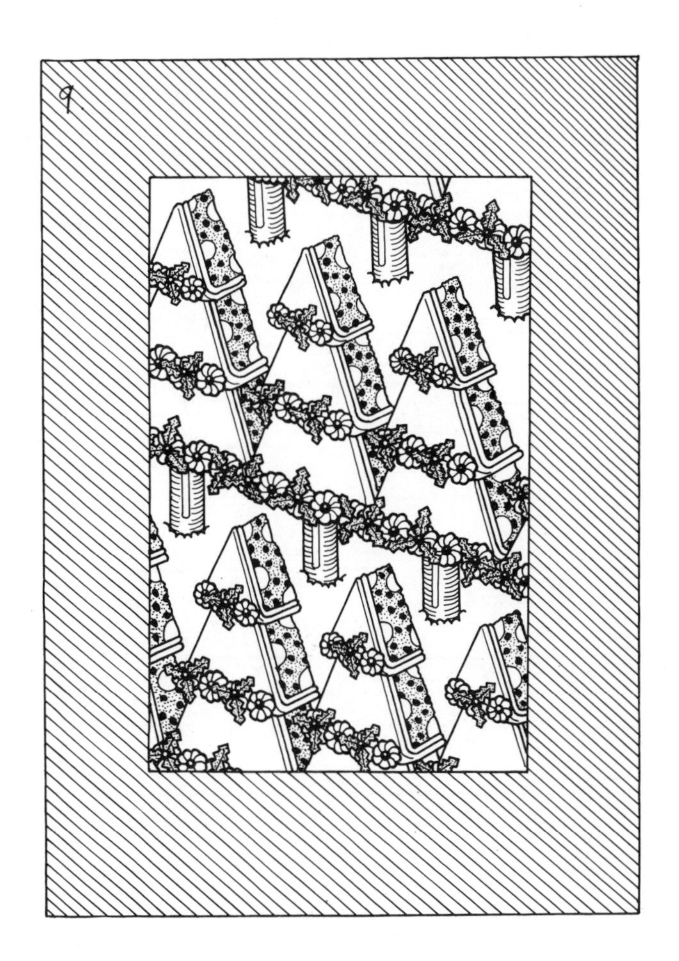

Notes on the sheets of drawings

All the ideas for solutions shown on the nine sheets of graphics were drawn on A4 sheets of paper in black ink with the aid of technical pens. Notes were added with a black biro pen. All the drawings were made freehand with the occasional use of a ruler to assist in drawing straight lines. The final sheet of graphics which was submitted for printing is shown on page 108. The artwork (as such sheets of graphics are called) was coloured with the aid of an air-brush but small details were added with a fine paint-brush. The artwork was produced on a sheet of good quality art board and the painted surface was covered with a sheet of clear, transparent acetate sheet to protect the paint surfaces from being damaged during the printing processes.

The flow diagram given on page 8 is included here as a revision of the graphic design process explained in previous pages. Examine the flow diagram carefully. Has this process been followed during the design of the Christmas card?

GRAPHIC DESIGN

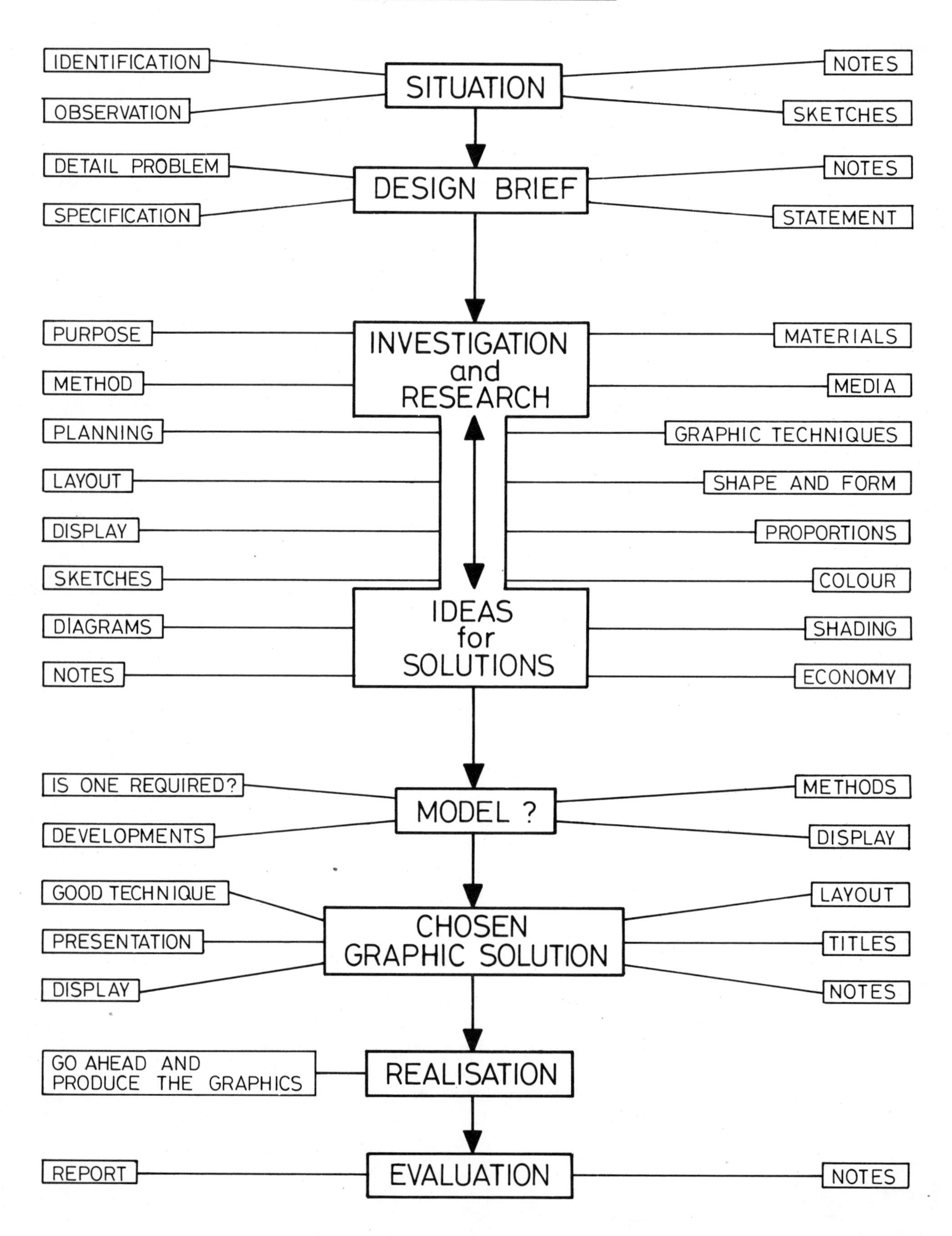

Design drawing

Using lines in graphic design drawings

Lines in design drawings are commonly drawn with pens or with pencils. The illustrations on the next few pages show a variety of methods of showing shape and form using pen or pencil linework. All lines in the accompanying illustrations have been drawn freehand.

Thin, thick and curved lines
Lines can be thick or thin, curved or sloping. The actual position of lines – horizontal, vertical or sloping – can indicate direction upwards, downwards or on a varying path up or down.

Parallel lines
Groups of thick or thin parallel lines can indicate a flow of movement from side to side horizontally or up and down vertically, or may be drawn so as to indicate an area, a space or a shape.

Sloping lines
Sloping lines, whether thick or thin, can indicate direction towards points or away from points; they may also be drawn to form a pattern of lines, whether the pattern consists of all thick lines, all thin lines or a mixture of thick and thin lines.

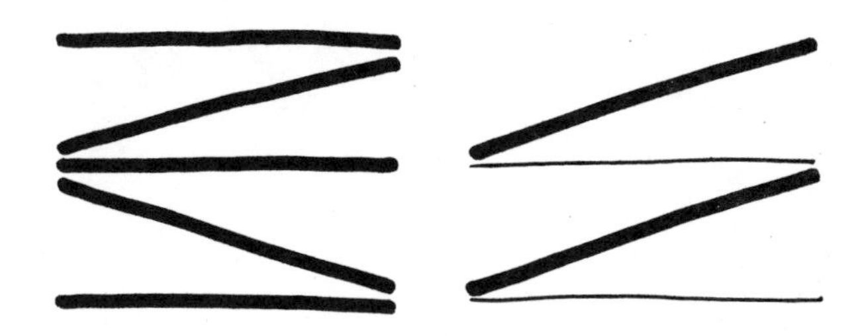

Rhythm
Wavy parallel lines, thick or thin, can indicate a rhythmic movement – horizontally, vertically or along a slope.

Direction
Lines, thick or thin, or thick and thin, can indicate direction towards a distant point.

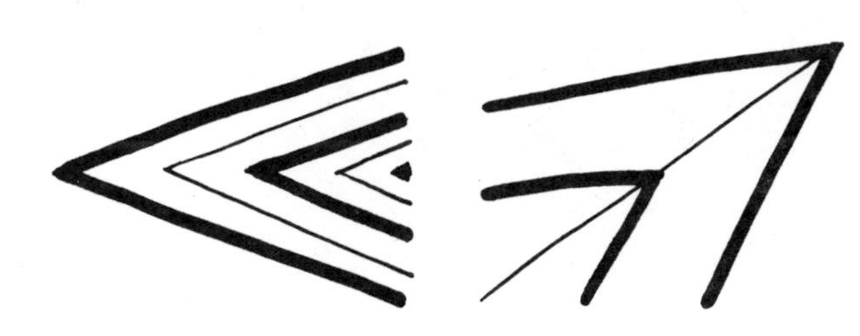

Focal points
Thick or thin lines can be drawn as if moving to focal points – left, right, above, below or central.

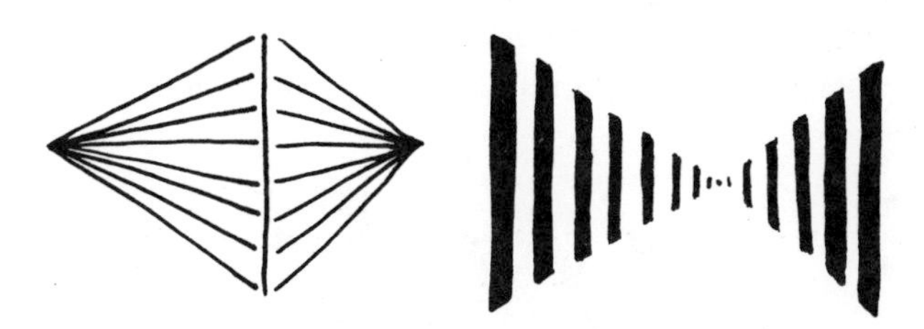

Pattern
Combinations of lines can be drawn to produce patterns of a simple or complex nature.

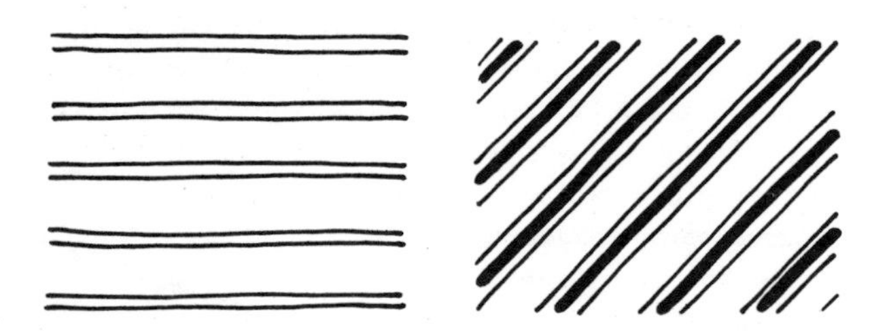

Surface and texture
Two examples are given of lines drawn to show a shiny surface and a cloth textured surface.

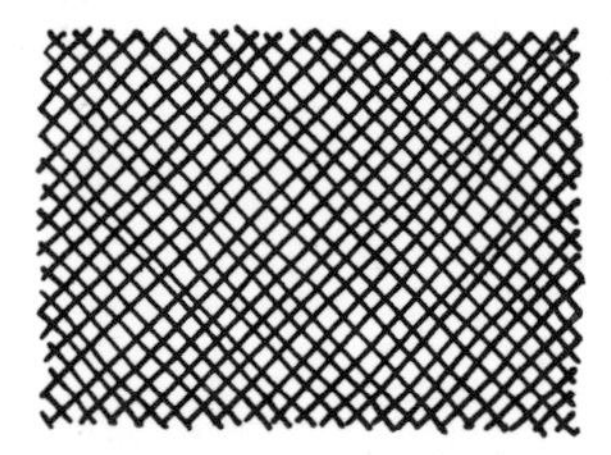

Movement
Lines can be drawn to show movement in any direction.

Illusion
Some combinations of lines produce optical illusions. In the example given the thick lines are parallel but they appear to diverge and converge towards each other. Take care when drawing some lines to avoid causing such optical illusions unintentionally.

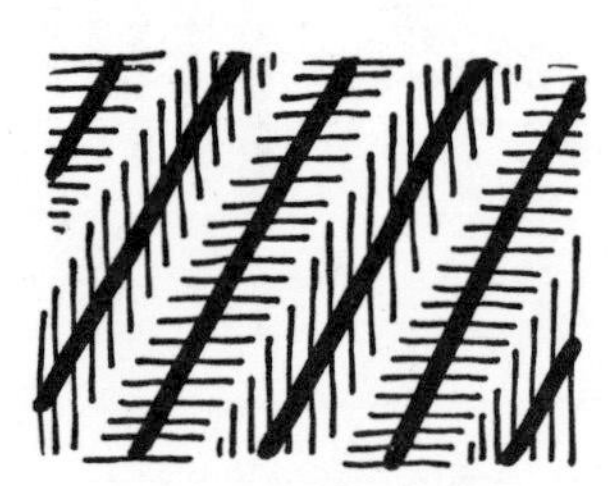

Point of interest or a central point
A series of lines drawn towards a point or towards the centre of a circle draw the attention to the point or the centre as a clear focal point.

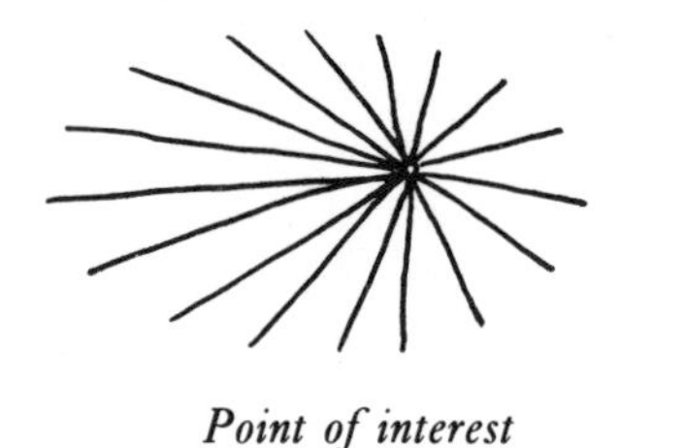

Point of interest

Central

Shapes shown by lines in graphic design drawings

Many of the shapes found in design drawings are geometric in outline. Methods of drawing accurate geometric shapes are described later (pages 37 to 46). A number of examples of shapes are given here. Different

Triangle *Square*

Rectangle

Pentagon

Hexagon

Octagon

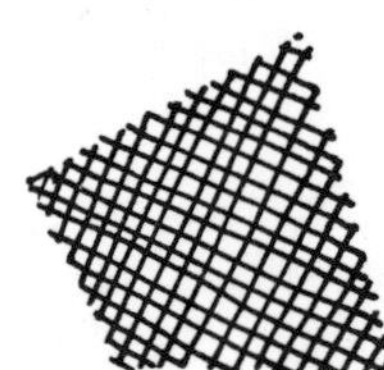

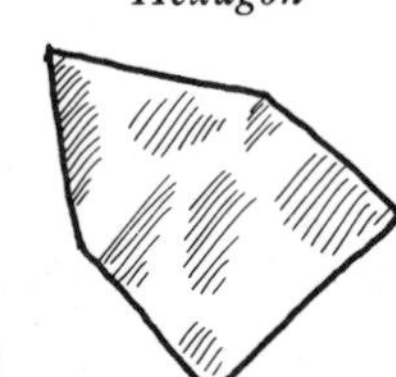

Irregular plane figures

Circle

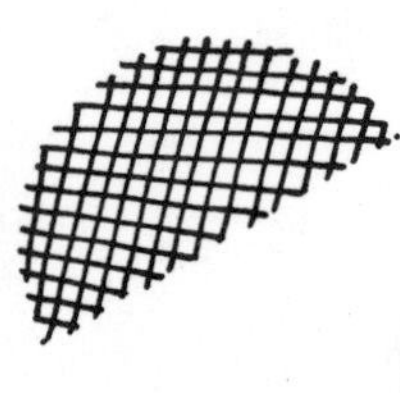

Semicircle

Arcs

types of linework can be used to emphasise the geometric shapes.

Geometric shapes

The illustrations show thick line outlines for a triangle, a square, a rectangle and a regular pentagon; thin parallel lines for the surface of a regular hexagon; thick parallel lines for the surface of a regular octagon. An irregular quadrilateral surface is formed by crossing thin lines; an irregular pentagonal surface is outlined in thin lines with thin line shading to indicate a shiny surface; an arc, outlined with thick lines, has its surface emphasised with concentric arcs. A circle outlined with thick lines, a semicircle made up with crossing thin lines and a series of concentric arcs emphasised with radial lines make up another group of shapes.

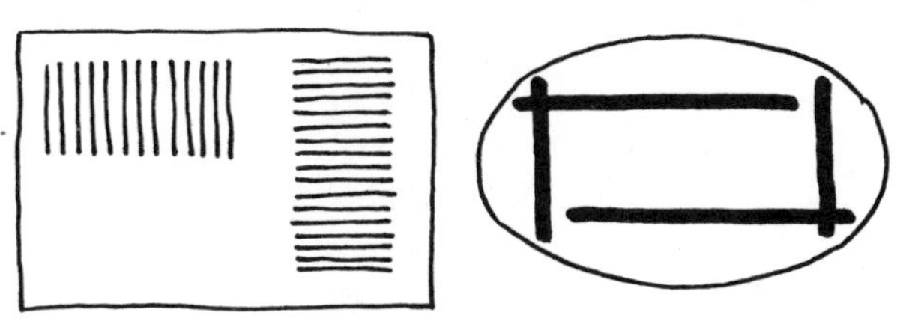

Shapes and lines

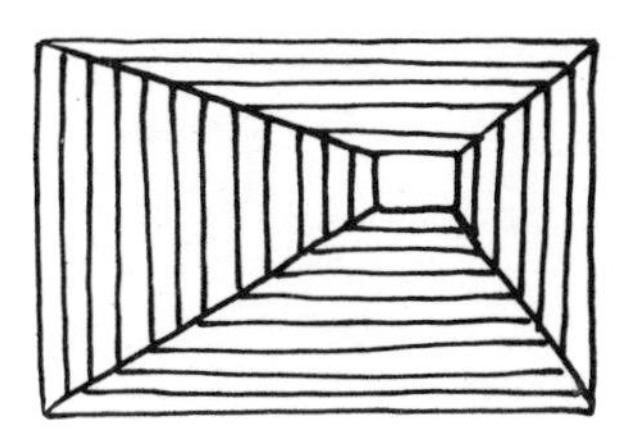

Shapes and lines

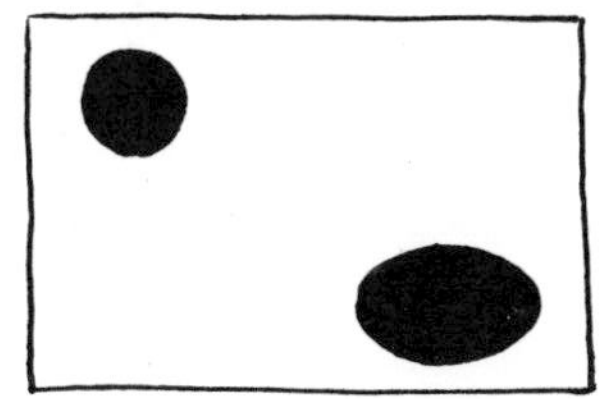

Combined shapes

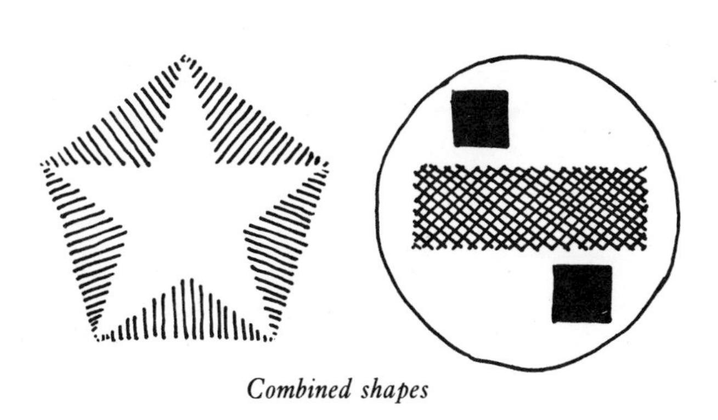

Combined shapes

Combinations of geometric shapes

Six illustrations show combinations of shapes and lines. These are: a pair of rectangles made up of thin parallel lines within an outlined rectangle; crossing thick lines forming a frame pattern within a thin outline of an ellipse; a small rectangle within a larger one, with lines between the two rectangles which appear to indicate a tunnel or a passageway; a rectangle outlined with thin lines within which a circle and an ellipse have been drawn: a star within a regular pentagon formed with thin parallel lines; a thin circle outlining a pattern of a rectangle and two squares.

Patterns

Lines can be drawn to show patterns. Two patterns are shown. The first is simply based on light and dark squares, the shading of the dark squares being represented by sloping parallel lines. The second is a simple *tessellated* pattern. The *tesserae* of such patterns are identical in shape and interlock with each other.

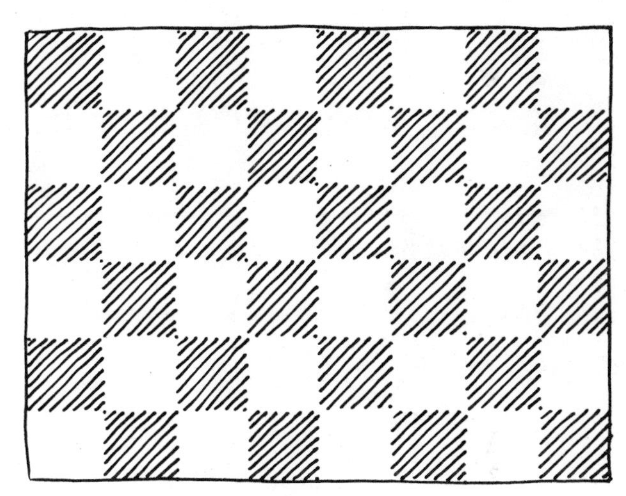

Simple pattern

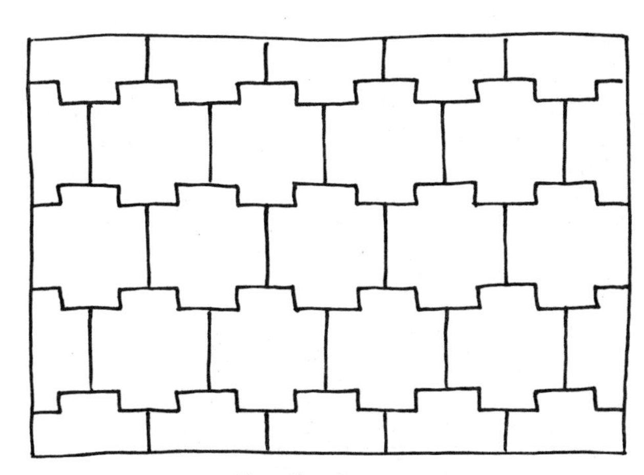

Tessellated pattern

Patterns drawn with the aid of a template

Care is necessary in designing a pattern template to achieve a repeat pattern, even one as simple as that shown. The pattern is again based on squares. After designing its shape a template was cut to shape with a sharp knife. The template was then placed on the drawing sheet and its internal shape copied by running a pencil around its internal edges. In this example, after each square with its internal pattern had been drawn, the template was turned through 90° to draw the adjoining square of pattern.

Pattern drawn with the aid of a template

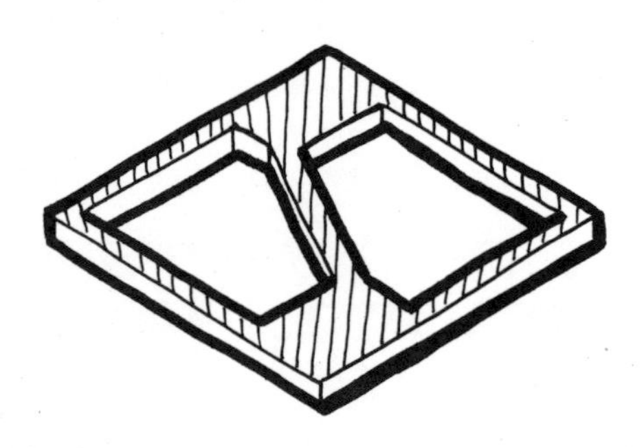

Pattern template

Shapes selected from jumbles of lines

Draw crossing straight or curved lines at random on a sheet of paper. Select suitable shapes from among those found in the jumble of lines.

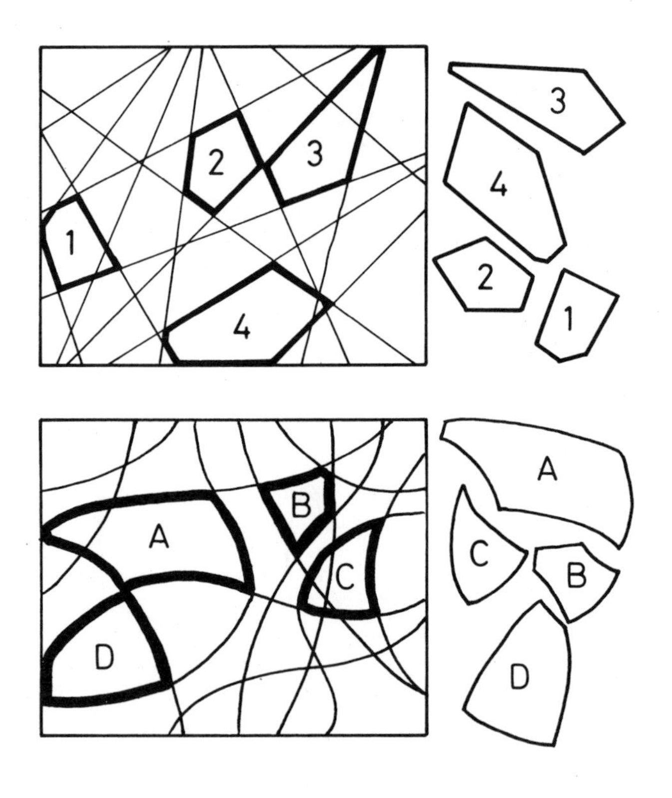

Shapes selected from a jumble of lines

Shapes from nature

The shapes of natural objects or animals often provide ideas for shapes for design. Eight examples are given – a seal, vine leaves and a bunch of grapes; a mushroom; some fern leaves; a lizard; a parrot; a fish; a starfish.

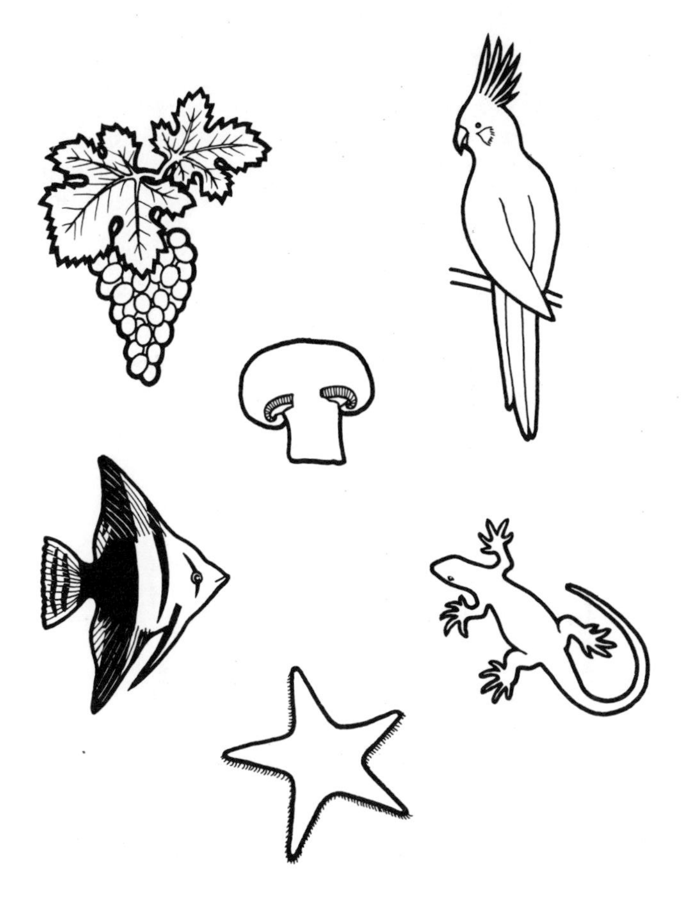

Shapes from nature

A simple filograph

Filographs are shapes produced by drawing straight lines between a number of selected points in order to produce a pattern or design. The given example is based upon a pair of circles divided into 12 equal parts.

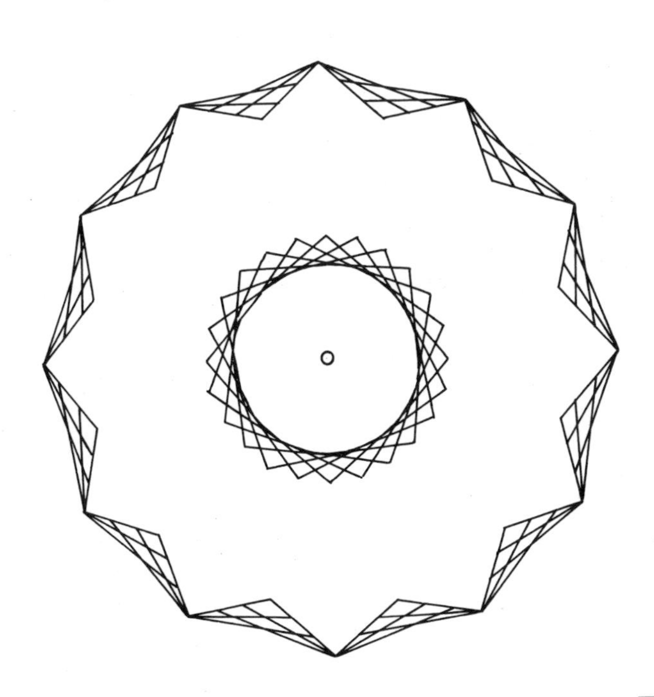

Toning and shading

Sixteen examples of tones and depths of shading which can be applied to drawings are shown. The sixteen are as follows.

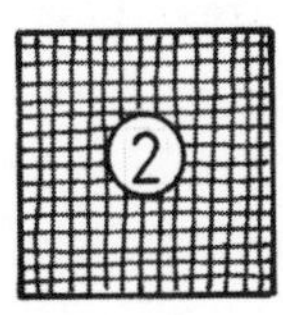

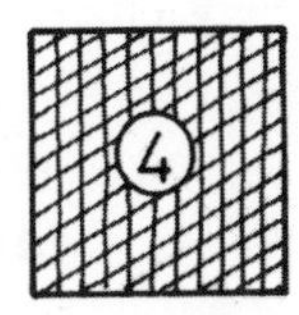

1. Thin parallel lines. Variation in the width between the lines will give a variation of depth of tone.
2. Crossing thin lines.
3. A herring-bone pattern of thin lines. Variation of spacing and angle of the lines will provide different tones.
4. Thin lines crossing at an angle other than 90°.

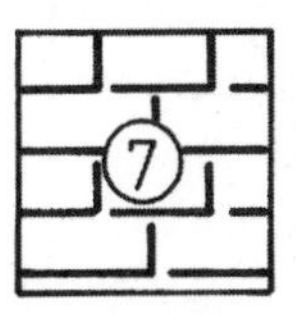

5. Thin lines to show a shiny surface.
6. Wood grain pattern formed with thin lines.
7. Brick pattern formed from thick lines.
8. Textured surface formed from small touching ellipses.

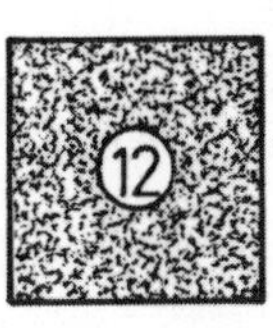

9. 10% Letratone dot shading – from dry transfer sheet.
10. 20% Letratone dot shading – from dry transfer sheet.
11. 30% Letratone dot shading – from dry transfer sheet.
12. A Letratone surface texture shading – from dry transfer sheet.

13. Light pencil shading – minimum pressure with H grade.
14. Medium pencil shading – HB grade.
15. Medium to dark pencil shading – heavier pressure with HB grade.
16. Dark pencil shading – B grade pencil.

Forms in graphic design drawings

A convention commonly adopted when referring to shape and form in graphic design work is to refer to two-dimensional outlines (2D) as shapes and to three-dimensional items (3D) as forms. This convention is adopted here.

Methods of showing forms

Lines used for drawing 3D forms
Three examples are given for 3D forms – a cube, a pyramid and a block – in which the form of each solid is obtained by drawing thin parallel lines on the sides and varying the directions taken by the sets of parallel lines.

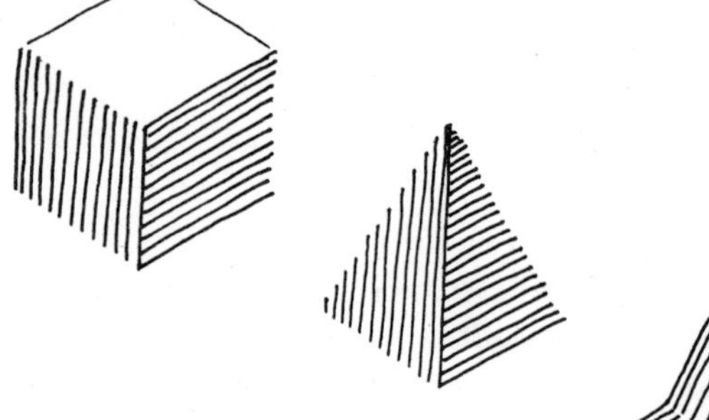

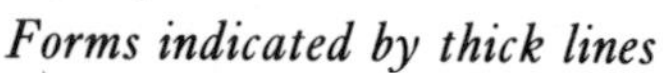

Forms indicated by thick lines
A rectangular prism, a cylinder and a pyramid are represented by thick lines drawn along edges and on the vertical surface of the cylinder.

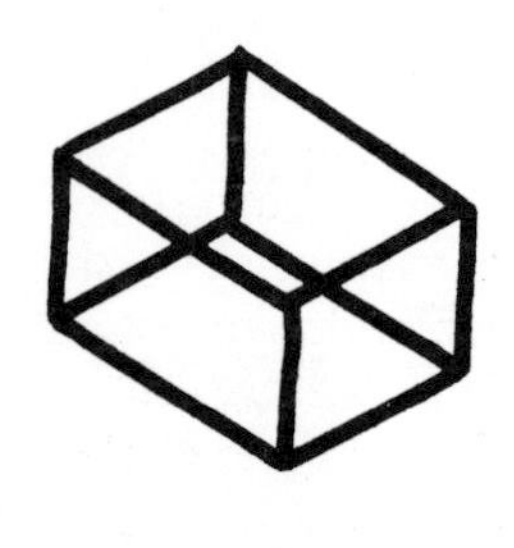

Variation of line thickness to emphasise 3D form
The outlining of 3D forms with thick lines to give emphasis to form is a practice very commonly adopted in graphics. All other lines in such drawings are thin.

Line shading of solid forms
Two examples of shading by adding thin lines are given – a letter cube and a bottle. Note that the cylindrical form of the bottle is emphasised by the shading becoming darker towards the outsides of the drawing.

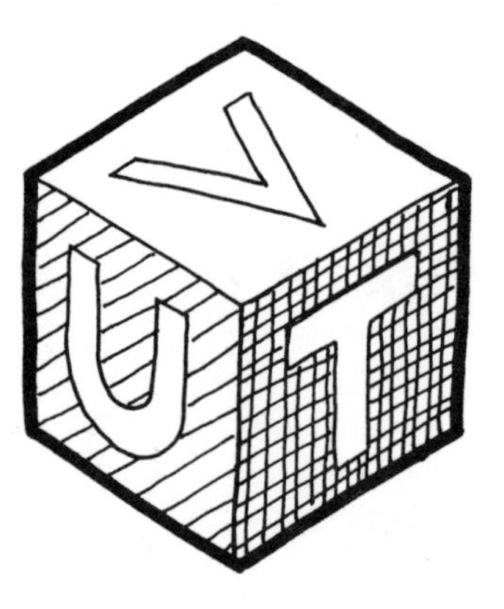

Line shading of cylinders
Three examples given – thin line shading, the lines becoming closer together towards the sides of the drawing; thick line shading; thin line shading in which the shading lines follow the curve of the cylindrical surface.

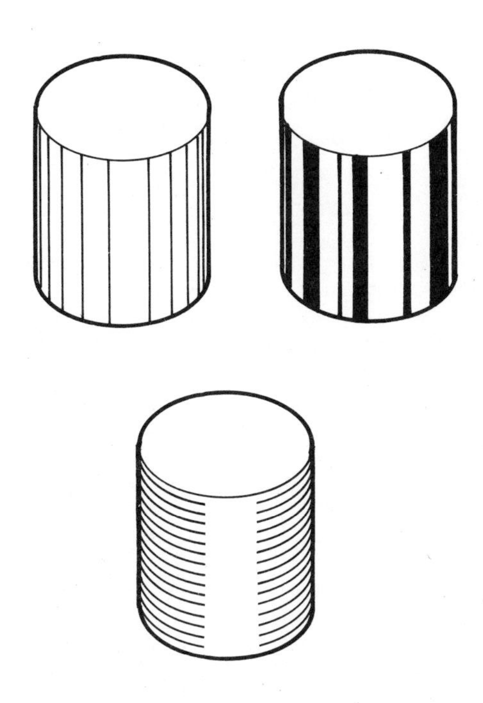

The three examples shown are suitable for shading any curved surface to give it a three-dimensional appearance (3D). Methods for the colour shading of curved surfaces are shown on pages 95 to 96.

Geometric method of drawing shadows

Assume the light source is at L (selected in any desired position by the person making the drawing).

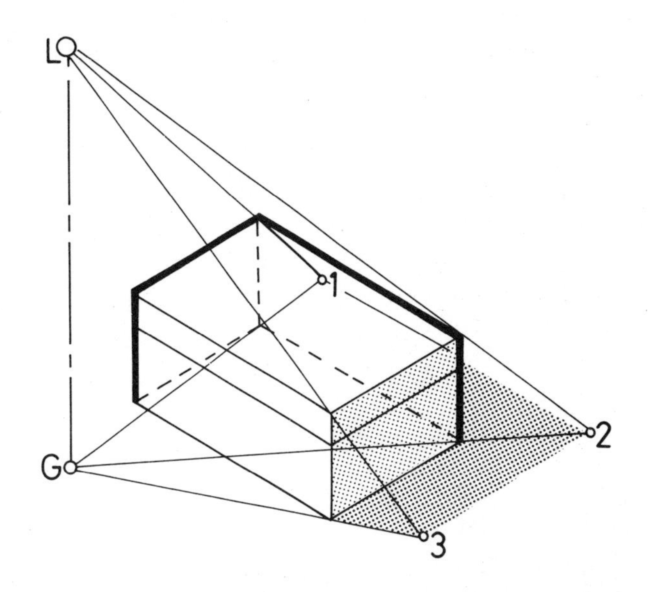

1. Select a point G vertically below L and presumed to be on the surface on which the box is standing.
2. Draw lines from G through the corners of the box where they touch the surface on which the box stands.
3. Draw lines from L through the upper corners of the box.
4. Where the pairs of lines – from L and from G – meet are corners of the required shadow.
5. Join the shadow corners as shown.

Mini graphic design projects

Two mini projects which demonstrate the use of linework in graphic design are included in the following four pages (pages 16 to 19). These projects are the result of working to the following design briefs.

1. Design a hanging 'mobile', suitable as a decorative feature in a room – pages 16 and 17.
2. Design a logotype which can be used as a symbol representing the Craft, Design and Technology department in a large school – pages 18 and 19.

The two projects follow the graphic design process outlined in the flow diagram on page 8. Space does not allow for the inclusion of an appraisal here. The reader is advised to attempt writing appraisals for the two designs as a valuable exercise. All four sheets of graphics were produced on A3 size drawing sheets.

Note the following

The sequence: design brief (given above); investigation; solutions; chosen solution; realisation is followed. A full realisation of the hanging mobile design could only be completed when the design has been realised.

The reader is advised to study the contents of the two projects with care. Note how various ideas for solutions have been included, how some have been rejected and others carried forward for further development towards a final solution. All the drawings have been made freehand, with the occasional aid of drawing instruments. Note the use of a variety of shadings to produce required graphic effects.

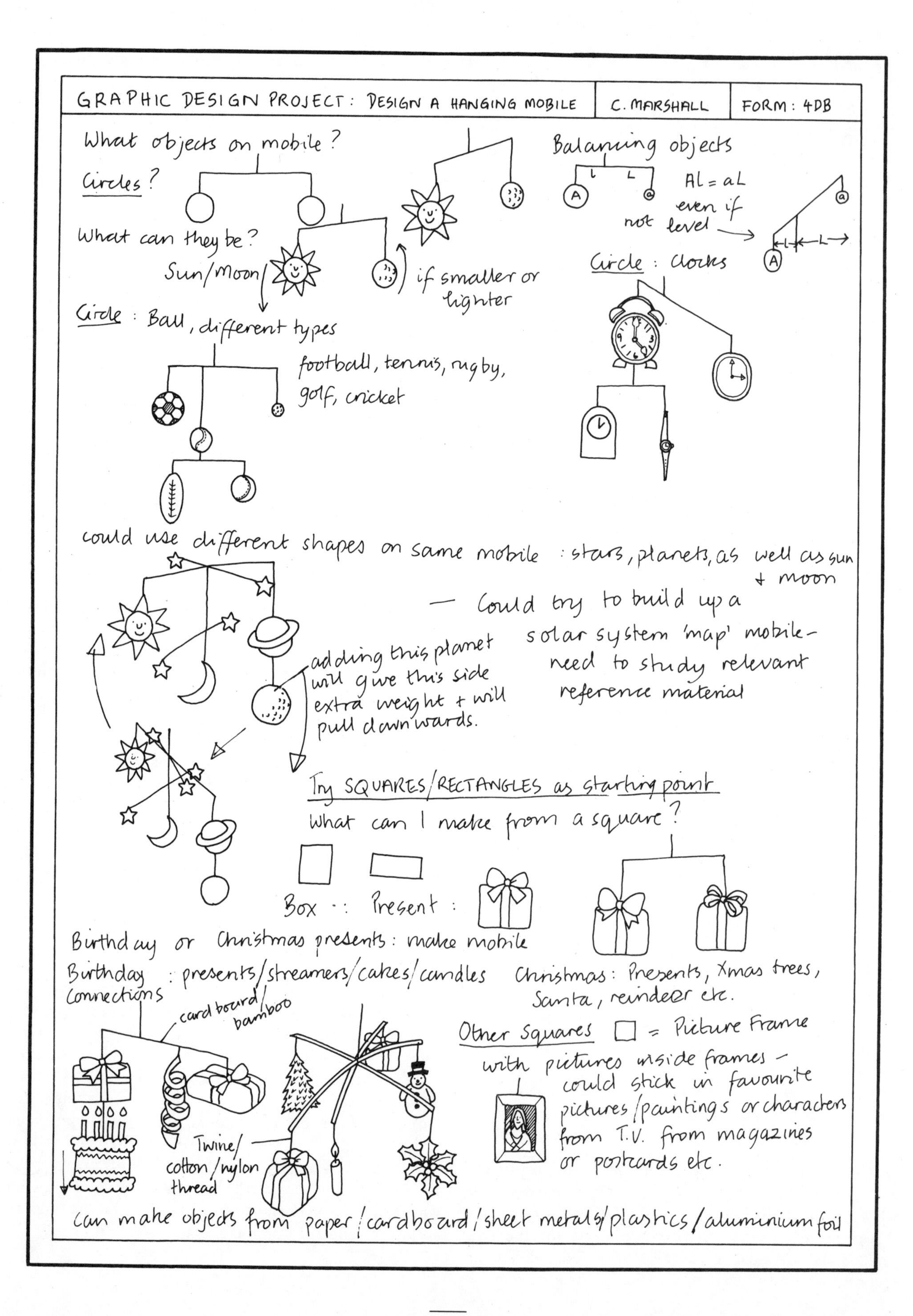
GRAPHIC DESIGN PROJECT: DESIGN A HANGING MOBILE
C. MARSHALL
FORM: 4DB
What objects on mobile?
Circles?
What can they be?
Sun/Moon
if smaller or lighter
Circle : Ball, different types
football, tennis, rugby, golf, cricket
Balancing objects
AL = aL even if not level
Circle : Clocks
could use different shapes on same mobile : stars, planets, as well as sun + moon
— Could try to build up a solar system 'map' mobile - need to study relevant reference material
adding this planet will give this side extra weight + will pull downwards.
Try SQUARES/RECTANGLES as starting point
What can I make from a square?
Box : Present :
Birthday or Christmas presents : make mobile
Birthday : presents/streamers/cakes/candles
Christmas: Presents, Xmas trees, Santa, reindeer etc.
Connections
cardboard/bamboo
Twine/cotton/nylon thread
Other Squares = Picture Frame
with pictures inside frames - could stick in favourite pictures/paintings or characters from T.V. from magazines or postcards etc.
can make objects from paper/cardboard/sheet metals/plastics/aluminium foil

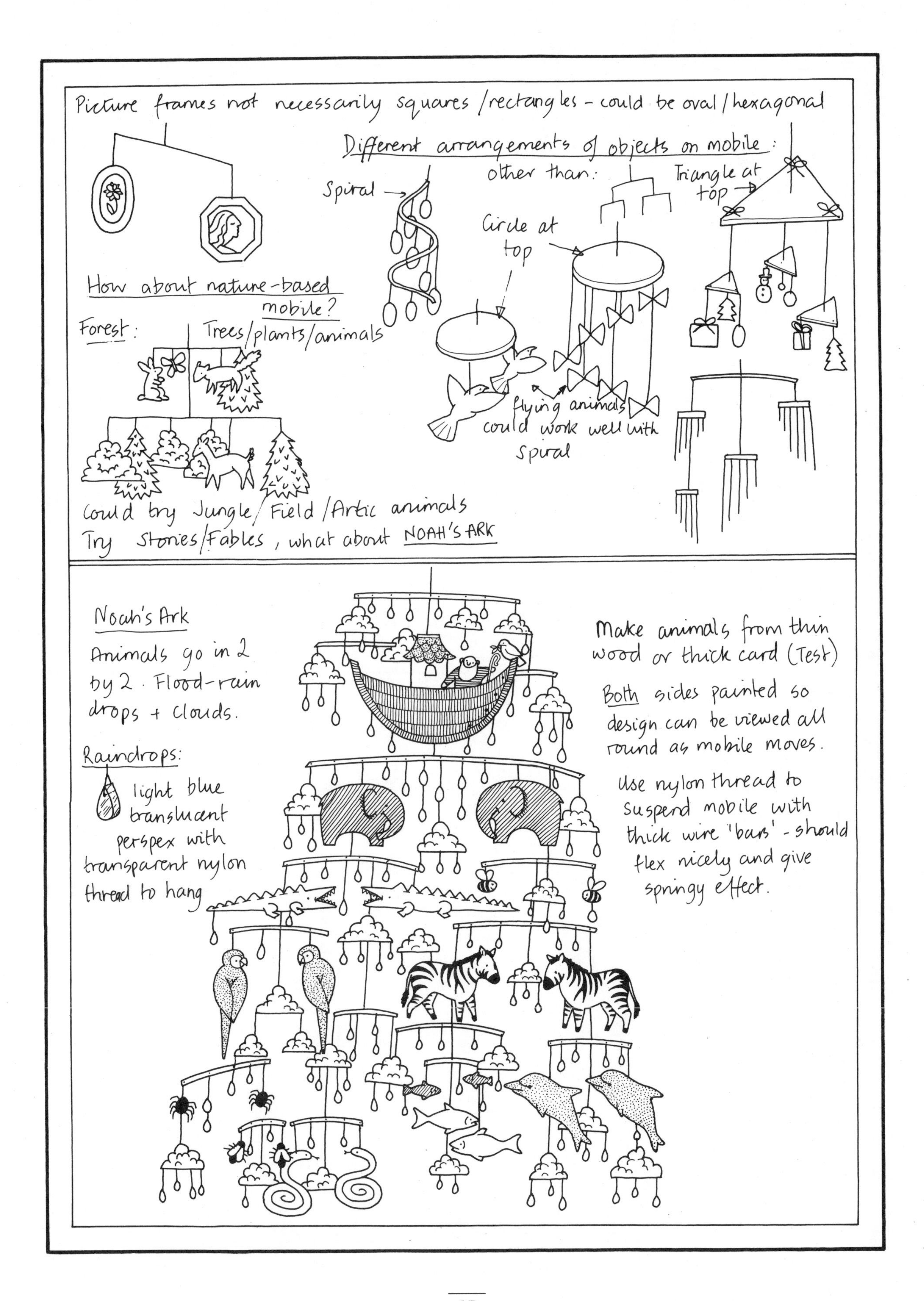

Picture frames not necessarily squares/rectangles – could be oval/hexagonal
Different arrangements of objects on mobile:
other than:
Triangle at top
Spiral
Circle at top
How about nature-based mobile?
Forest: Trees/plants/animals
flying animals could work well with spiral
Could try Jungle/Field/Artic animals
Try Stories/Fables, what about NOAH'S ARK
Noah's Ark
Animals go in 2 by 2. Flood-rain drops + clouds.
Raindrops:
light blue translucent perspex with transparent nylon thread to hang
Make animals from thin wood or thick card (Test)
Both sides painted so design can be viewed all round as mobile moves.
Use nylon thread to suspend mobile with thick wire 'bars' – should flex nicely and give springy effect.

A Logotype needs to represent both the letter forms and the subject (CDT)
At the very least be aware of what CDT involves in order to assist in symbolising it, otherwise design may look alright but at best be irrelevant and at worst inappropriate.

Therefore i) Examine the subject of Craft, Design + Technology
ii) Examine letters, typefaces

Also consider where logotype is to go (letterhead, on classroom door etc) and how it is to be reproduced and to what materials it will be applied

i Examine the subject of Craft, Design and Technology

(Look for useful graphic representations of elements of CDT)

CDT includes: Structures, Metalwork, Woodwork, Electronics, Design, Drawing...

Wood

Metal

Structures: Meccano?

Plastics

Tools:

Electronics - has some useful graphic representations

Drawing / Design

pencil

rapidograph

paper

What can typify each element of CDT? CRAFT: WOOD DESIGN: GRAPHICS/DRAWING
TECHNOLOGY: ELECTRONICS

ii Examine letterforms:

"C": can be semi-circle "(" or almost full circle "C" usually curves, but could be represented as "[" "C"

"D": can also be semi-circle "D" or full circle "O"
or straight lines: "[]" "D"
lower case: d d

"T": Always straight lines (upper case) or straight blocks "T" "T" "T"
lower case can curve or straight lines "t" "t"

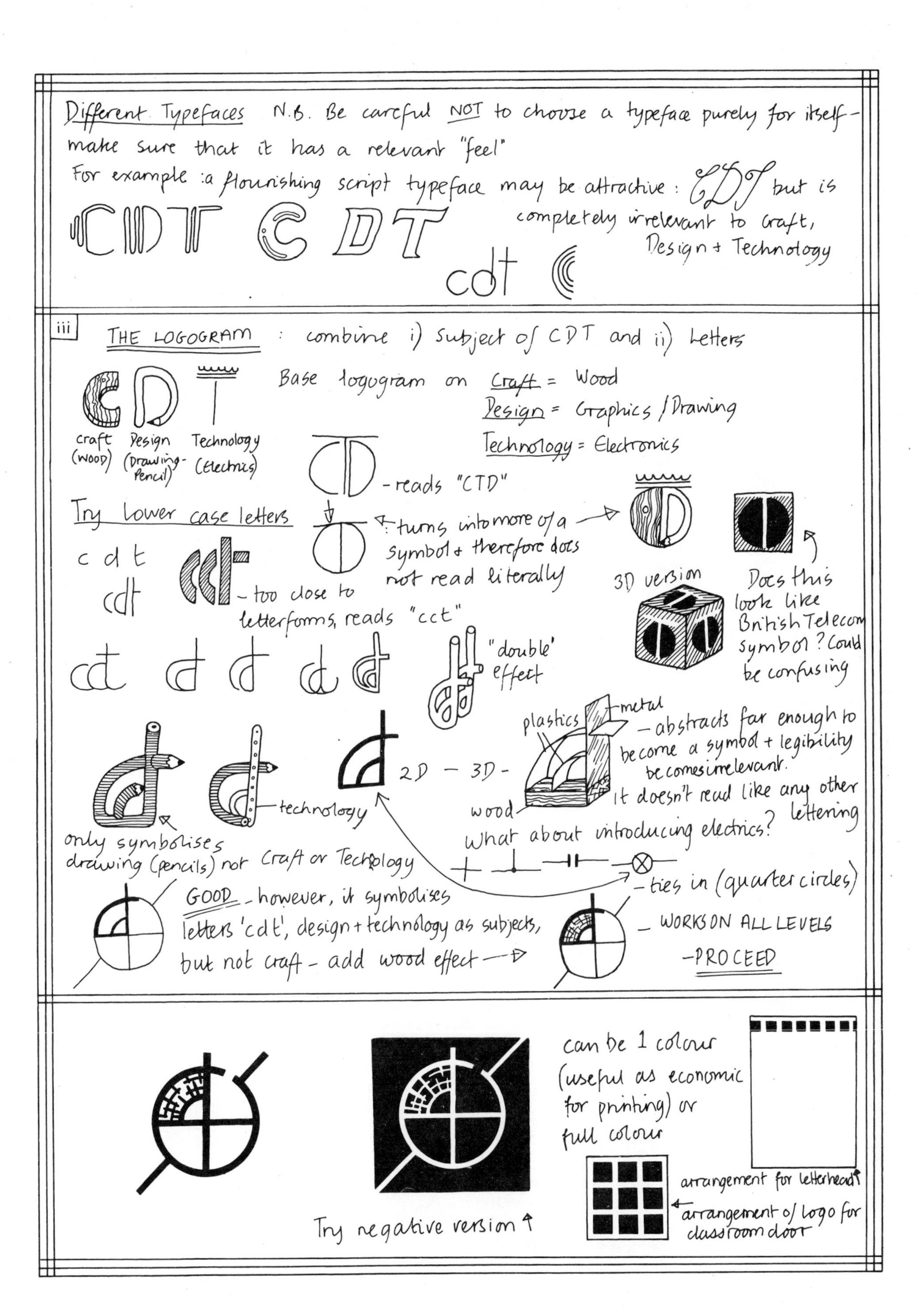

Different Typefaces N.B. Be careful NOT to choose a typeface purely for itself - make sure that it has a relevant "feel"
For example: a flourishing script typeface may be attractive: CDT but is completely irrelevant to Craft, Design + Technology
CDT
CDT
cdt
iii
THE LOGOGRAM : combine i) subject of CDT and ii) letters
Base logogram on Craft = Wood
Design = Graphics / Drawing
Technology = Electronics
Craft (wood)
Design (Drawing-Pencil)
Technology (Electrics)
- reads "CTD"
Try lower case letters
turns into more of a symbol + therefore does not read literally
3D version
Does this look like British Telecom symbol? Could be confusing
c d t
cdt
- too close to letterforms, reads "cct"
"double" effect
plastics
metal
- abstracts far enough to become a symbol + legibility becomes irrelevant.
2D - 3D -
wood
It doesn't read like any other lettering
technology
only symbolises drawing (pencils) not Craft or Technology
What about introducing electrics?
- ties in (quarter circles)
GOOD - however, it symbolises letters 'cdt', design + technology as subjects, but not craft - add wood effect
- WORKS ON ALL LEVELS
- PROCEED
can be 1 colour (useful as economic for printing) or full colour
Try negative version
arrangement for letterhead
arrangement of logo for classroom door

Discussion topics

The following discussion topics are based on statements from various Design and Communication GCSE syllabuses. They are included here as a basis for discussion between pairs of students, groups of students, for general class discussion, or as a basis on which students can compile notes about designing.

1 Consider the following:

(i) the restraints of costs, personal skills, resources and time in connection with the design process;
(ii) the preparation of time and progress sheets to forecast how a project will be developed;
(iii) the aesthetic, technical, economic and moral factors connected with designing.

2 Discuss the following:
Designing requires skills such as problem identification, searching for information, selection, analysis and synthesis of collected information, the development of discrimination in order to evaluate designs.

3 Communication skills are central to design, making and evaluation.

4 A project in Design and Communication must show evidence of identification of problems, analysis and synthesis relating to form, method, sequence, assembly and procedure and the ability to evaluate a completed design.

5 Discuss the following:
The sequence of designing follows an order such as: recognition of problems, drawing up of a specification, research, analysis, synthesis, evaluation.

6 Designing requires: the development of an understanding of graphics for representing design concepts; the ability to produce design specifications which take technical and aesthetic factors into account; the ability to generate ideas to solve problems; the ability to propose solutions to a design problem; the ability to compare and evaluate the design of an artefact.

7 The design process follows this sequence:

(a) Specify the chosen design assignment.
(b) Research problems connected with the assignment.
(c) Analyse and synthesise the results of the research and find solutions from this analysis and synthesis.
(d) Choose a suitable solution from the ideas.
(e) Develop the chosen solution.
(f) Produce models or prototypes of the design solution.
(g) Evaluate the finished design.

8 Models and mock-ups to simulate the principles and functioning of ideas are a necessary part of the design process. From such models a designer can recognise and test geometric principles of action, true shape and form. Models also serve to test structural principles, to test visual aspects of ideas and to test form, colour and texture.

EXERCISES

Note: To copy drawings from those given in the exercises, proceed as follows: please do *not* shade the drawings in this book.

(a) Lay a piece of clean tracing paper over the drawing you wish to copy. Hold it in position with one hand.
(b) Trace the outlines of the drawing onto the tracing paper with a pencil.
(c) Turn the tracing paper over and blacken its rear surface by shading with a grade HB or B pencil.
(d) Place the tracing paper in position the right way up on your drawing sheet and pencil over the traced lines. The drawing will then be copied from this book on to your drawing sheet. The dimensions given are in millimetres.

1. A cube, a cylinder, a cone and a sphere are given in drawing 1. Copy the drawings freehand with the aid of tracing paper. Shade your copies to produce a 3D appearance in the drawings.

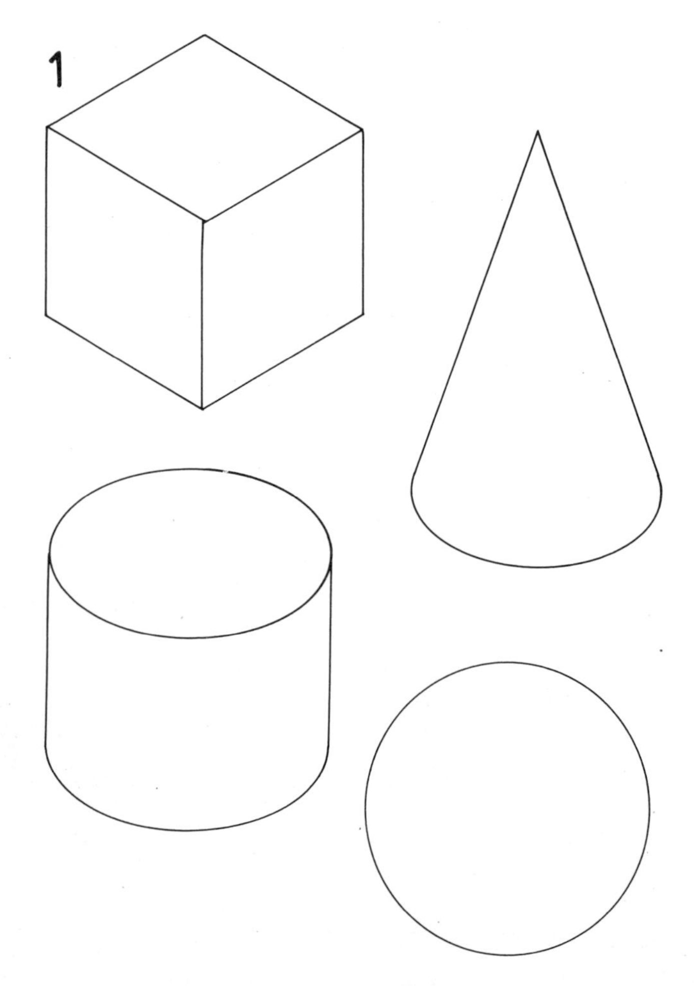

2. Drawing 2 shows a wall brick. Copy the drawing to the given sizes with the aid of a ruler. Emphasise the outlines of your drawing with thick lines. Shade its surface as if light is falling on the brick from above, the light source being from front left.

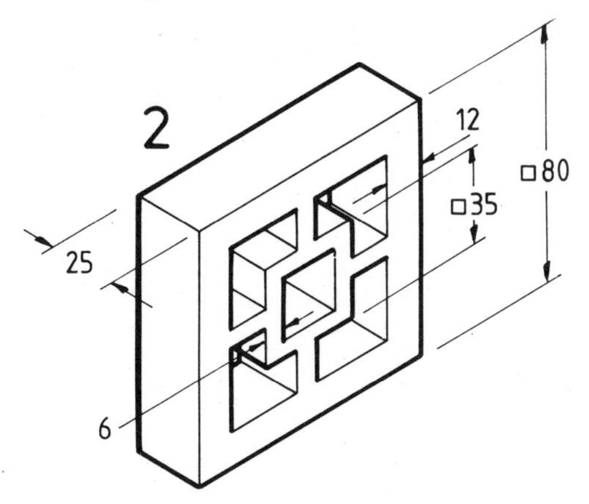

3. Drawing 3 shows three sets of grid lines. Draw similar sets of lines at the spacings shown with the aid of a ruler and a 30°, 60° set square. Shade parts of your grids to form interesting patterns. The wavy lines can be drawn by cutting a piece of card to the wavy line shape and then using the card as a template to draw lines with a pencil.

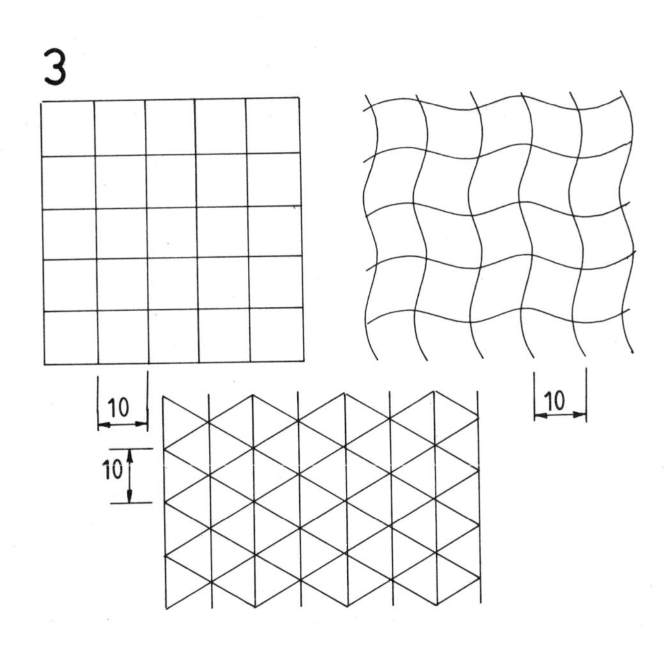

4. Using the geometric method of drawing shadows shown on page 15, copy drawing 4 to the sizes given and add shadows as if the light were from a source above the object, from behind and to the right.

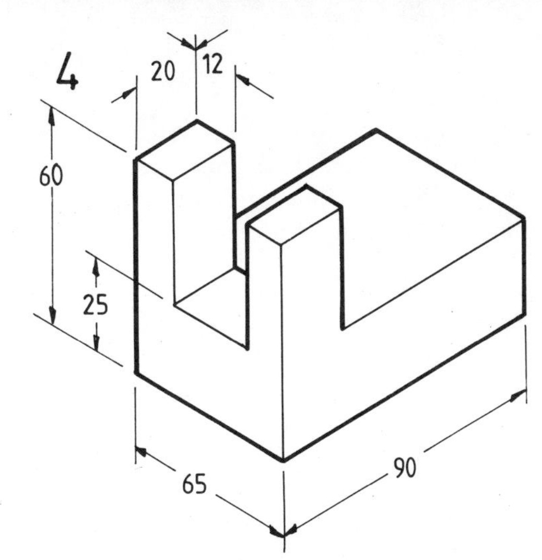

5. Drawing 5 shows a brickwork pattern surface. Copy the drawing to the sizes given and add lines and shading designed to emphasise the pointing between the bricks and the surface texture of the bricks.

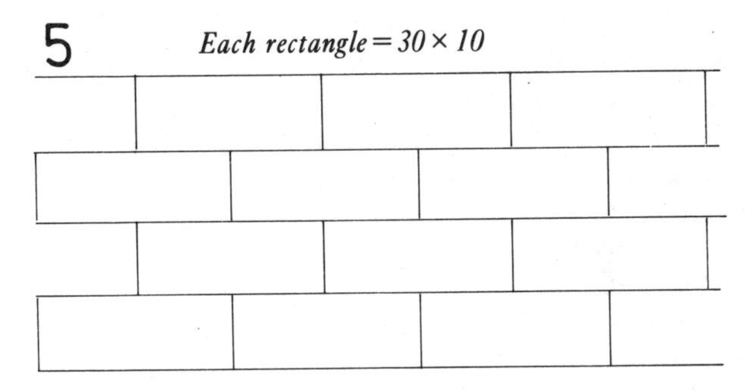

6. Drawing 6 shows outlines of two containers which have been designed to hold beauty treatment lotions. Copy the given outlines with the aid of tracing paper and, either by line or pencil shading, shade your drawings so as to give a 3D effect. Design labels for the containers which indicate the preparation each is to hold.

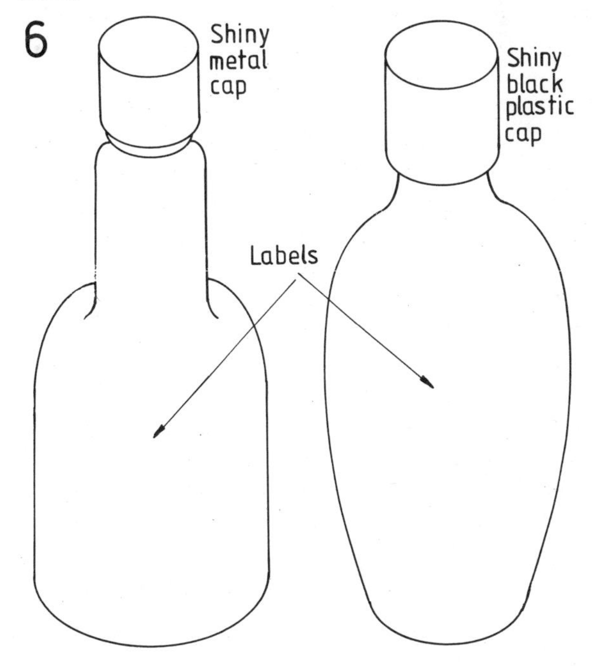

7. Copy drawing 7 to the sizes given with the aid of a ruler and a set square. Divide each line into 7 mm spacings. Design a filograph by joining the spacing points with ruled straight lines.

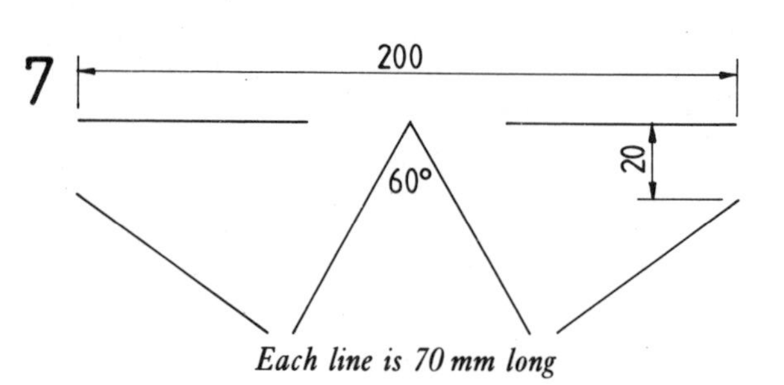

Each line is 70 mm long

8. Copy the circles, ellipses, pentagons and stars of drawing 8 with the aid of tracing paper onto different pieces of coloured papers. Cut out the shapes with scissors. Design a variety of combinations of shapes by moving the cut-out shapes into different positions on a table top. Attempt this with other materials: acetate sheet, polythene sheet, metal foil, coloured cardboards.

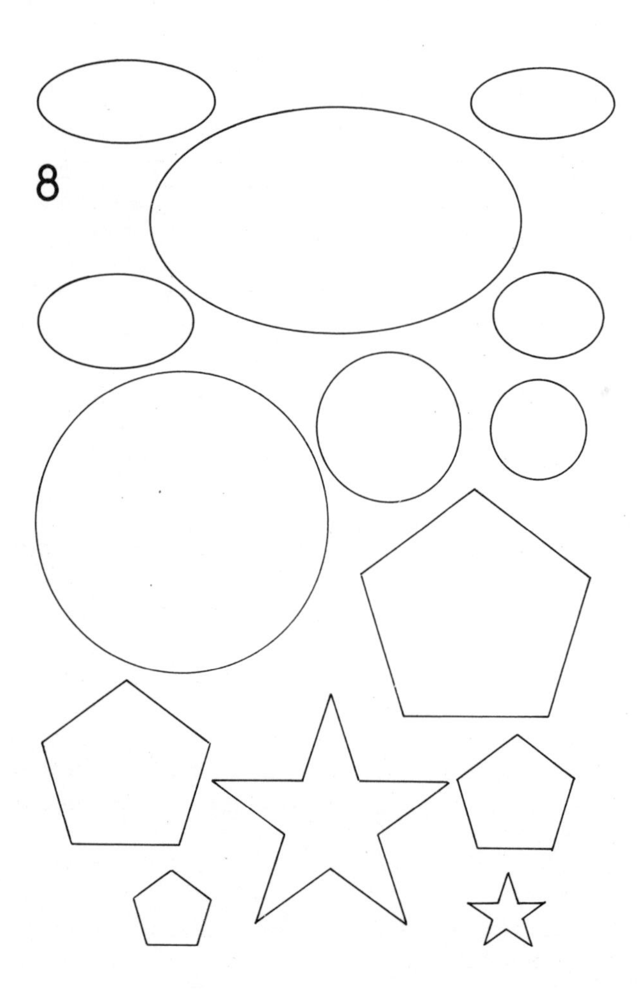

9. Drawing 9 is the outline of a cat, the shape of which was suggested by a photograph of a cat. Attempt similar styles of drawings suggested by photographs of other animals, for example a dog, a horse, an elephant, a snake, a butterfly.

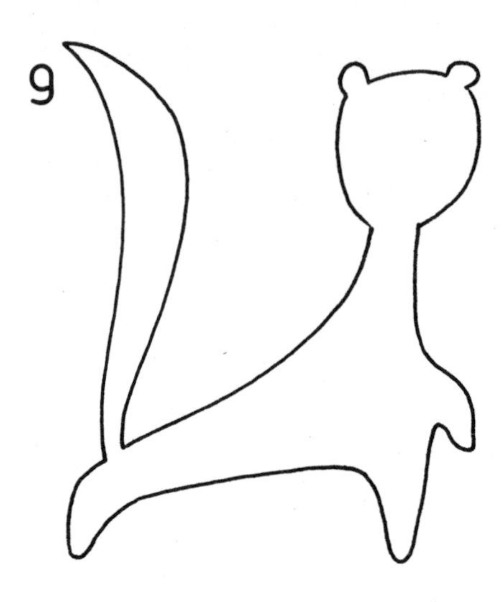

10. Using the method of selecting a shape from a jumble of either straight or curved lines as given on page 12, design outlines suitable for the following:
(a) a number plate for a house;
(b) a name plate for a house;
(c) a letter plate suitable for the letter B (for bedroom) for placing on a room door.

Drawing sheet layouts

Borders and titles

Depending upon the requirements of any particular item and form of graphics, a format may be preferred in which the long edges of drawing sheets are horizontal – a *landscape* format – or in which the long sheet edges are vertical – a *portrait* format. After selecting the most appropriate format, landscape or portrait, common practice is to draw margins around drawing sheets. The drawings of margins produces borders to drawings, which have several advantages.

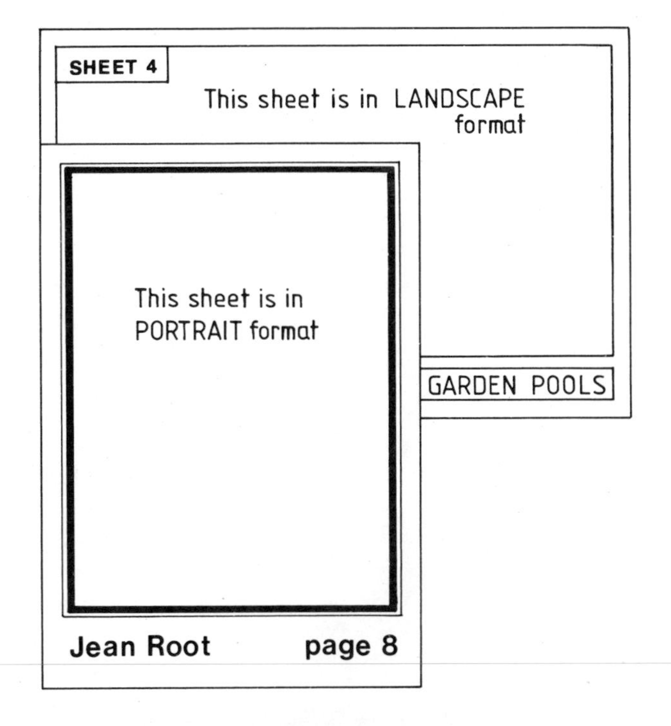

1. The borders form a frame to the graphics on the drawing sheet, so enhancing their appearance.
2. The area outside the actual graphics on the sheet may become frayed or otherwise damaged as sheets are sorted out for filing, as sheets are passed between people for design evaluation or when drawing sheets are used in workshops. A border contains such drawings, allowing the area of the sheet on which the actual graphics are drawn to remain undamaged.
3. Margins can be designed to enhance and emphasise various features on a sheet of drawings.

Engineering drawings

When drawings are produced for the engineering industries, it is customary to draw margins of the type shown. Schools and colleges have generally adopted this method of straight, ruled margins. Included within the borders is a title in a title block and other printed information:

1. the name of the school or college;
2. the name of the pupil or student;
3. the form or study group of the pupil/student;
4. the date the drawing was completed;
5. the scale to which the drawing has been made;
6. the name or title of the article the drawing describes;
7. the dimensioning unit;
8. if the drawing has been made in orthographic projection, the angle of projection employed.

Borders and layouts for design drawings

When producing drawings for design graphics, many of the sheets of drawings may well follow the pattern set for engineering drawings. There is no real necessity, however, to follow such a pattern too rigidly. Many

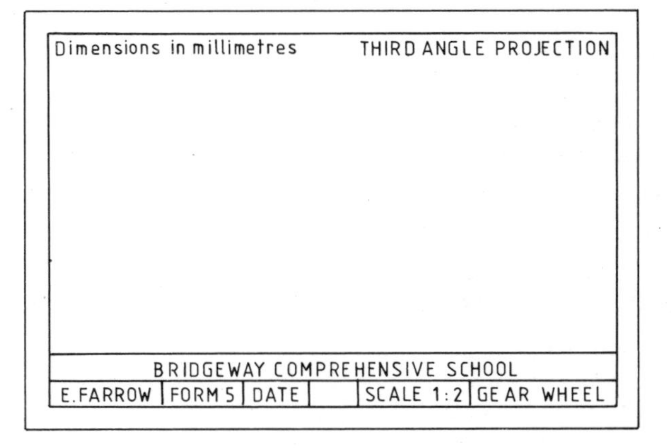

other types of margins and sheet layouts can be designed. Eight different borders/layouts are shown here. These include:
1. margin lines meeting at corners at right angles;
2. margins radiused at corners;
3. margins joined at corners by quarter ellipses;
4. margins drawn with pairs of parallel lines;
5. thick line margins;
6. margins designed from alternating thick and thin lines;
7. margins designed to divide a sheet into convenient smaller areas.

These examples do not, by any means, exhaust the possibilities. Students are advised to design their own margins, borders and page layouts, remembering to relate them, if possible, to the design of the contents of the sheets. From the given suggestions note the following details.
1. Titles can be placed in any convenient position on the sheet.
2. Each sheet should have a sheet number. Numbers should be clear and placed in any position.

3. Printing in titles and headings may be:
   (a) freehand;
   (b) drawn with the aid of lettering stencils;
   (c) added to the drawing from dry transfer sheets;
   (d) drawn in italic script;
   (e) drawn all capitals or capitals and lower case lettering.

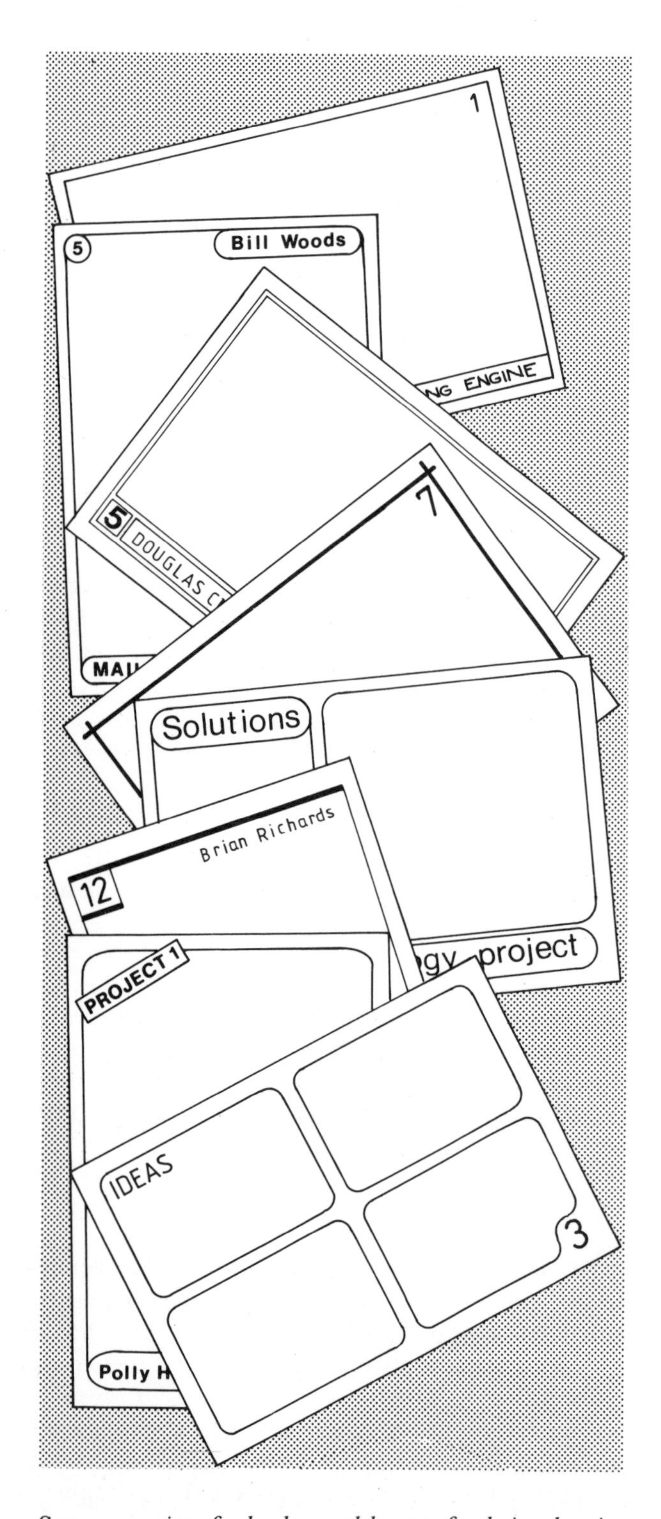

Some suggestions for borders and layouts for design drawings

Suggestions for sheet layouts

Eight drawings showing suggested sheet layouts are given. Many sheets of drawings in design graphics will take the form of an orderly set of sketches interspersed with explanatory notes. Some of the drawings may overlap each other. Occasions will arise when the content of a sheet of graphics look better if designed beforehand. Some suggestions on which the sheet design layout can be based, are:

1. using the 'golden mean' proportions – a rectangle of sides in an approximate ratio of 8:5;
2. dividing either the vertical or the horizontal areas of a sheet into thirds;
3. selecting points towards which the sheet design moves – perhaps from a broad outline of ideas to a final conclusion;
4. designing the layout on spirals or within areas bounded by curves to represent a movement or a flow of ideas;
5. avoiding the temptation to place a main drawing exactly in the middle of a sheet.
6. Titles and other prominent details of lettering do not necessarily have to be placed at the top or at the bottom of a sheet. They do not even have to be placed horizontally.

The illustration shows:

1. a set of four drawings (or photographs) placed centrally on a portrait format with notes to the left or right of each item of graphics;
2. drawings (or photographs) staggered down a portrait format, with accompanying notes against each item;
3. drawings placed within notes written as if bounded by curved lines;

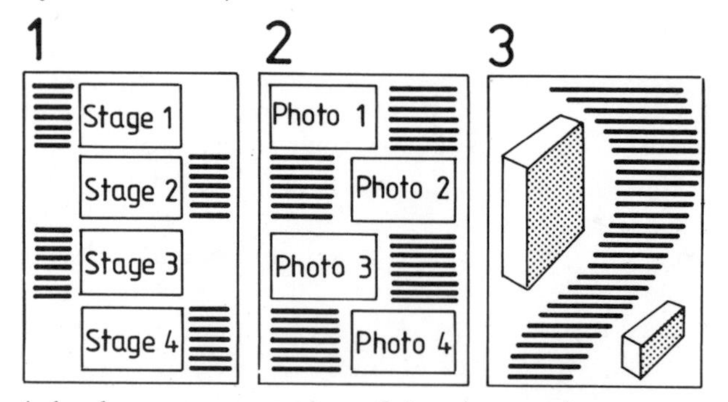

4. landscape presentation of drawings and notes;
5. another landscape design with both graphics and notes seemingly moving towards a vanishing point;

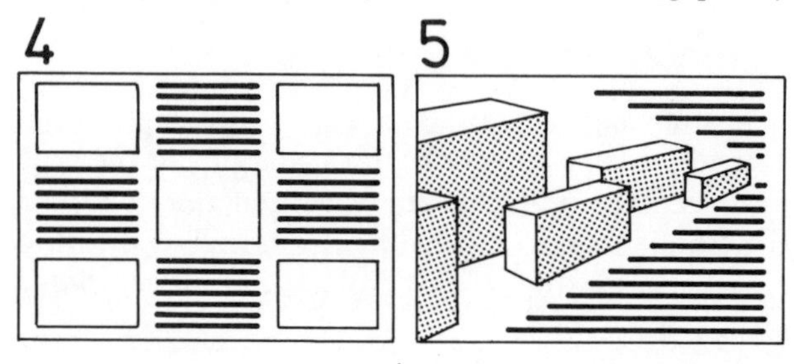

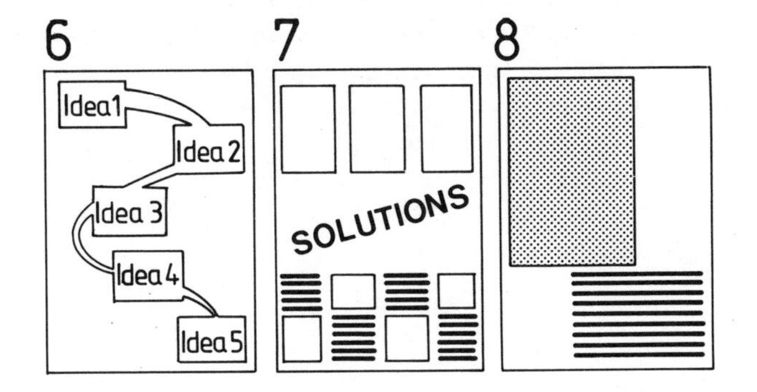

6. a flow of ideas based on a spiral curve;
7. a series of graphics with a central large title;
8. a photograph, with notes, based on a sheet divided both vertically and horizontally into thirds.

Photographs

Photographs can be used in the design process. Some possibilities for the inclusion of photographs on design sheets are:

1. to record a realised design;
2. to record possible solutions to a design brief;
3. to show how a design can meet a variety of situations;
4. to illustrate stages in the making of a design;
5. to illustrate stages in construction processes.

In school and college design work, the best photographer is the student or one of a group of students engaged in the designing process. Modern photographic equipment is so varied and comparatively easy to use, that photographs of a reasonable quality can be produced with a minimum of photographic experience. The processes of developing film and printing from a developed film are relatively cheap. Photographs taken from magazines and other periodicals, or photographs from firms and organisations, do have a place in the design process. There is no place for slavish copying of designs taken from photographs of other people's work, but such photographs can suggest starting points from which ideas can be developed, or can be used as illustrations of solution ideas other than your own, or for comparing other people's designs with your own.

Types of photographs

There are three types of photographs suitable for inclusion in a design folder:
1. black and white photographs;
2. colour photographs from colour negative films;
3. colour transparencies or colour prints made from colour transparencies.

Each type has advantages and disadvantages.

1. Film for black and white photographs can be developed without much difficulty at school or at home. The negatives produced by development can then be placed in an enlarger to print enlargements. In particular, selected areas of negatives can be enlarged. One main advantage of black and white film is that it can be used with any of the various forms of lighting – natural or artificial.
2. Colour negative film, from which colour photographs can be produced, can only be processed (developed and printed) by firms specialising in photographic processing. The processes are too complicated for the school/college design student. Prints can, however, be obtained at reasonable prices.
3. Good colour transparencies have a depth of colour and tone that is difficult to reproduce from either black and white or colour negative film. Transparencies, however, can only be seen with the aid of a viewer or on a screen from a projector. Prints can be obtained from transparencies, but at greater expense than from colour negative film.

Taking photographs

Cameras

For students wishing to take photographs, cameras using 35 mm film are probably the easiest to handle. Either *single lens reflex* (SLR) or automatic focusing cameras are suitable. Most modern 35 mm cameras have automatic exposure devices incorporated in their mechanisms.

Lighting

Care must be taken when considering the lighting of an object which is to be photographed. The object can be lit in one of four ways.

1. *Natural lighting out-of-doors*
Provided the day is not too dull or too bright, photographs taken out-of-doors under natural light can produce good results. If too dull, the tones of the resulting photograph will be dulled, if too bright,

unpleasant harsh shadow effects will hide some of the details in the photograph. Black and white, colour negative or transparency films are suitable.

2. *Flash lighting*
Flash 'guns' can be purchased for any 35 mm camera. Some are sold with a 'gun' incorporated in the camera body. Be careful when taking photographs with flash. Unless precautions are taken (see below) the resulting picture can be rather flat in appearance with very harsh shadow effects. Use the same films as for natural lighting.

3. *Neon tube lighting*
Many rooms in schools and colleges are equipped with abundant overhead neon lighting. This will provide excellent, almost shadow free, general lighting for photographic purposes. However, only black and white films or colour films made for 'tungsten' lighting can be used, although other colour films can be used providing suitable filters are fitted over the lens.

4. *Tungsten lighting*
Precise and accurate lighting can be obtained from lights such as 'photofloods' or high wattage lights such as those used for stage lighting. Only black and white films and tungsten colour film can be used with this type of lighting, although tungsten filters can be fitted to a camera, enabling other colour films to be used.

Taking photographs under artificial lighting

Tungsten lighting

A plan showing the positions of the object being photographed, the camera and the lights for artificial lighting is given. Note the following.

1 *Two lights*
Main light (say 500 watts) at the camera position; 'filling-in' light (say 250 watts) to one side. The 'filling-in' light eases the blackness of shadows and 'models' the object, giving it more of a 3D effect.

2 *Camera*
Its distance from the object depends upon the size of the object and the type of camera lens.

3 *Background*
Many a photograph is ruined by objects intruding in the background. The background shown in the given plan consists of a sheet of cloth fixed to a wall with masking tape and allowed to drape against the wall and onto the floor so that no hard corners can be seen in the resulting photograph. Stage canvas, painted white or grey with emulsion paint, forms an excellent background material. Other backgrounds can be made up from sheets of plywood, hardboard or chipboard or from sheets of cloth showing an open weave. The colour of the background material may be varied according to the colour of the object being photographed.

Flash lighting

A similar set up is used as for tungsten lighting, but with two flash guns in place of the tungsten lights – the 'fill-in' flash being less powerful than the flash on the camera. Another method of avoiding harsh shadows and flatness in flash photography is to point the flash gun overhead at a white ceiling or wall so that the light comes from the flash and is also reflected (bounced) from the wall or ceiling. Another method is to dull the flash by placing muslin or the equivalent over the flash gun. Some flash guns are fitted with covers specially made for this purpose.

Tripods

If possible use a tripod with your camera. The camera is held steady and you can spend more time composing the picture in the view finder and be sure that the camera will not move away from your composed picture when the photograph is taken.

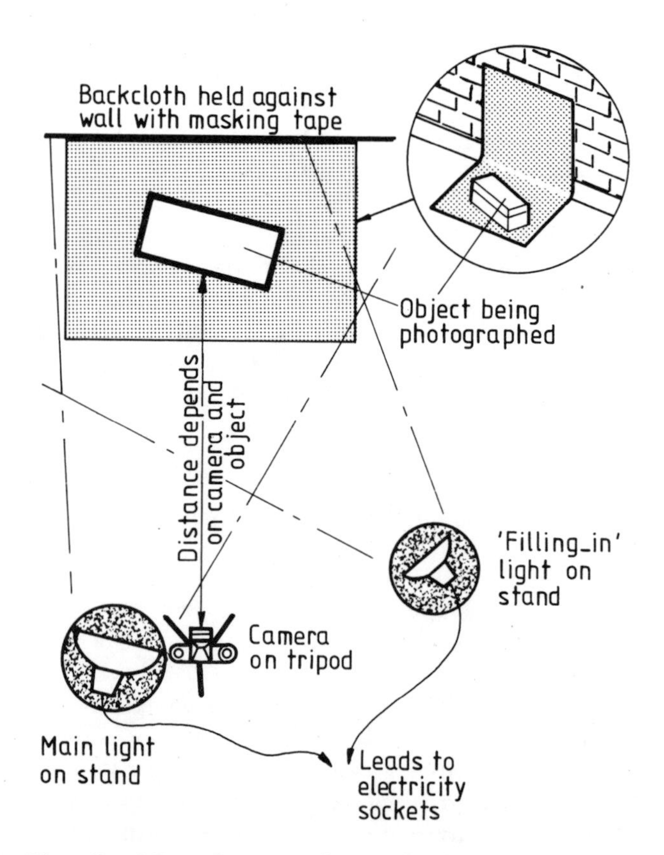

Plan of positions of camera, object and lighting

Mounting photographs in a design folder

Several types of adhesive are suitable for mounting photographs (or any other items of graphics) on a background or on an area of a design sheet.

1. *PVA glue*
White liquid PVA wood glue, sold in plastic squeegee containers, is a good, clean and cheap adhesive. Apply sparingly and evenly to the back of the photograph and smooth the photograph in position with hand pressure.

2. *Stick adhesive*
This is also a PVA glue but in stick form. Apply as for the liquid form.

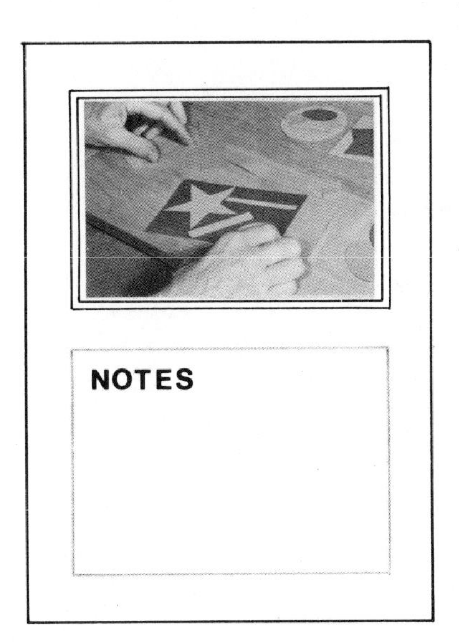

3. *Rubber adhesives*
These are sold in tubes or cans. Apply to the back of the photograph with the spreader supplied with the glue.

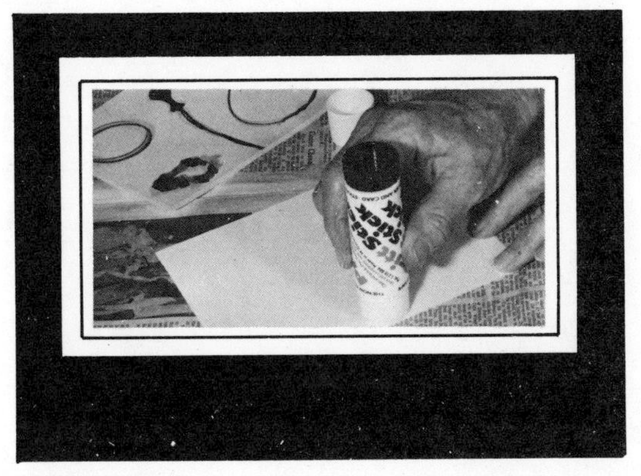

4. *Spray mount*
This is supplied in an aerosol. It is possibly the easiest and cleanest of all mounting adhesives, but is rather expensive. Place photograph upside down on clean newspaper; lightly spray its back with spray mount; position the photograph on its mounting; place clean paper over the photograph and smooth down with hand pressure; fold the newspaper with its now sticky side inwards and throw it away. Spray mount allows re-positioning if necessary.

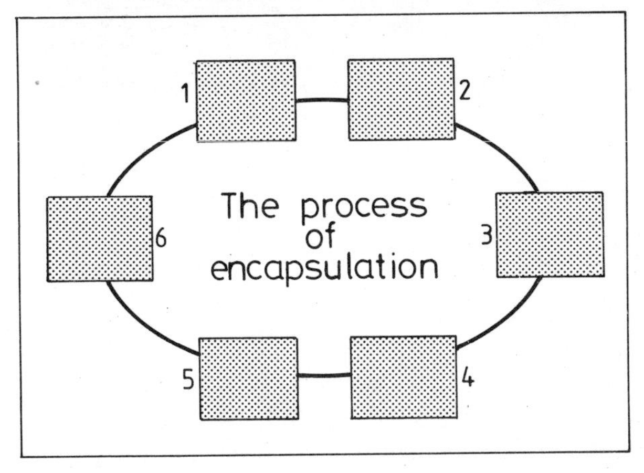

A suggestion for a layout of a series of photographs

SECTION 2 Background knowledge

Drawing papers

Paper is made from the fibres found in plant materials such as wood, straw, esparto grass, cotton and rag waste. Other materials are included but to a lesser extent. Cheap papers such as newsprint are all wood fibres, held together with lignin from wood cell walls. Better quality papers, such as those for graphics work, usually contain other plant fibres, for example from cotton.

The A range of drawing sheets

Most papers used for graphics are sold in sheets of the internationally recognised A range of sizes. Sheets in the A size range are rectangles with sides in the proportion of $1:\sqrt{2}$. An A0 size sheet has an area of 1 square metre and each sheet in the range is exactly half the area of the next larger sheet.

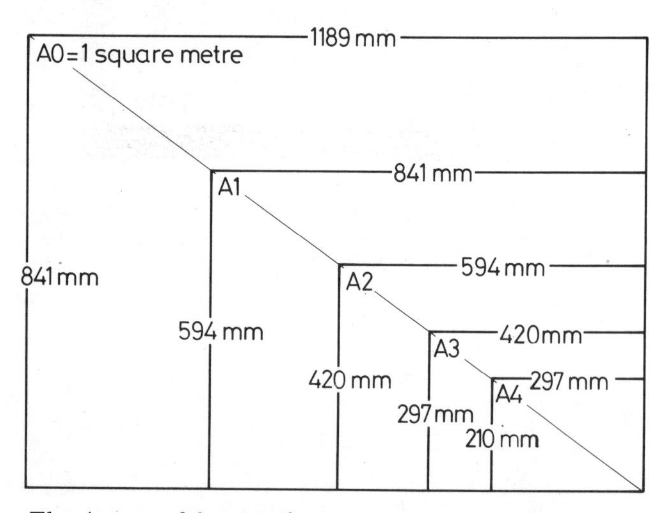

The A range of drawing sheets

Types of paper

Some of the papers used in graphics are described below. Papers are often sold as being of a weight of so many grams per square metre (gsm or g/m^2). Paper weight gives a good idea of paper thickness.

Cartridge paper – A good quality drawing paper suitable for pencil work. Takes colour wash well. Common weights are from 60 gsm to 120 gsm. Can be purchased in A size sheets or in rolls.

Bleedproof paper – A paper for use with marker pens. A non-absorbent paper usually supplied in A4 or A3 size pads with 50 sheets to the pad.

Tracing papers – Semi-transparent papers that can be used for tracing drawings for repetition elsewhere. Another use is for placing over a drawing as a cover sheet through which the original drawing can be seen. The cover sheet protects the drawing from damage and from becoming dirty. 'Natural' tracing paper is sold in rolls or in A size sheets in weights of 38 gsm, 63 gsm and 90 gsm. 'Natural' tracing papers take pencil, ink and colour washes well.

Detail paper – A thin but good quality rag-based paper which takes pencil lines well.

Bristol board – A heavy weight paper/board mainly for ink and colour work, with hard, smooth, white surfaces.

CS10 – One of a series of papers made by a firm specialising in artists' papers. Its surfaces are particularly suitable for line work with inks. White, hard and tough surface.

Grid papers – Square, isometric and perspective grid papers can be purchased. Grid papers allow you to make accurate drawings very quickly.

Low-tack masking film – Can be cut and placed in position to mask areas which are not to be coloured when air-brushing. Its back surface is coated with adhesive, allowing the film to be placed in position on a drawing sheet and peeled off when colouring is completed without damage to the drawing surface.

Plastic tracing films – Made from various plastics: polyesters, vinyls and polycarbonates. Much more expensive than natural tracing paper but considerably stronger. Will take either pencil or ink drawings.

Drawing equipment

Basic items of equipment are a drawing board, a Tee-square and a pair of set squares. Some schools and colleges are equipped with drawing boards with a parallel motion device which provides a straight edge, enabling a Tee-square to be dispensed with. Drawing and design offices are usually equipped with draughting machines.

Drawing boards
Drawing boards are commonly made from wood. Some drawing boards are faced on one or both sides with plastic sheets of a type which allows pencil and pen work to proceed smoothly.

Fixing drawing sheets to a drawing board
Drawing sheets are best fixed to drawing boards with strips of masking tape at each corner. Sellotape can also be used. Purpose made steel clips are another method. As a last resort use pins, but these cause considerable damage to drawing boards. The corners of the board become pitted with holes after repeated pinning of sheets to it.

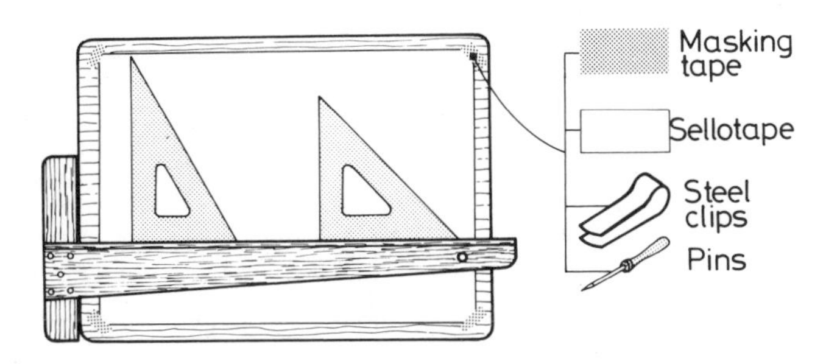

Tee-squares
These are made from hardwoods or from plastics or a combination of wood and plastic.

Set squares
These are made from sheet plastics, either clear transparent or coloured semi-opaque. Three types are common – 30°, 60° squares, 45° squares, and adjustable squares.

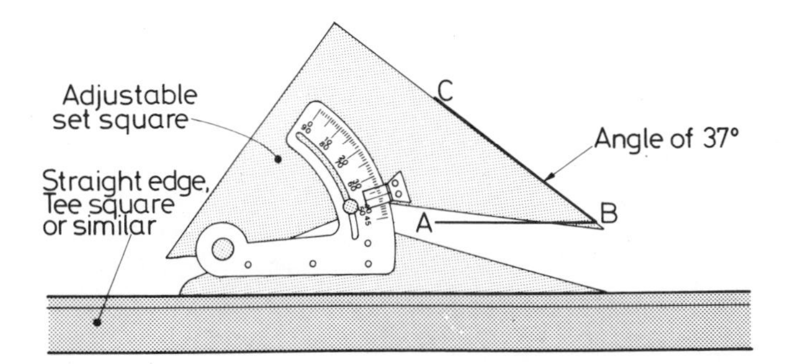

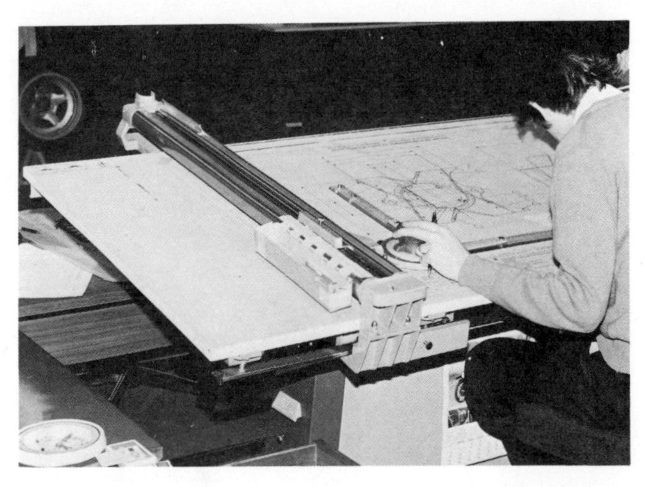

A draughting machine and table in use in a drawing/design office

Rules
Rules made from plastic materials are very suitable. One measuring up to 300 mm in 10 mm and 1 mm intervals will cover most graphics work.

Pencils
The common wood pencil or a clutch pencil are both suitable. Clutch pencils are designed to hold lengths of pencil 'lead' which can be released from the clutch mechanism by pressing a button on the top of the pencil. Although seventeen grades of pencil, from 9H to 6B can be purchased, those of grades 2H, H, HB, B and 2B are most suitable for work in graphics.

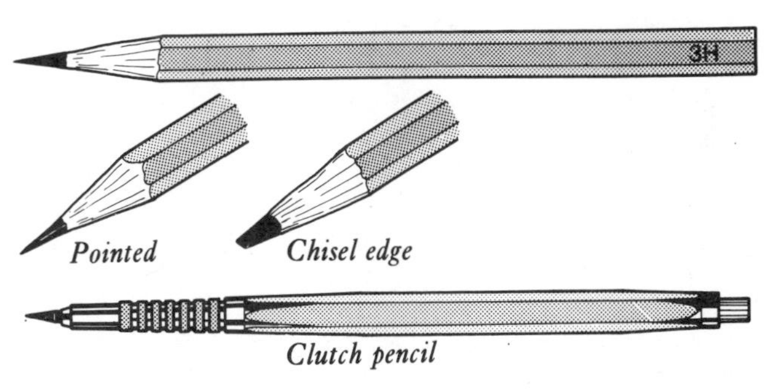

Technical pens
Technical pens have nibs which are designed to allow a flow of ink from a reservoir to the working point of the nib. A technical pen can be used continuously over a long period of time until the ink in its reservoir is used up. The greatest value of technical pens is that they will draw lines of exact widths. The width of the line each nib will draw is printed on it and on its cap. Colour strips on the nib and its cap also identify the line thickness it will draw.

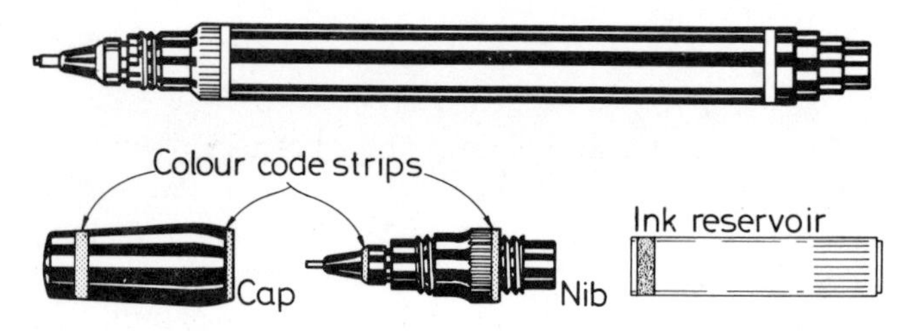

Technical pen and its parts

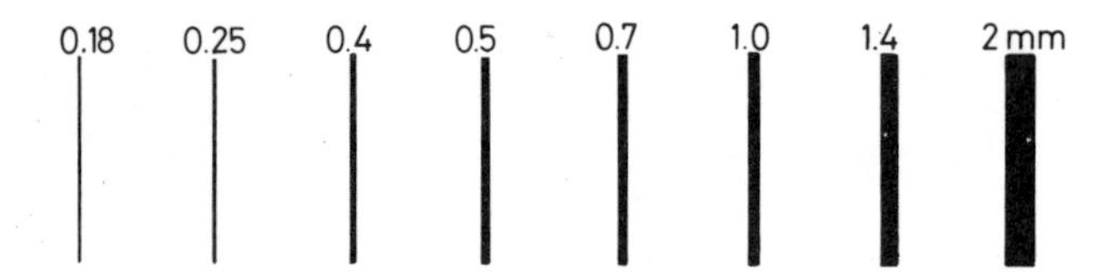

Line widths of one range of technical pens

Compasses
Most graphical work requiring compasses can be carried out using a spring bow for small circles or arcs and a pencil compass for larger work. When working with technical pens, a technical pen compass is needed. For drawing very large arcs and circles, a beam compass is advisable.

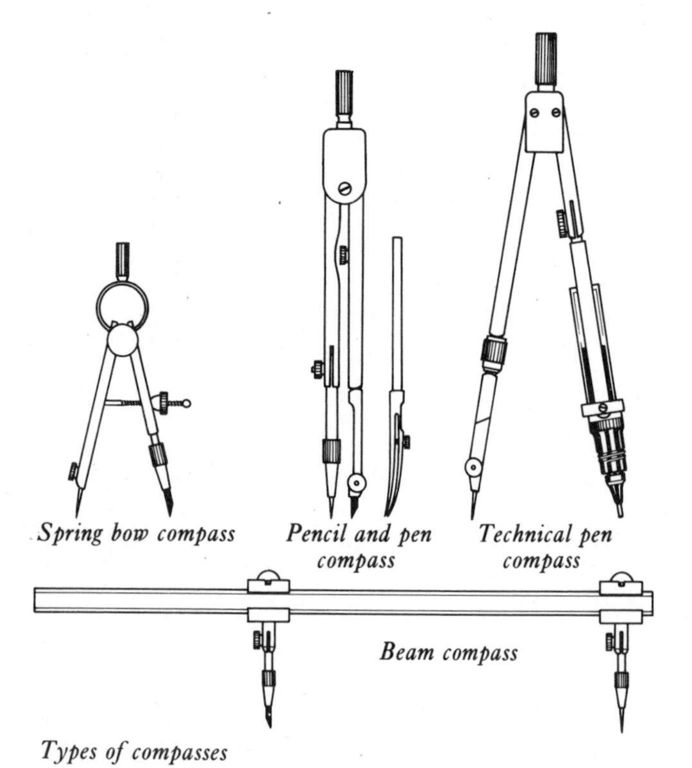

Types of compasses

Drawing aid templates

Drawing aids are made in a very large array of shapes and sizes to assist in drawing the curves and outlines required in graphics. They are suitable for the drawing of curves, ellipses, circles, radii, triangles, squares, hexagons, nuts and bolts, electronics and electrical circuit symbols, pneumatic and hydraulics symbols, flow chart symbols and many other outlines. They allow accurate drawings to be made quickly. Only a few of these aids are shown here.

A set of three French curves

French curves
This is the most common aid for drawing curves. A typical use is shown in drawing a number 7. The number is first sketched freehand on a grid of lines (drawing 1). Drawing 2 shows how an accurate outline can be completed with the aid of a French curve.

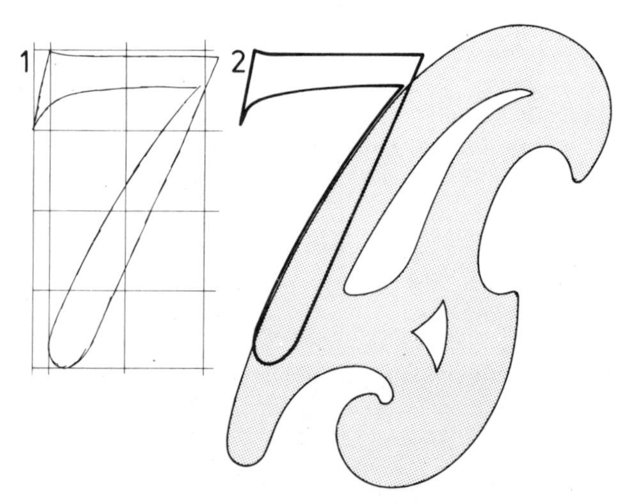

Drawing a curve with the aid of a French curve

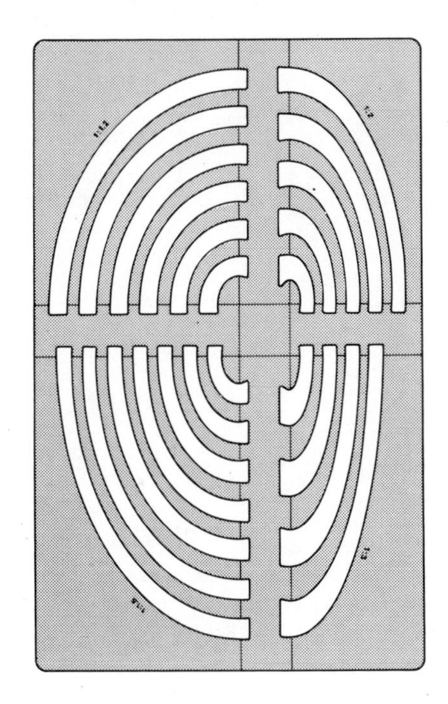

Ellipse templates
A large range of ellipse templates are made, one of which is illustrated. This particular template allows 44 different ellipses to be accurately drawn.

Nut and bolt template
This allows you to draw hexagonal nut outlines from M5 to M25 ISO metric screw thread sizes. M stands for ISO metric thread form, the figures following the M giving the outside diameter of the screw thread in millimetres.

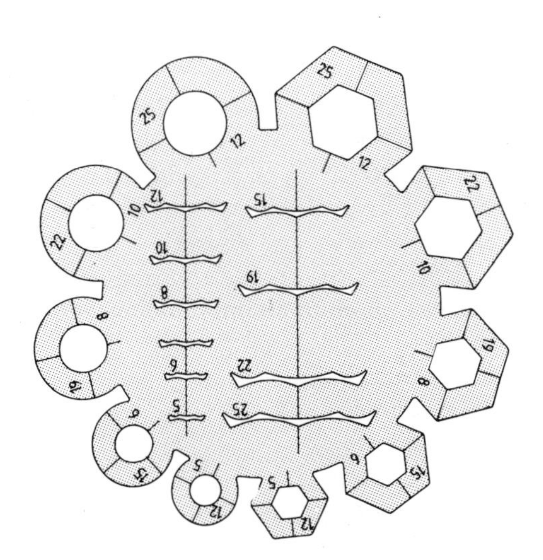

Dry transfer tools

When applying dry transfer letters, or cutting masking films to shape, cutting knives and a burnisher will be of value. Throw-way knives are so cheap that they can be thrown away when the cutting edge becomes blunt. Artist's knives have interchangeable blades for renewal when one becomes blunt. A burnisher is of value for rubbing down dry transfers firmly onto a drawing.

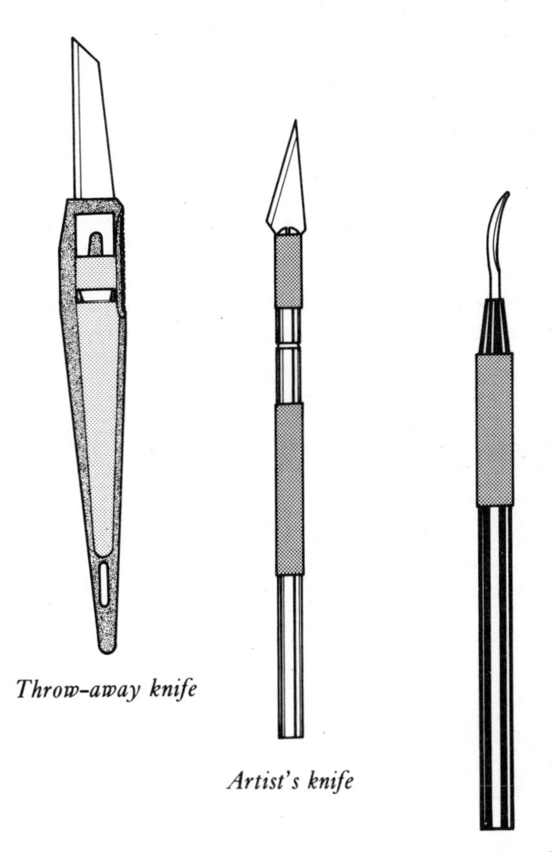

Throw-away knife

Artist's knife

Dry transfer burnishing tool

Care and maintenance of drawing equipment

Drawing equipment, instruments, aids and tools should be stored in clean and dry boxes or drawers. Wood faced drawing boards may have to be sanded smooth, flat and clean occasionally. The edges of Tee-squares may need to be straightened by planing if they become damaged. Set squares and drawing aids can be washed in warm water in which some drops of detergent have been placed. Technical pens must be thoroughly washed under running water to clean their parts of ink from time to time. If this is not carried out every two or three months the nibs of technical pens may become blocked with hard dried ink. When not being actually used, place the nib cap back over the nib to prevent drying out of the ink.

Lines, symbols and lettering

British Standards

Some of the drawings we produce should be drawn to the standards recommended by the British Standards Institute. Over 4000 British Standards have been published, some dealing with the graphics for industries such as engineering and building. Others deal with symbols for drawings in public and commercial undertakings. British Standards can be borrowed from public libraries. Parts of some British Standards are published in shortened form as PDs or PPs for use by students in schools and colleges.

Engineering drawings

Reference: British Standard BS 308: Parts 1, 2 and 3 *Engineering drawing practice*: PP 7308: *Engineering drawing practice for schools and colleges.*

Lines

Type of line	Application
Ⓐ Thick line	All drawing outlines
Ⓑ Thin line	Imaginary lines of intersection; dimension lines; projection lines; leader lines; Hatching lines
Ⓒ Thin line / Thin line	Limits of interrupted views or of partial views
Ⓓ Thick lines or Thin lines	Hidden detail lines. Note–in the U.K. common practice is to use thin hidden detail lines
Ⓔ Thin lines	Centre lines or lines of symmetry
Ⓕ Thick & thin lines	Section cutting plane lines
Ⓖ Thick lines	Surfaces which require special treatment (drawn adjacent to the surface)
Ⓗ Thin lines	Outlines of adjacent parts and alternative positions of moveable parts

Types of lines used in engineering and in building drawings

All lines should be black and dense whether drawn with pencil or with pen. Thick lines should be about twice as thick as thin lines. Note that in the United Kingdom it is common practice to use thin hidden-detail lines.

Scales

Recommended scales are:
Full size 1:1
Smaller than full size 1:2; 1:5; 1:10; 1:10; 1:50; 1:100; 1:250; 1:500 and so on.
Larger than full size 10:1; 5:1; 2:1.

Dimensioning

Some of the methods of dimensioning used on engineering drawings are shown below. Note the following.
1. Projection lines are thin with a 3 mm gap between the line and the drawing.

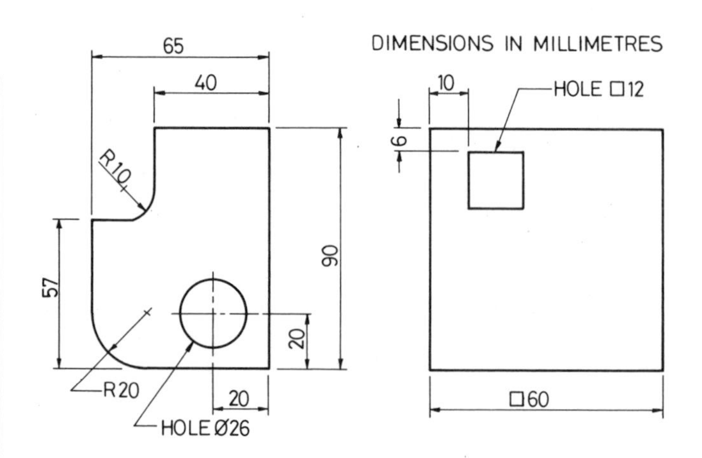

2. Dimension lines are thin and end in arrows 3 mm long.
3. Place dimensions above dimension lines to read from the sheet bottom or from the right.
4. Do not use centre lines as projection lines.
5. R, S, Ø and □ are placed in front of dimensions of radius, sphere, diameter and square.

R, S, Ø and □ are the abbreviations most often used with dimensions on engineering drawings. Others not so common are:
SR—sphere radius
SØ—sphere diameter
CSK—countersunk
CBORE—counterbore

Abbreviations of the units of dimensioning must be placed after the dimensions. Examples are:
35 mm—35 millimetres
10 m—10 metres
5 km—5 kilometres

An example of an engineering drawing

An engineering drawing is shown which includes all the types of line shown earlier on page 32. Leaders from each type of line end in circles (called 'balloons'), each enclosing a letter. A is a thick outline line; B is a thin hatching line, C is a thin limiting line; D is a thin, broken hidden-detail line; E is a thin centre line; F is a section plane cutting line; G is a thick chain line showing that the handle is to be bright polished; H is an alternate handle position line.

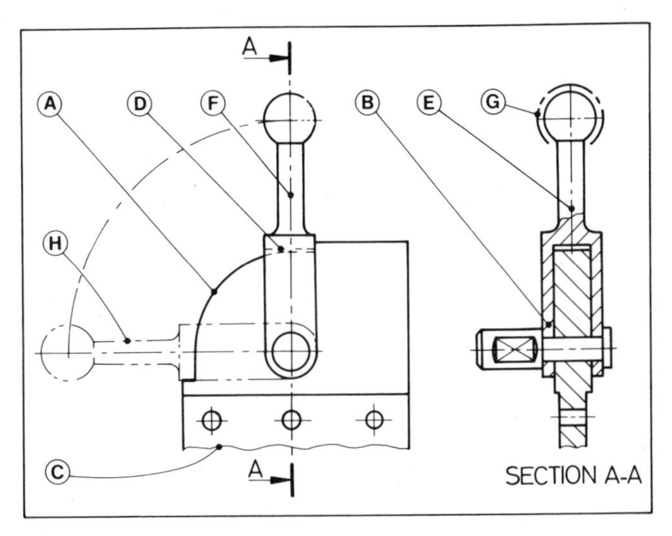

An engineering drawing showing use of lines

An example of a building drawing

The given drawing is the *site plan* of a bungalow. The encircled letters, with leader lines pointing to types of lines, should be compared with those given in the table on page 32.

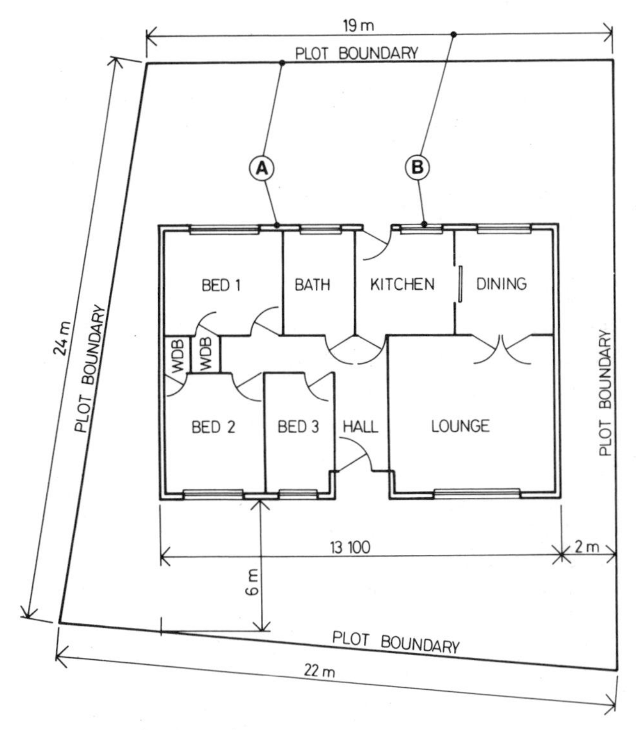

A site plan of a bungalow

Building drawings

Reference: British Standards BS 1192 *Construction drawing practice.*

Lines in building drawings should generally be of the same type as those for engineering drawings as given on page 32. Dimensions are, however, usually of a different type as shown in the examples.

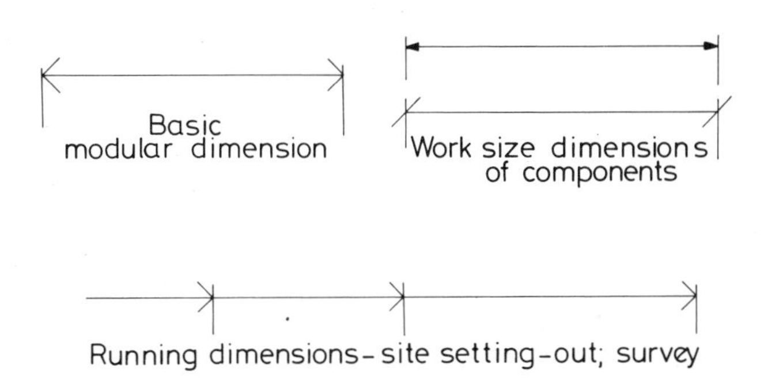

Symbols for electrical and electronics circuit diagrams

Reference: British Standard BS 3939: *Graphical symbols for electrical power communications and electronics diagrams.* The two tables show symbols used in electronics and in electrical circuit diagrams.

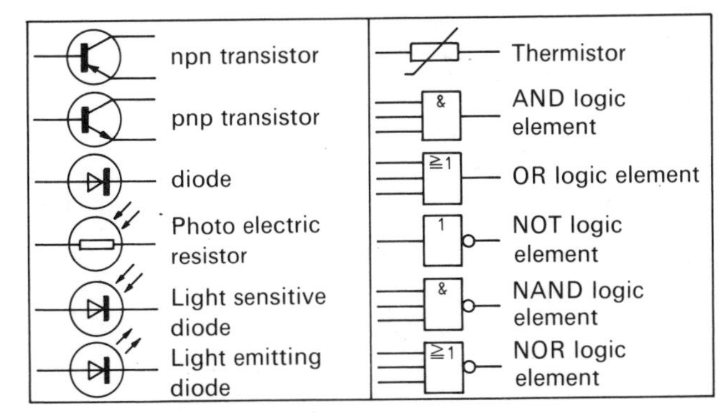

Symbols used in electronics circuit drawings

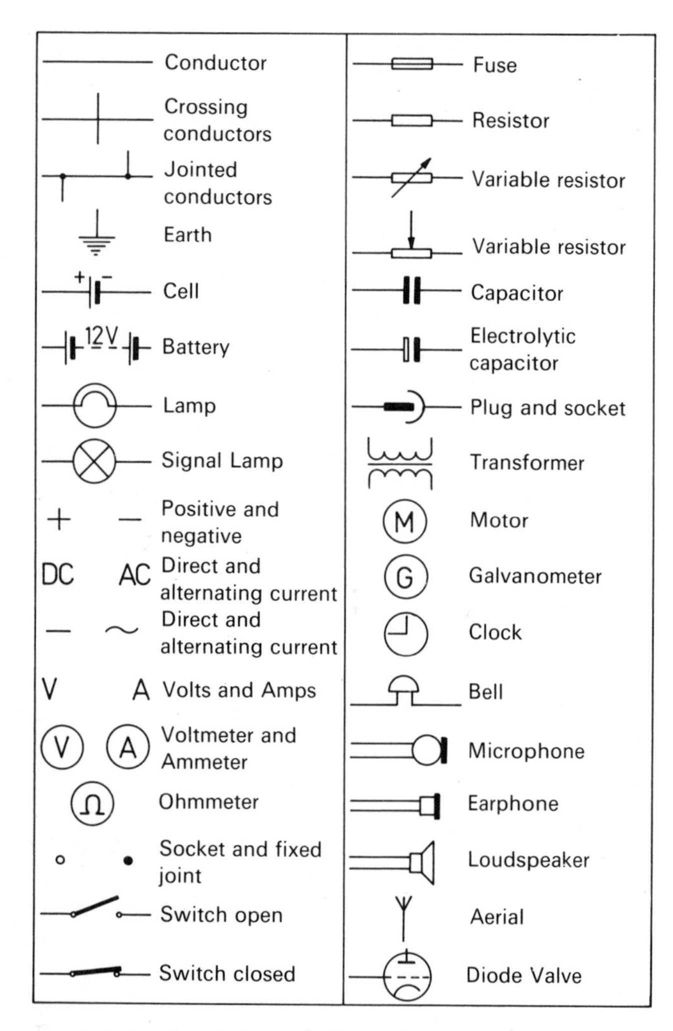

Symbols for electrical circuit diagrams

An example of an electronics circuit diagram

The given circuit diagram shows a circuit containing a battery, a switch, two npn transistors, a variable resistor, a photoelectric resistor, a fixed resistor and a light emitting diode (LED).

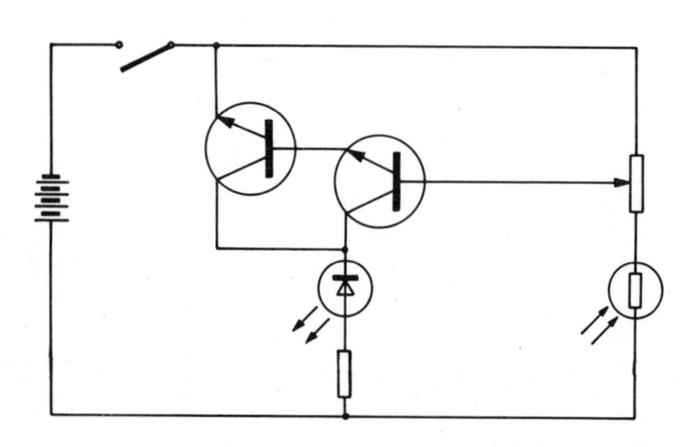

Flow chart diagrams

Reference: British Standard BS 4058: *Data processing flow chart symbols, rules and conventions.*
Some symbols from BS 4058 are given here. These symbols would be used in a flow chart describing a sequence of operations in a computer program. The flow diagram shows the symbols for Terminal/Interrupt, Process and Decision, used in a flow chart that describes the flow of operations involved in unpacking and checking the working of an electric washing machine.

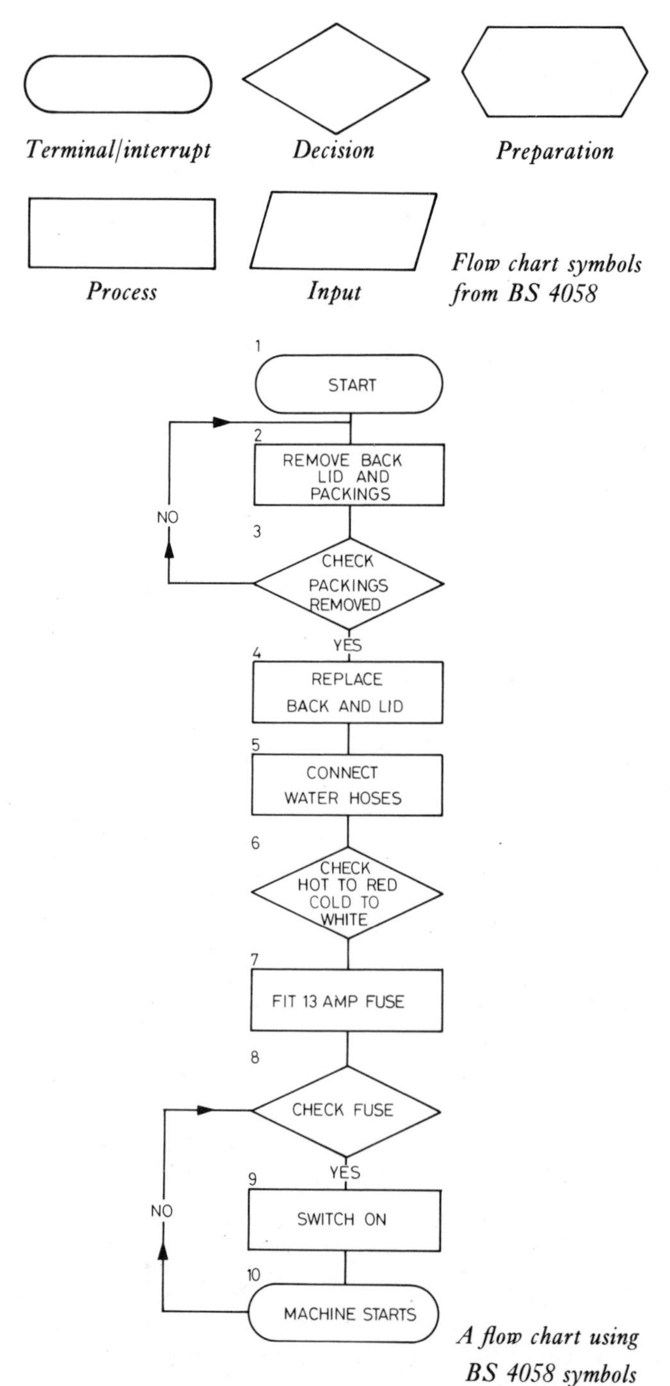

Flow chart symbols from BS 4058

A flow chart using BS 4058 symbols

Information symbols

Geometrically drawn symbols for conveying information are in common use throughout the world. They are displayed to attract attention, to advise, to instruct, to warn or to give an order. Information symbols must be clear in the message they give and should be designed with an economy of line and shape to make a quick, clear impact. Words should usually be unnecessary although they may be added to avoid confusion.

A small selection is shown here.

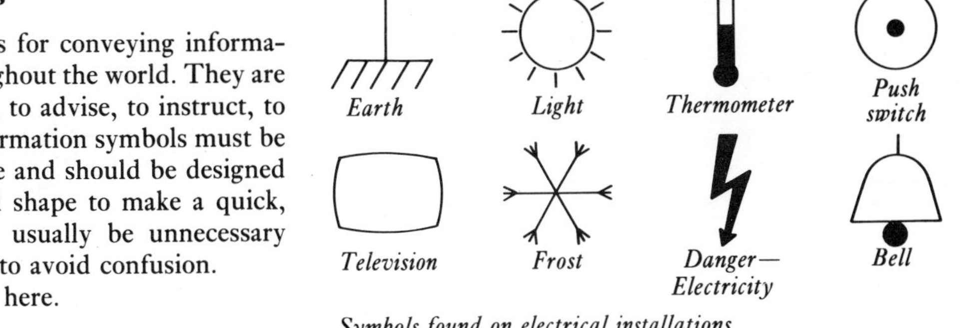

Symbols found on electrical installations

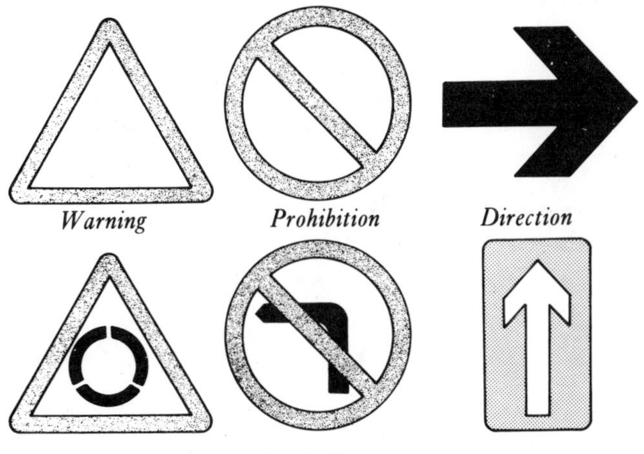

Warning and prohibitory symbols

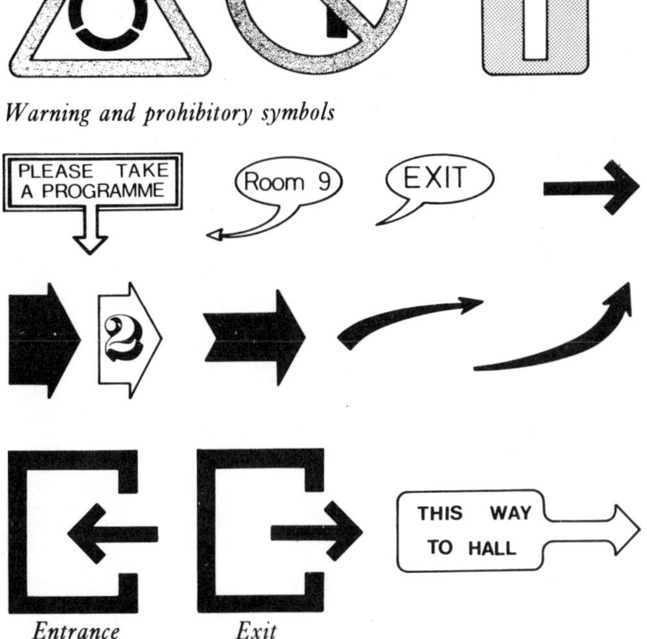

Some examples of arrow symbols

Cigarette lighter

Choke

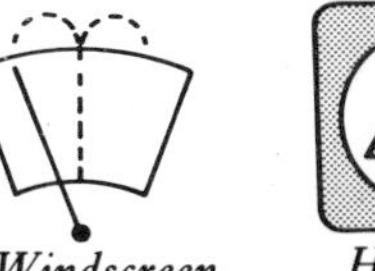
Windscreen wash

Hazard

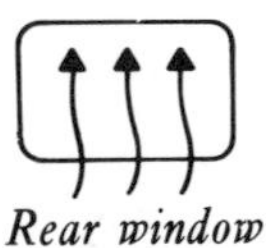
Rear window de-mist

Battery

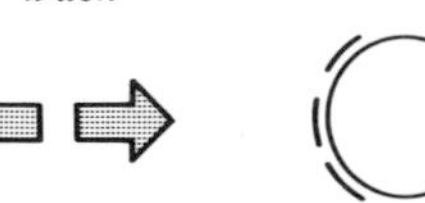
Direction indicators

Brake wear

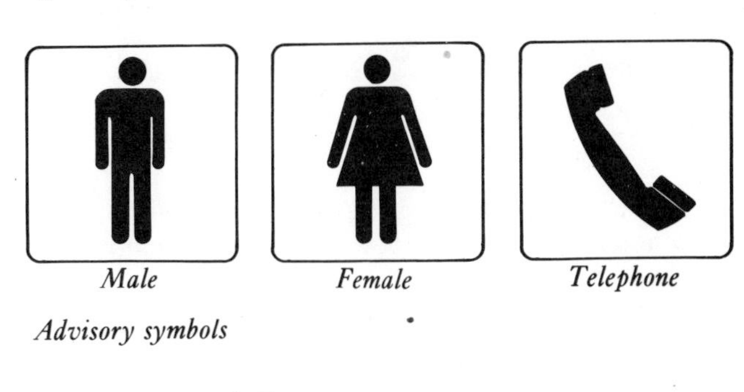

Advisory symbols

Letters and figures

1. *Freehand* – with practice, good, neat lettering can be achieved.
2. *With lettering stencils* – The photograph shows letters being added to a drawing with a technical pen with the aid of a lettering stencil.

Adding lettering to a drawing with the aid of a Rotring letter stencil

LETTERING BY STENCIL 1234567890

Another example A third example 1234567890 And a fourth 1234567890

LETTERING WITH ROTRING STENCILS 567890

Graphical communication 1234567 A SMALLER LETTER

and smaller (!:;+=-&?%) A DIFFERENT STYLE 01234567% Same style but smaller 1234567

Drawing a circuit diagram with the aid of a template

3. *From dry transfer sheets* A very large range of figures and letters in a variety of typefaces and heights can be obtained from sheets. The photograph shows letters being added to a drawing from such a sheet.

Applying dry transfer letters with a ballpoint pen

Dry transfer sheets

A number of firms publish dry transfer sheets. Three main groups are produced as aids to those engaged in producing graphics. These are of great value in helping to achieve good quality graphics. The three groups are:
1. *Letters and figures* – Some examples are shown.
2. *Tones and shadings* – Some drawings in this book have been shaded with dry transfer tones.
3. *Architectural and other symbols* – Some examples of architectural symbols from the Letraset range are given. Others such as arrows, border patterns, electrical, electronic, pneumatic and hydraulic circuit symbols are produced. In addition, human figures in different poses and drawn to a variety of scales can be added to sheets to give reality to drawings.

PRINT and print

LETTERS figures

TRANSFER ~ ~ Horatio

DRY TRANSFER .

THORNE 456789

COMPUTER graphics 123456

DRY TRANSFER Helvetica medium; Figures 123456789

HELVETICA LIGHT; graphical communication; 123456789

Examples of different typefaces from dry transfer sheets

Examples of Letraset architectural symbols from dry transfer sheets

Dry transfer sheets – letters, tones and symbols – are available in colours other than black.

Basic geometry

Lines

Bisection of a line

Draw AB. Set a compass to about $\frac{2}{3}$ of AB. With the compass centred first at A, then at B, draw two sets of crossing arcs C and D. Draw a line between C and D. E is the centre of AB. AE = EB.

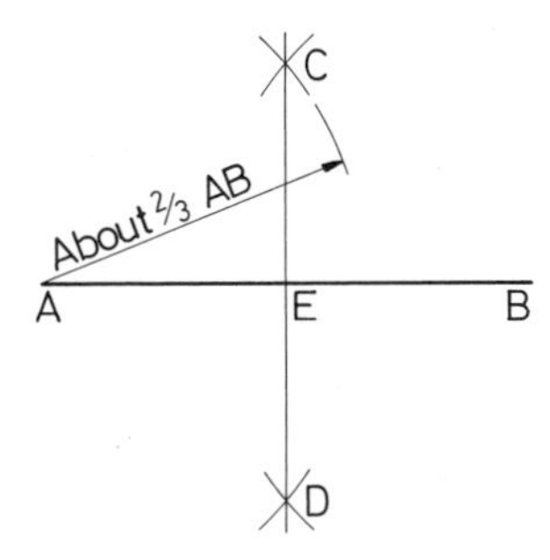

Bisection of a line

Division of a line into four equal parts

Draw FG. Bisect FG to give J. Bisect FJ to give H and bisect JG to give K. FH = HJ = JK = KG.

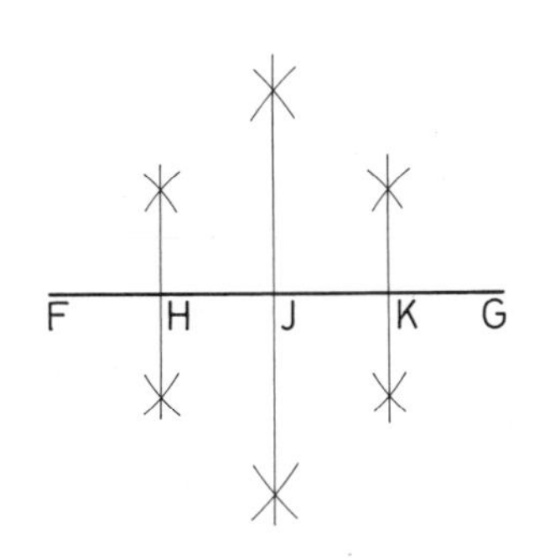

Division of a line into four equal parts

Division of a line into any number of equal parts

Stage 1 – Draw AB. From A draw AC at any convenient angle to AB. With a compass set to any convenient size mark off five equally spaced points 1, 2, 3, 4 and 5 along AC. Draw line B5. With the aid of a set square and straightedge draw lines parallel to B5 through points 4, 3, 2 and 1.
Stage 2 – The completion of dividing AB into five equal parts.

Stage 1

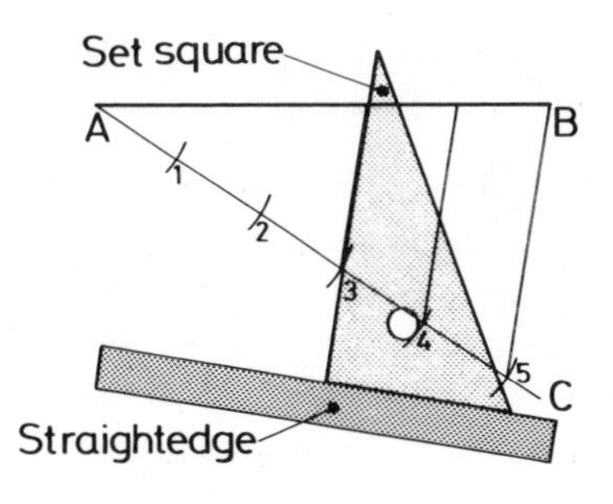

Stage 2

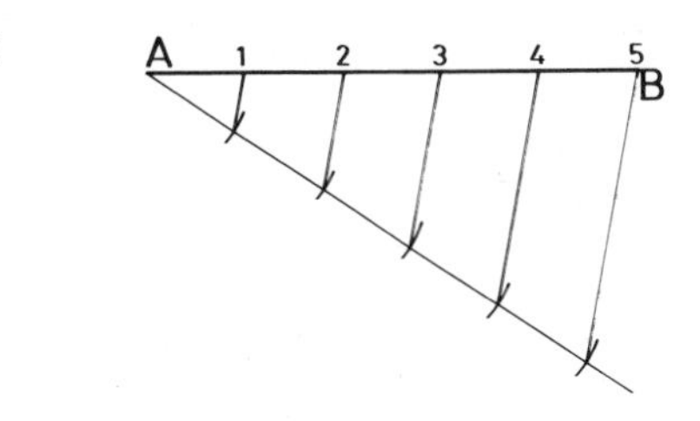

Division of a line into equal parts

Scales

Two scales are given – 1:5 (one fifth size) and 50 mm ≡ km (1:20 000). In both cases a dimension taken from the scales is shown.

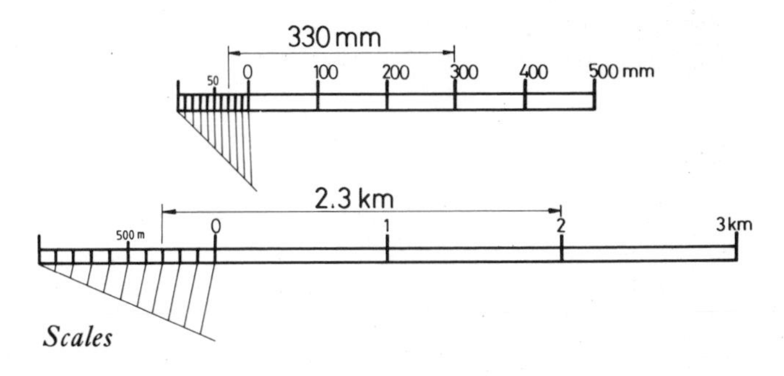

Scales

Angles

Angles of less than 90° are known as *acute* angles. Those of between 90° and 180° are called *obtuse* angles. Note how angles are dimensioned.

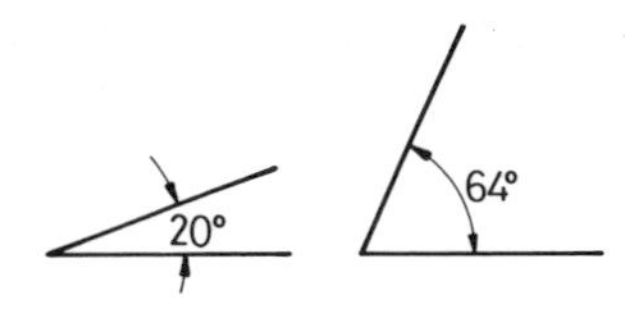

Acute angles (less than 90°)

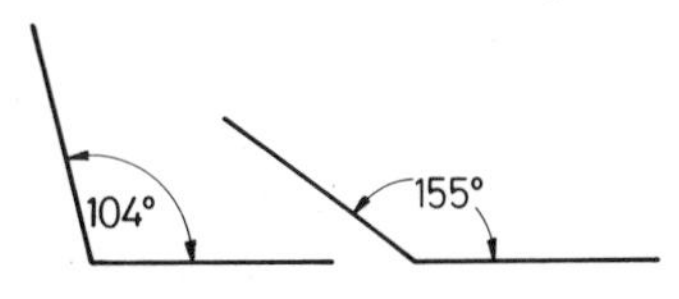

Obtuse angles (between 90° and 180°)

Construction of an angle with the aid of a protractor

Any angle can be drawn with the aid of a protractor. An example (42°) is given. Draw AB. Set the protractor with its base line exactly over AB with the centring mark of the protractor exactly on A. Draw a pencil mark against the required angle dimension – in this example 42°. Remove the protractor and draw a line from A through the pencil mark to obtain angle BAC of 42°.

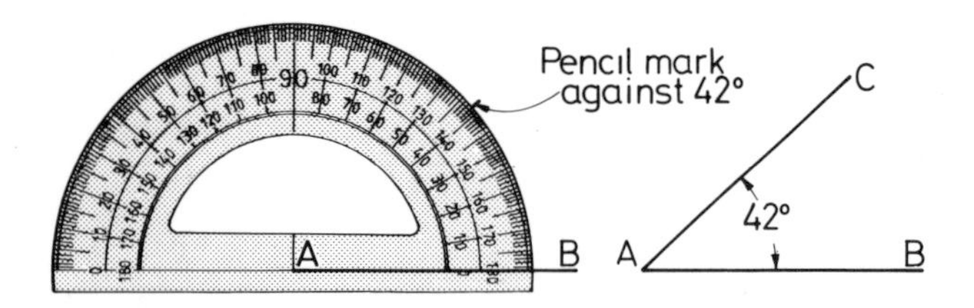

Angles drawn with the aid of set squares

Angles of 30°, 45°, 60° and 90° can be easily and speedily drawn with the aid of the two common set squares. Obtuse angles of 120°, 135° and 150° can also be drawn with their aid. If two set squares are employed, placed edge to edge, angles such as 15°, 75°, 105° and 165° can also be drawn using set squares.

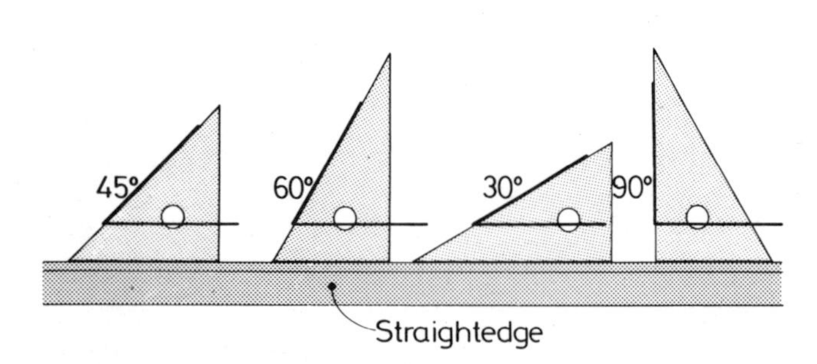

Construction of 60° angle

The construction of a 60° angle follows on from the facts that there are 360° in a circle and the radius of a circle can be stepped off six times around its circumference.

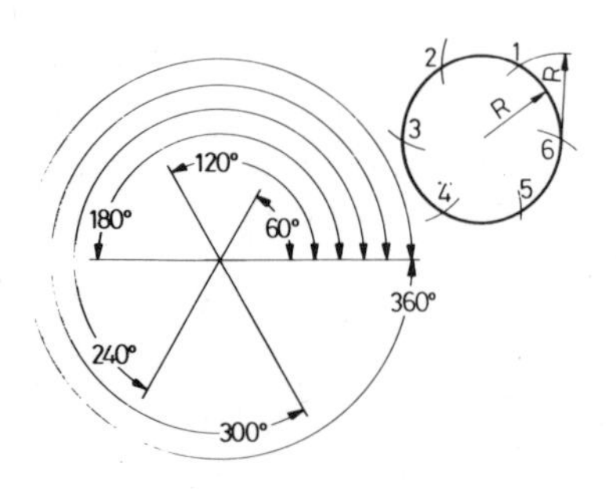

Angles in a circle

Draw AB. Set a compass to any convenient radius (greater than 25 mm). With the compass centred at A draw an arc of radius AE. Without altering the compass and with its centre at F, draw a second arc to cross the first arc. Draw AC through the point at which the arcs cross.

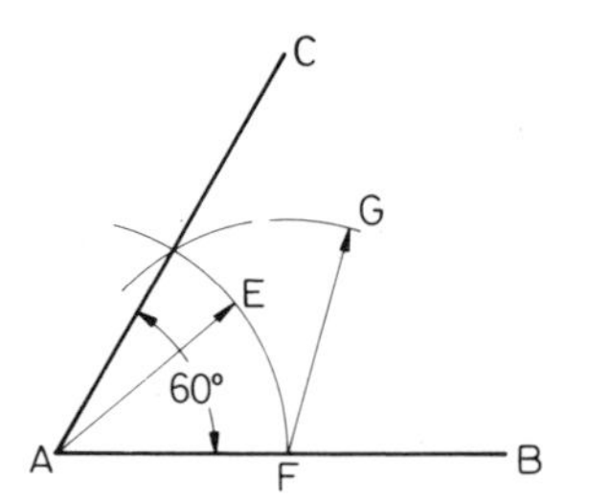

Construction of a 60° angle

Bisection of an angle

Draw the angle BAC. Set a compass to any radius greater than 25 mm. With the compass centred at A draw an arc to cross the arms of the angle at E and F. The compass can now be re-set if it is thought necessary. With centre E draw an arc; without resetting the compass and with centre F draw an arc crossing that centred at E. This gives G. Draw AD through G. Angle BAD = angle CAD.

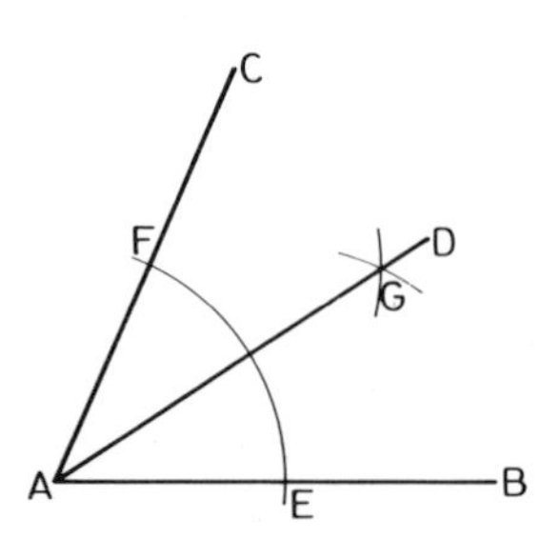

Bisection of an angle

Construction of 30° and 90° angles

To construct a 30° angle bisect a 60° angle.

To construct a 90° angle bisect the angle formed by the construction of 60° and 120° angles.

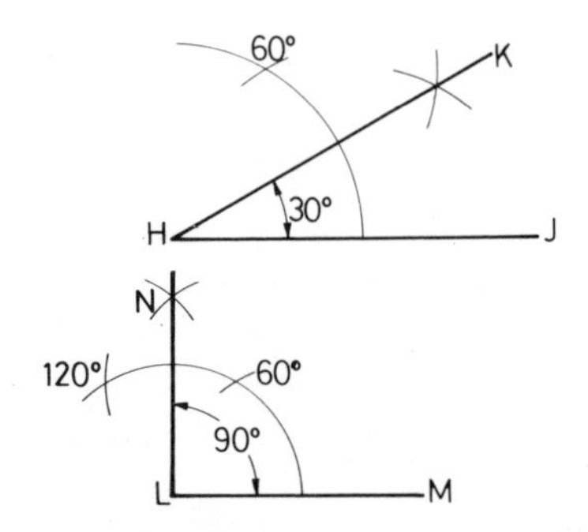

Triangles

The three angles of any triangle add up to 180°. The four types of triangle are:
Equilateral – All three sides of equal length. Each angle is 60°.
Isosceles – Any two sides are equal in length. The two angles opposite to the equal sides are equal in size.
Right-angled – One angle is a right angle (90°).
Scalene All angles are of different sizes. All sides are of different lengths.

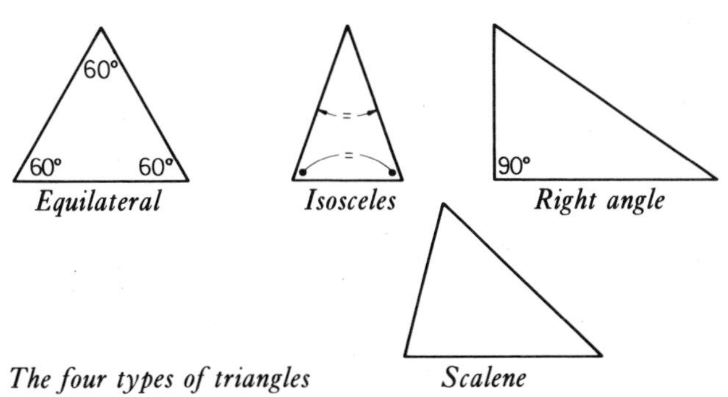

The four types of triangles

Construction of triangles

Only the most commonly used methods are shown.

Equilateral
Any size of equilateral triangle may be constructed with the aid of a 30°, 60° set square.

Isosceles
If the length of the two equal sides is known, draw the base AB. Set a compass to the length of one of the equal sides. With the compass centred first at A, then at B, draw crossing arcs to give C.

If the two equal angles are known, draw AB. Construct the angles at A and at B with the aid of a protractor.

Right-angled
Can be constructed with the aid of a set square.

Scalene
Draw CD. Set a compass to the length of side CE and, with the compass centred on C, draw an arc. Set the compass to the length of side DE and, with the compass centred on D, draw a second arc crossing the first at E. Complete the triangle.

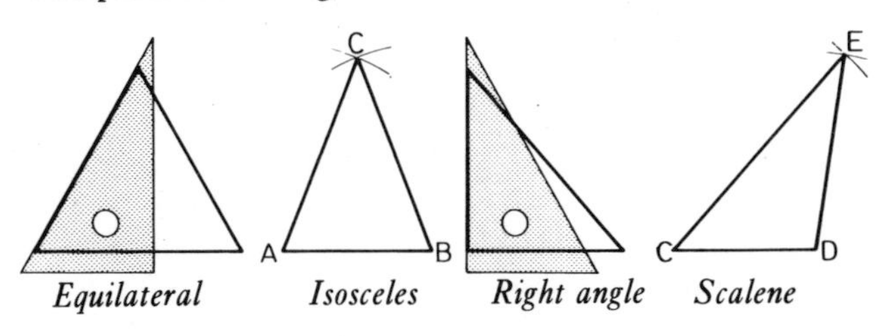

Construction of circles to triangles

To draw the *inscribed* circle to any triangle, bisect two of its angles. The *in-centre*, I, is at the point where the bisectors meet.

To draw a *circumscribed* circle to any triangle, bisect any two sides. The *circumcentre*, C,* is at the point where the bisectors meet.

The circumcentre of a right-angled triangle is at the bisection point of its *hypotenuse*.

Note: The angle within a semicircle is always a right angle.

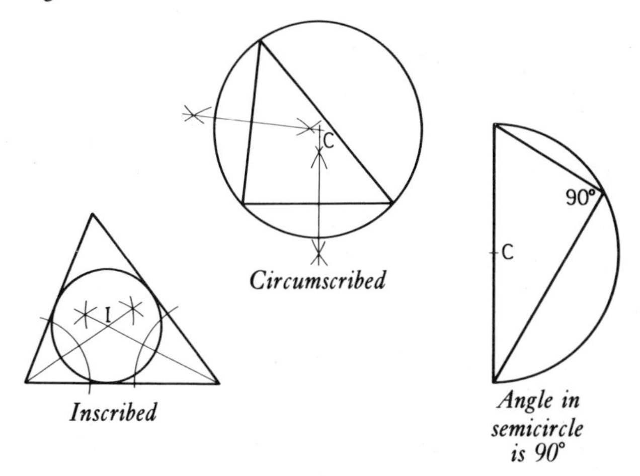

Construction of circles to triangles

Polygons

A polygon is a figure bounded by lines. The lines may be straight or curved. Here we are only concerned with straight-sided polygons. Polygons may be either *regular* or *irregular*. Regular polygons are those with all sides of equal length and also with all angles of equal size. Irregular polygons are those with sides or angles not all equal in length or size. Note that *both* conditions of equal side length and equal angle size must be met for a polygon to be classed as regular.

Parallelograms

Parallelograms are four-sided polygons (quadrilaterals), with straight sides in which opposite pairs of sides are parallel. Squares and rectangles are special forms of parallelogram in which all angles are right angles. Only the square, of the four polygons shown, is a regular quadrilateral.

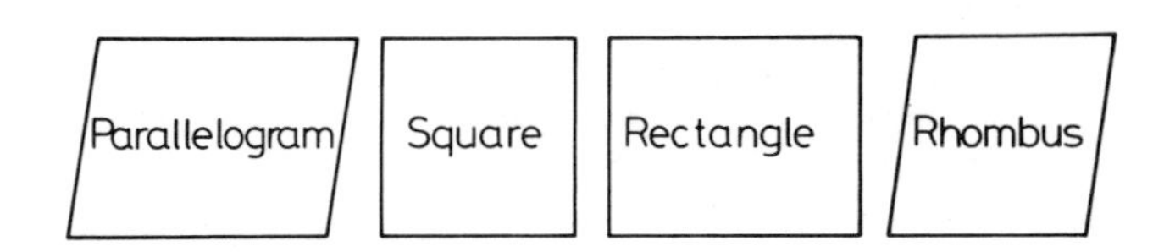

Regular polygons

Pentagon (five sides) – Each angle is 108°.
Hexagon (six sides) – Each angle is 120°.
Octagon (eight sides) – Each angle is 135°.
Heptagon (seven sides) – Each angle is $128\frac{4}{7}$°.

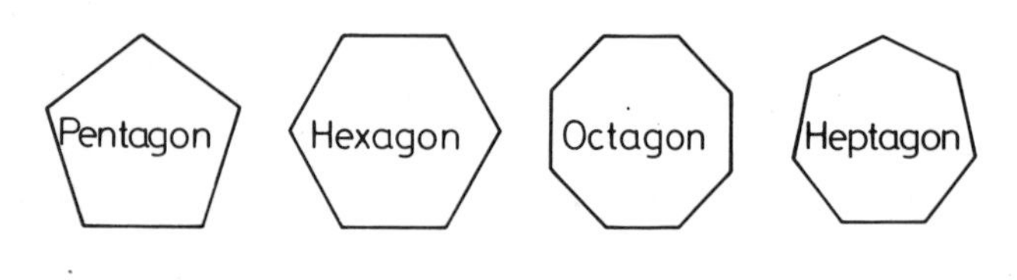

Irregular polygons

Four irregular polygons are shown.

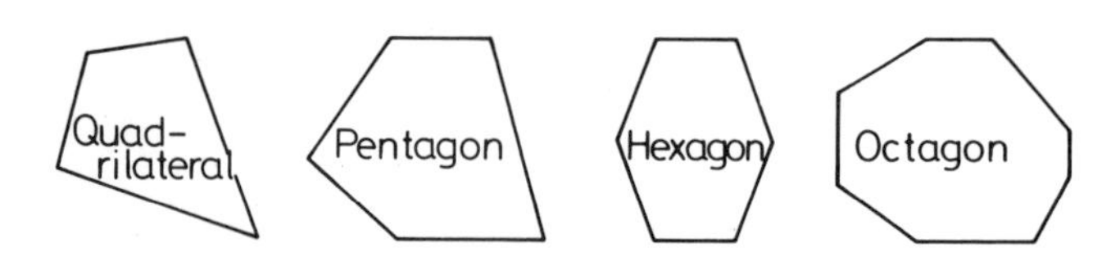

Construction of parallelograms

Squares and rectangles can be constructed with the aid of set squares to obtain the 90° angles, or the angles can be constructed with a compass. Side lengths are then marked off with the aid of a compass. Equal side lengths of a square can be obtained with the aid of a 45° set square.

Parallelograms can be constructed by first drawing one of the angles with the aid of a protractor and then stepping off side lengths with a compass.

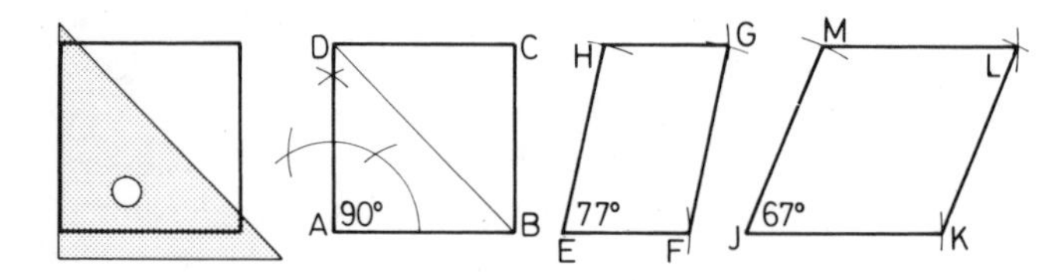

Construction of a regular pentagon

Method 1, with a protractor
Draw base AB. Draw angles of 108° at both A and at B. Set a compass to AB. With the compass, from A mark off E and from B mark off C. From C and from E mark off D.

Method 2, with set squares
Draw base FG. Draw 60° angles to FG at F and at G. to meet at H. Draw 45° angles to FG at F and at G to meet at J. Bisect HJ to give K. With a compass centred at K and set to KF (or KG) draw a circle. Re-set the compass to base length FG and step off side lengths around the circle.

Method 3, within a circle
Draw a circle with centre P. Draw the diameter QPR. Find RT which is $\frac{2}{5}$ of QR. With a compass set to QR and centred first at Q, then at R, draw arcs to meet at S. Draw ST and extend to meet the circle at U. Then RU is one side of the pentagon. Step off the length RU around the circle to give V, W and X.

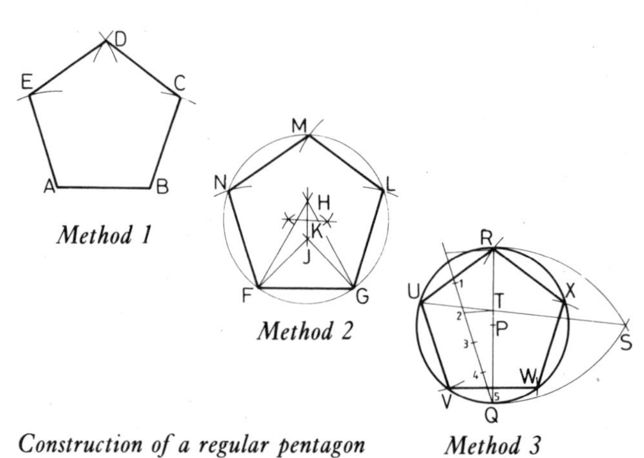

Construction of a regular pentagon

Construction of a regular hexagon

Method 1, within a circle
Draw a circle of radius equal to the side length of the required hexagon. Draw a diameter AOB. Without re-setting the compass and centring it first at A, then at B, draw arcs across the circle to give points 1, 2, 3 and 4. Now draw the hexagon.

Method 2, another method
Draw a circle of radius equal to the hexagon's side length. Draw a diameter COD. Complete the regular hexagon with the aid of a 30°, 60° set square.

Method 3, circumscribing a circle
Draw the circle. With the aid of a 30°, 60° set square draw a regular hexagon whose sides touch the circle. Note that in this example, the circle diameter is equal to the *Across Flats* (A/F) size of the hexagon.

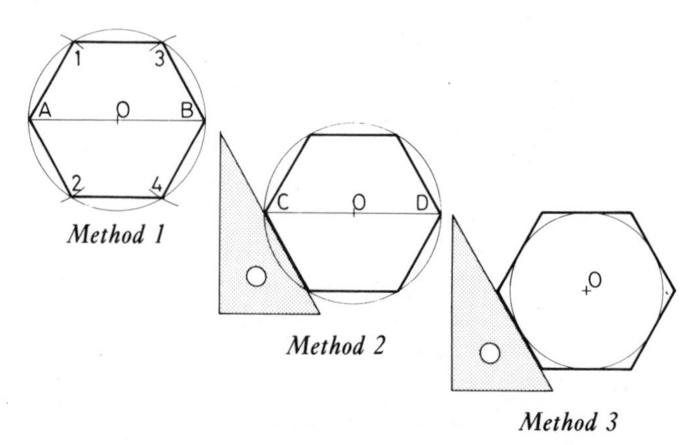

Construction of a regular hexagon

Construction of regular octagons

Method 1, within a square
Draw the square ABCD. Draw its diagonals AC and BD. With a compass set to AO (or to BO, CO or DO) draw arcs centred at A, B, C and D in turn. Complete the octagon as shown.

Method 2, circumscribing a circle
Draw the circle. With the aid of a 45° set square draw a regular octagon whose sides touch the circle.

Method 3, with a set square
Draw side AB. Construct the octagon with the aid of a 45° set square, stepping off the side lengths with a compass.

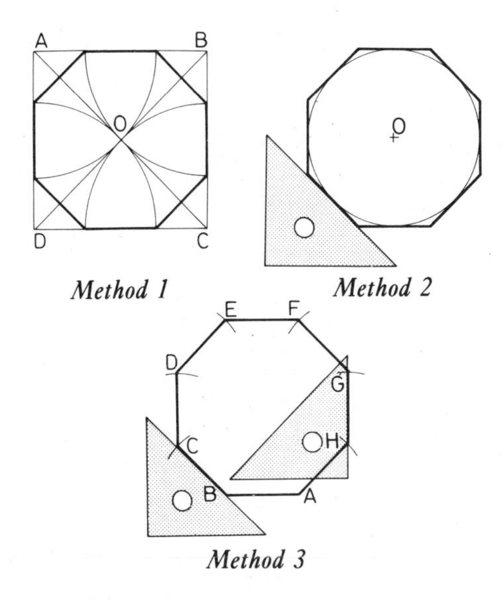

Construction of a regular octagon

Enlarging irregular polygons

Method 1
Draw the required polygon (shaded) of base side length AB. From A draw the diagonals to and through the vertices of the polygon. Mark off the new, enlarged side length AB_1. Draw lines 1, 2 and 3 parallel to the equivalent lines of the original polygon.

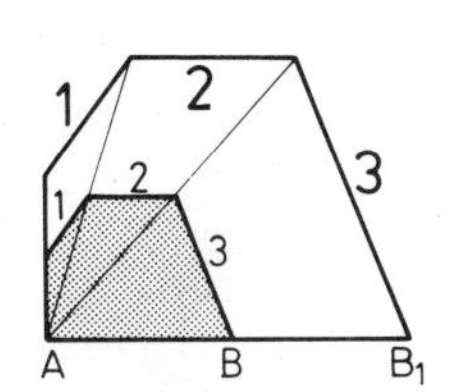

Method 2
Draw the polygon (shaded) of base CD. Draw the base C_1D_1 at the required enlarged scale. Select a *pole*, O. From O draw lines to and through the vertices of the original polygon. Draw lines from C_1 and D_1 and lines 1, 2 and 3 parallel to the equivalent lines of the original polygon.

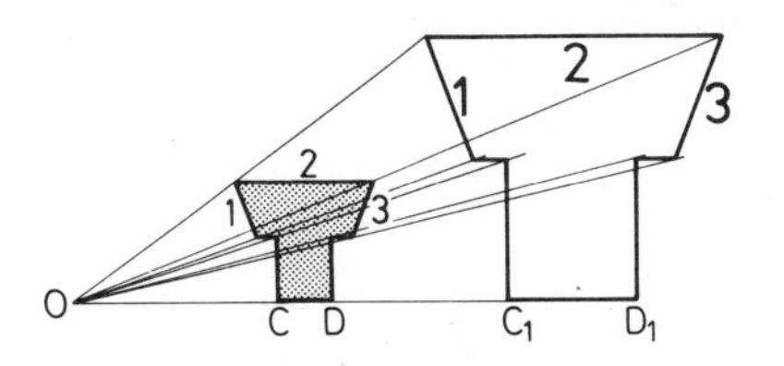

Applications of the two methods of enlargement (or reduction) of polygons are given.

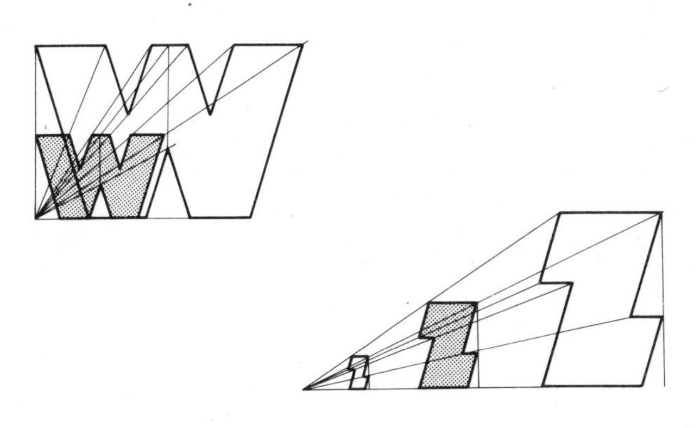

Applications of enlarging (or reducing) polygons

Circles

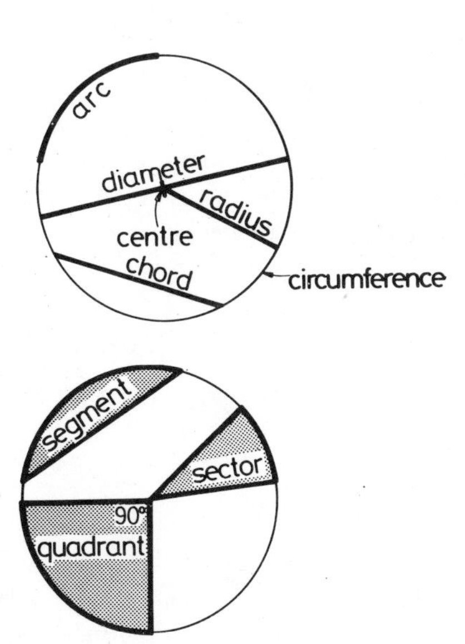

Parts of a circle

Constructions of straight line tangents to circles

At a point on the circle
T is the point at which the tangent is to be drawn. Draw the radius CT. At T construct a right angle to CT to give the tangential line TA.
Note: A straight line tangent touches a circle at a point where the tangent and the radius at the point form a right angle.

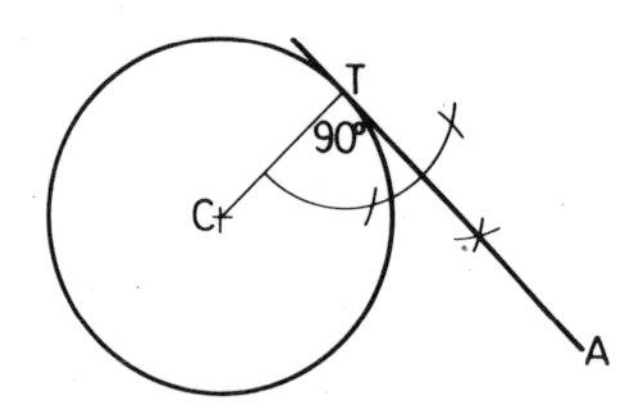

Constructions of tangents between arcs of circles

From a point outside the circle
Draw the circle. P is the point from which a tangent to the circle is required. Draw OP. Bisect OP to give S. Draw a circle on OP with centre S. T, where the semicircle crosses the circle, is the point at which the tangent PT touches the circle. Note the second tangent PT_1.

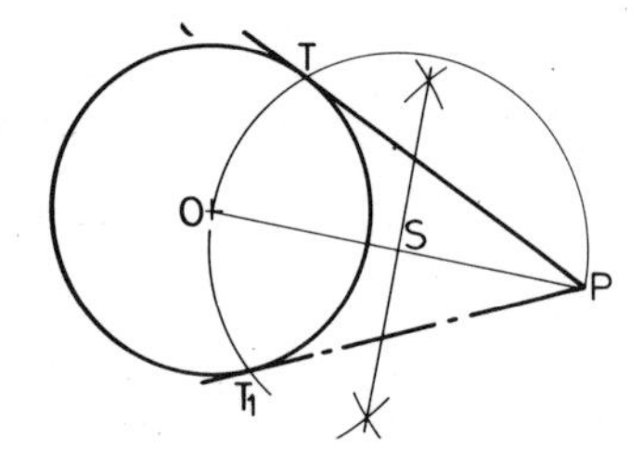

Constructions for joining arcs of circles

Arcs touching externally
Draw a circle of centre C. *Add* the radii of circle C and circle O. Draw a line through CT. With a compass centred at C, and of radius equal to the sum of the two circle radii strike an arc along CT to give O.

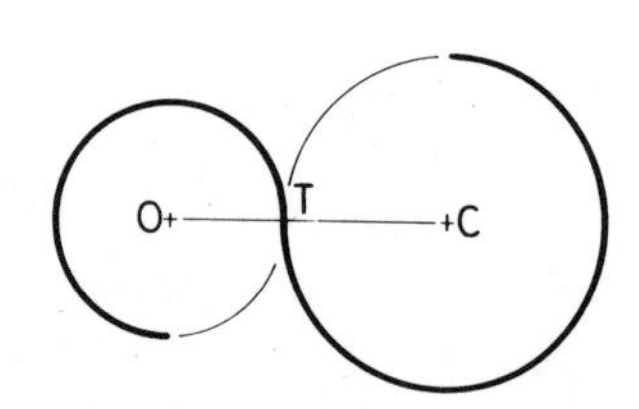

Note: To draw externally touching arcs, the radii of the two arcs must be *added.*

Arcs touching internally
Draw a circle of centre A. Draw the two lines AT_1 and AT_2. Set a compass to the *difference* between the two radii of circle A and circle B. With centre A and radius equal to the difference of the two circle radii, strike arcs along AT_1 and AT_2 to give B and D.

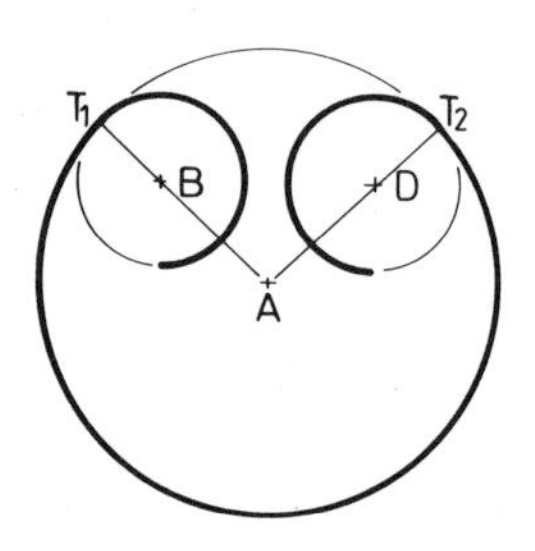

Note: 1. To draw internally touching arcs, the radii of the two arcs must be *subtracted.*
2. A line between the centres of tangential arcs will pass through the point of tangency of the tangential arcs.

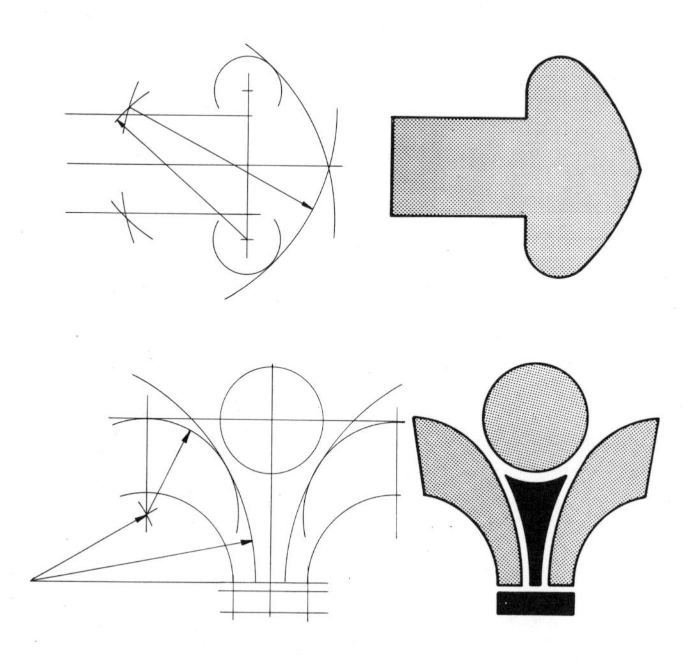

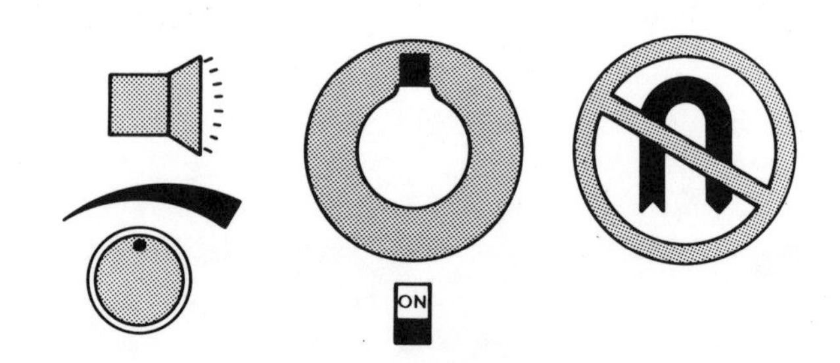

Some applications of circles and arcs to signs and symbols

Constructions for radiused corners

At a right-angled corner
Draw the right-angled corner. With a compass set to the required radius and centred at A, draw arcs across the arms of the right angle to give B and C. Without altering the compass, and with it centred first at B, then at C, draw crossing arcs to give D. D is the required centre of the corner radius curve.

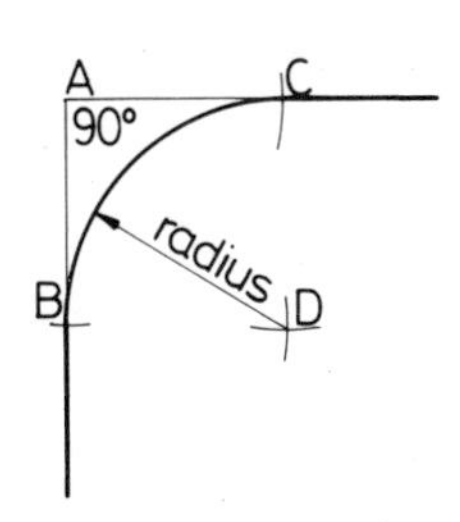

At obtuse or acute-angled corners
Draw lines 1, 2 and 3 parallel to and radius distance from the arms of the corner angles. Line 1 is parallel to EF; 2 is parallel to FG; 3 is parallel to GH. Where the parallels meet are the points for the radii centres.

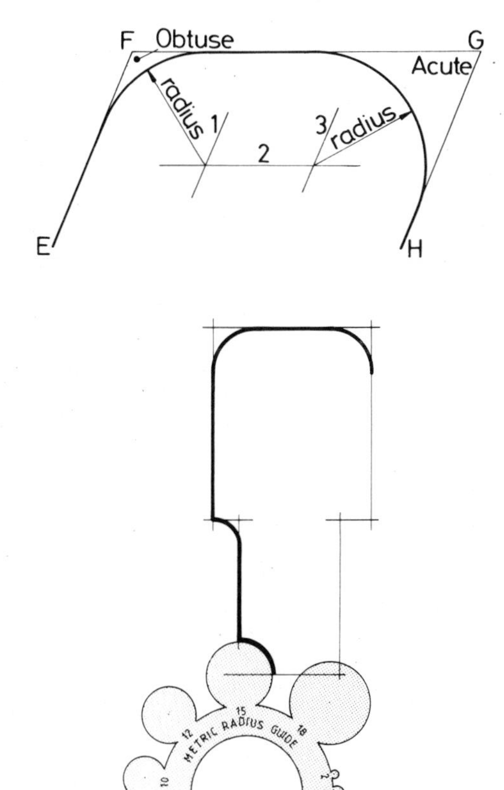

Drawing radius corners with the aid of a radius guide

Areas of plane figures

Arithmetical caclulations of areas of plane figures

The areas of polygons with straight sides and the areas of circles can be calculated by the use of simple arithmetic as shown.

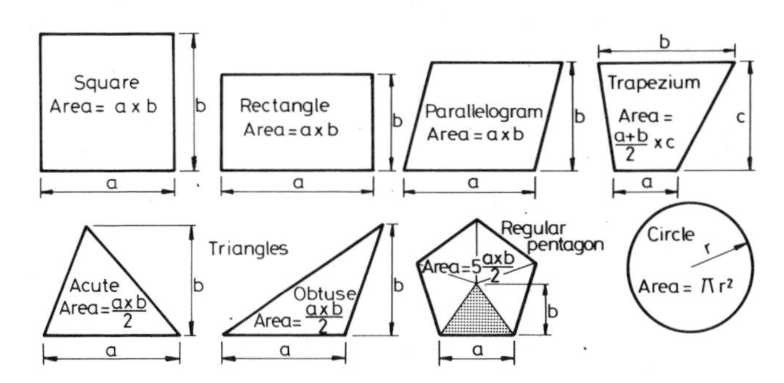

Arithmetical calculations of areas of plane figures

Note: 1. It is always the vertical height of the figure, as measured at right angles to its base, which is taken as one of the dimensions when finding the area of polygons with straight sides.
2. The area of any straight-sided polygon can be calculated by dividing it into the same number of triangles as the polygon has sides, and then adding the areas of the triangles so obtained. If the polygon is regular, it can be divided into triangles of equal areas by lines drawn from its vertices to its centre.

Calculations of areas of polygons by construction

Another method of finding the area of irregular polygons is to 'reduce' the polygon to a triangle of the same area and then calculate the area of the triangle. The method is shown. In the two given examples DE is parallel to CA, KL is parallel to JF and HM is parallel to JG.

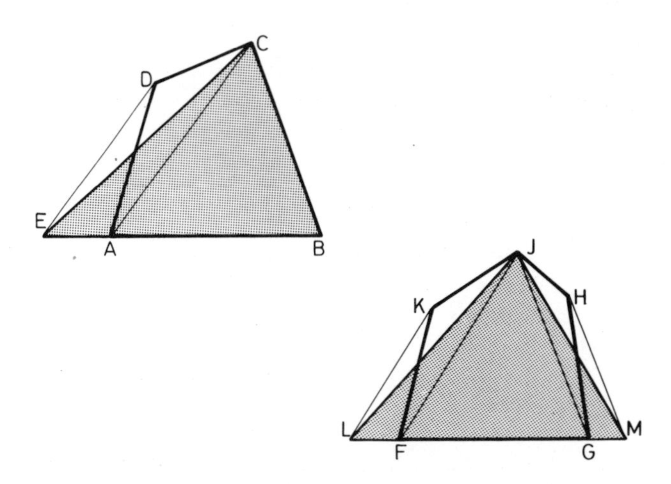

Calculations of areas of polygons by construction

Calculation of areas of irregular shape

By mid-ordinate method
Two examples are given. The mid-point of each centimetre is marked and *ordinates* are drawn at right angles to the base through the mid-points. The length of each ordinate across the figure is then measured in centimetres. These are added together to give the required area in square centimetres. In the first example the addition of the ordinate lengths is: $0.5 + 1.5 + 2.6 + 4 + 6 + 8 + 4.5 + 1 = 28.1$ square centimetre (cm^2).

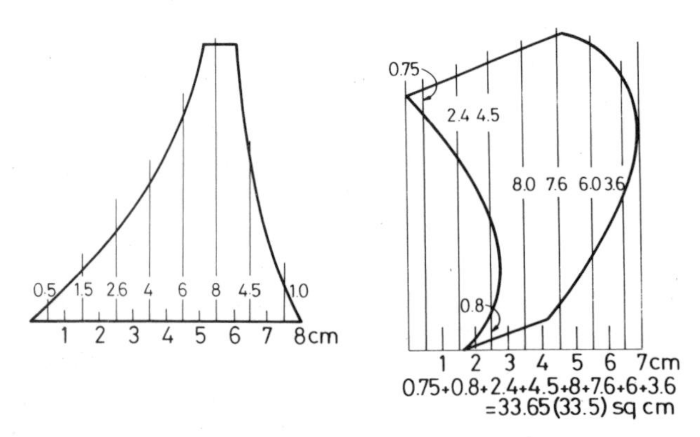

Calculations of areas by mid-ordinate methods

Note: The mid-ordinate method can only produce approximate results. In the given example the area could be quoted as 28 cm^2.

By adding squares
The square grid method of finding the area of irregularly shaped figures is shown by two examples. The figure is 'boxed' in a grid of squares. The edges of the squares may be of any unit length – centimetres, or scaled in metres, feet, kilometres or miles. The number of complete squares in the figure is then counted – in the first example it is 20. The remaining area of the figure outside the complete 20 squares is then found by estimating the area of each part square and adding the resulting fractions. In the first given example it is estimated that the fractional parts of squares add up to approximately 8.5. Thus the total area would be $20 + 8.5 = 28.5$ of square units. The calculation of the area of the second figure is left as an exercise.

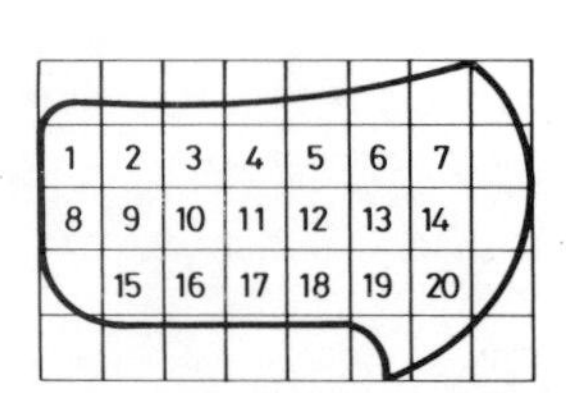

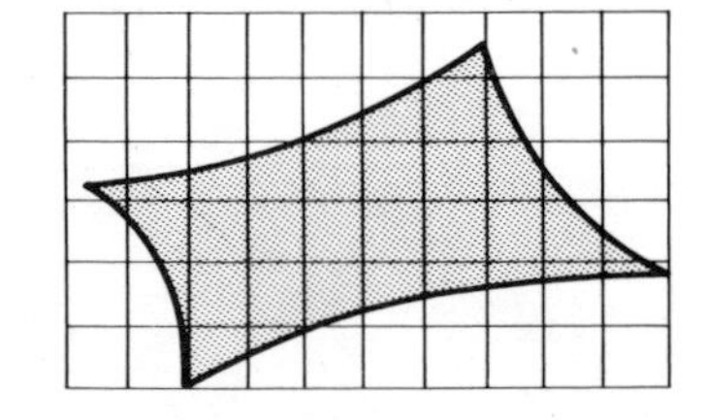

Calculation of areas from square grids

Conic sections

Ellipses, parabolas and hyperbolas are included in the group of curves known as conic sections.

Ellipses

Ellipses are centred on two *axes* at right angles to each other. The longest of the two axes is the *major* axis and the shorter is the *minor* axis. An ellipse can be defined as *the locus of a point which moves so that the sum of its distances to two fixed foci is constant and equal to the length of the major axis.* The two foci are named f_1 and f_2.

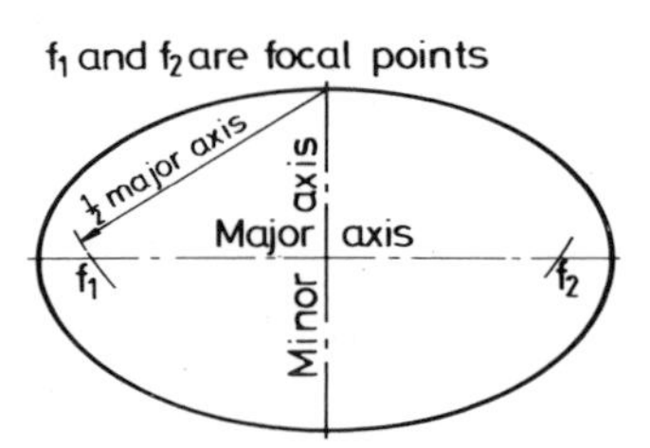

Parts of an ellipse

Construction of an ellipse: Method 1
Draw the major and minor axes. Draw two concentric circles on these axes as diameters. Draw a number of diagonals across both circles. One such diagonal is ACDB. At A and at B draw lines parallel to the minor axis. At C and at D draw lines parallel to the major axis. These lines meet at E and F. E. and F are points on the ellipse. Draw a good freehand curve through all the points so obtained.

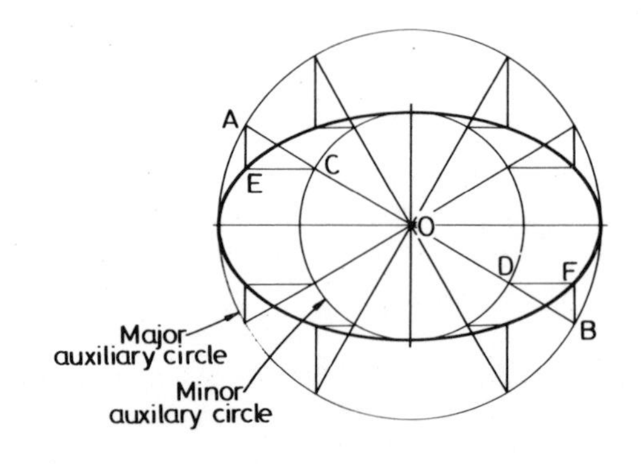

Construction of an ellipse: Method 2
Draw the major and minor axes. Mark f_1 and f_2. Make up a loop of twine of circumference equal to the major axis plus the distance f_1 f_2. Position pins at f_1 and f_2. With the loop of twine placed over the pins, an accurate ellipse can be drawn with a pencil moving within the loop, holding the twine taut as the ellipse curve is drawn.

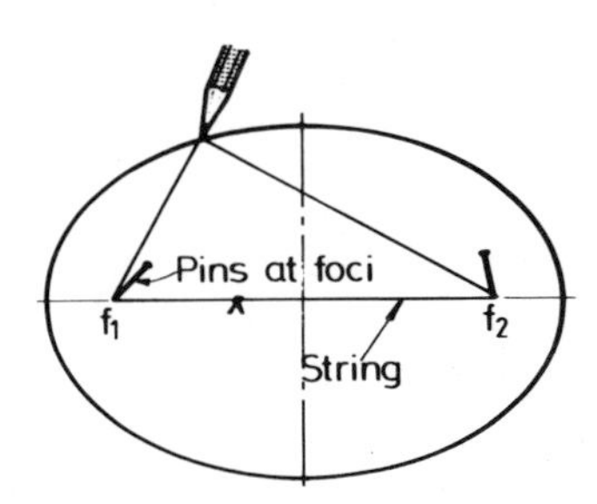

Construction of an ellipse: Method 3
This construction is also based on the two focus points. The arcs a, b, c, d, e and f are all such that:

$$f_1a + f_2a = \text{major axis} = f_1b + f_2b,$$

and so on.

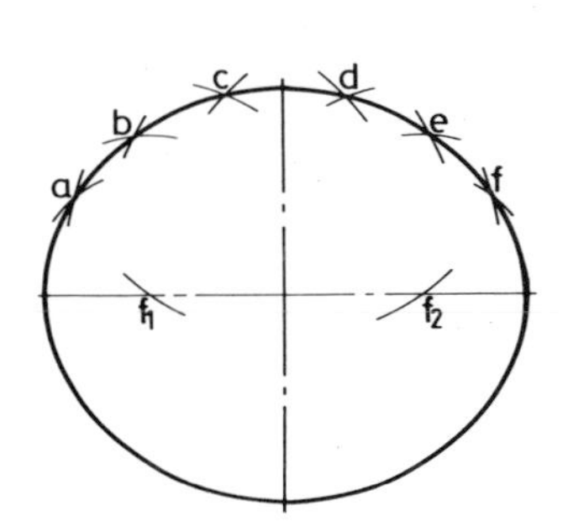

Parabola and hyperbola

A *parabola* can be defined as a curve which is the locus of a point which moves so that its distance from a fixed point (f) is always the same as its distance from a fixed line (DD_1).

A *hyperbola* can be defined as a curve which is the locus of a point which moves so that its distance from a fixed point (f) is always greater than its distance from a fixed line (DD_1) by a given ratio.

The following terms are applied to the two curves: directrix (DD_1), axis (at right angles to the directrix), focus (f), vertex (V).

Construction of a parabola given directrix and focus
Draw DD_1. Draw an axis at right angles to DD_1. Locate the focus f. Construct V, the mid-point of Af. Draw lines 1, 2, 3, 4 and 5 parallel to DD_1. Set a compass to the distance between line 1 and DD_1. With centre f, draw arcs crossing line 1. Repeat for each line 2, 3, 4 and 5. Draw a fair curve through the points so obtained.

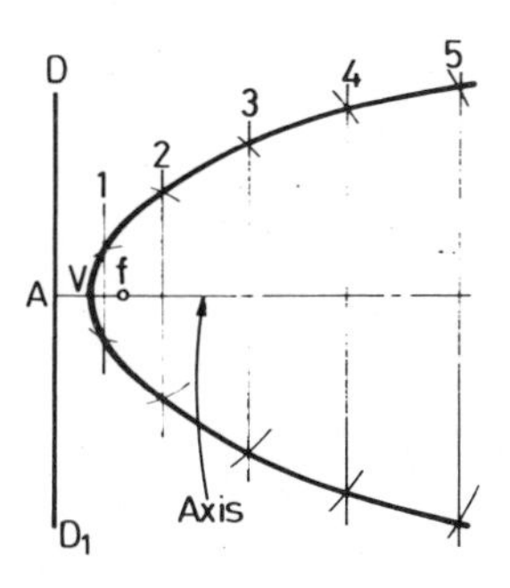

Construction of a parabola in a rectangle
Draw the rectangle ABCD. Find V, the mid-point of AB. Draw the axis at right angles to AB through V. Divide AD and BC into a number of equal parts (in the given example this is five parts). Divide AV and BV into the same number of parts. Draw lines as shown to obtain points on the parabola.

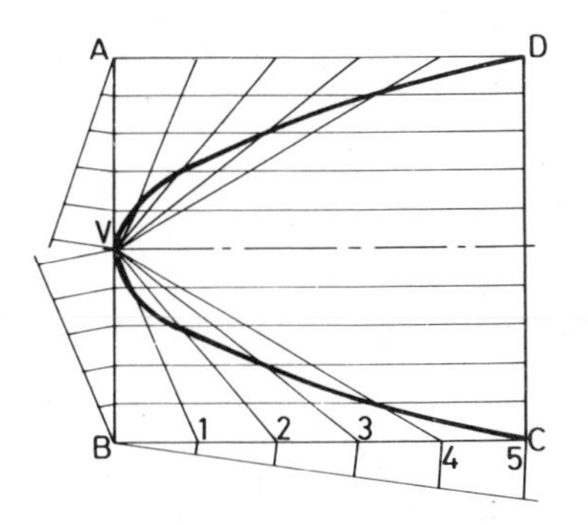

Construction of hyperbola

Follow the instructions given for the construction of a parabola, with the following differences.
1. The lines 1, 2, 3, 4 and 5 parallel to DD_1 are at set distances from DD_1.
2. The compass arcs centred at f are set to a given ratio (in the given example, 1.5) times the distance of each line from DD_1.

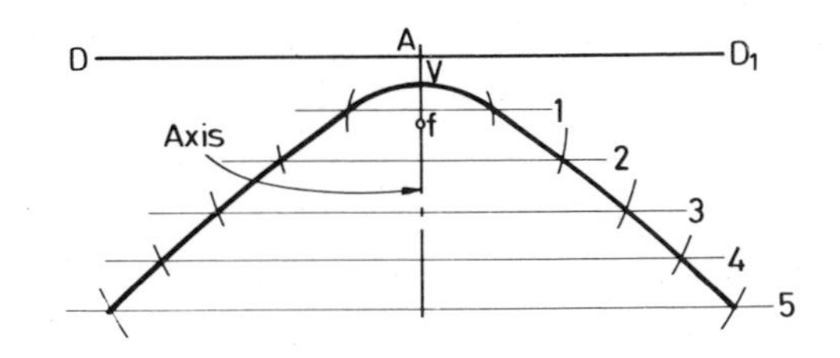

Construction of a helix

The drawing shows the constructions of a helix based upon the end and front views (see pages 47 to 49) of a transparent cylinder.

Divide the circle of the end view into 12 equal parts with the aid of a 30°, 60° set square. Draw a line showing the *pitch* length on the rectangle of the front view. Divide the pitch length into 12 equal parts. Project from 1 to 12 on the circle to the respective lines 1 to 12 on the rectangle to obtain points through which the helix can be drawn.

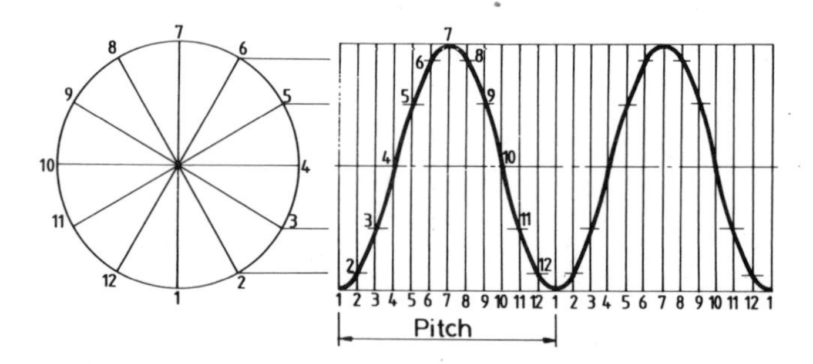

EXERCISES

1. Outlines of a pattern for a greetings card are given. Construct an accurate full size drawing of the outline.

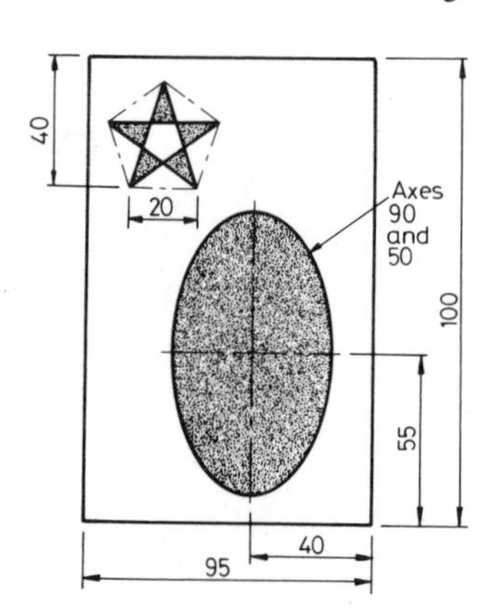

2. A symbol for a telephone is given. With the aid of instruments construct an accurate scale 1:1 copy of the symbol.

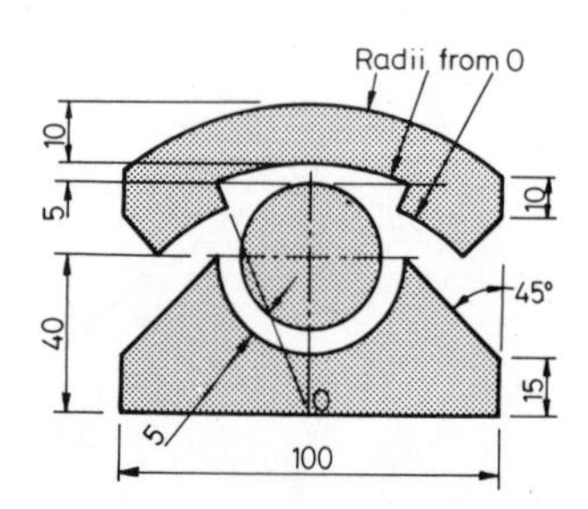

3. Make an accurate full size copy of the given ideogram.

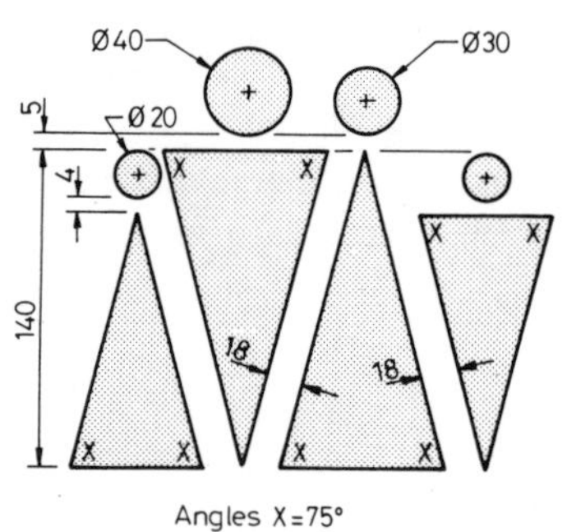

4. The cover for a ventilator from a window is based on a regular pentagon. Make an accurate full size drawing of the cover, showing how the constructions were obtained.

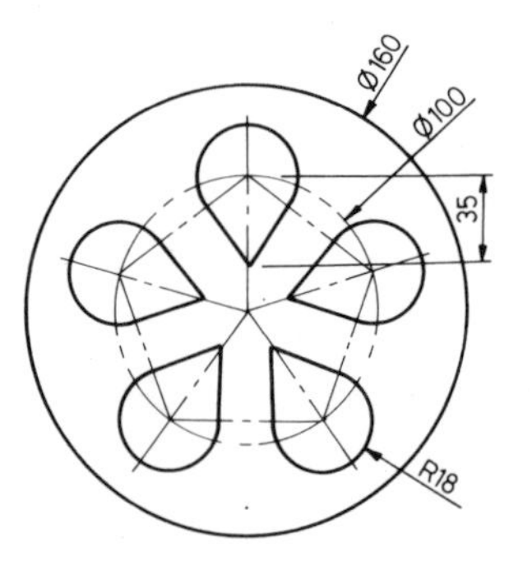

5. The given trade sign is to be drawn to scale 1:1. Draw the sign, showing all geometrical constructions.

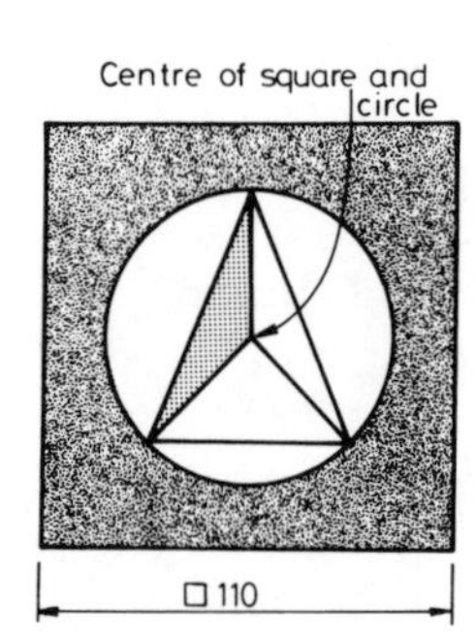

6. Make an accurate full size drawing of the given logo.

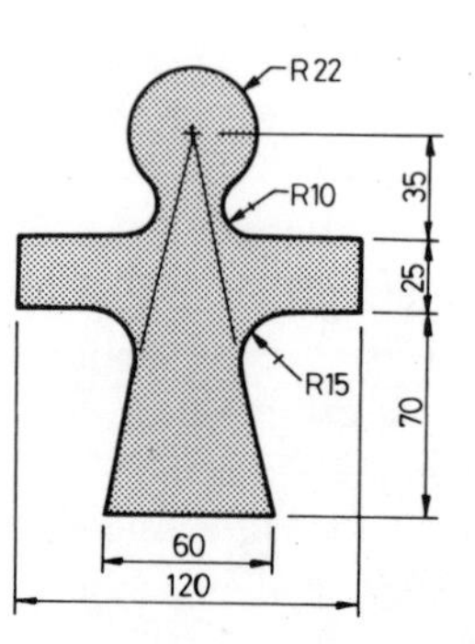

7. Draw the given 'No baggage' sign to a scale of 1:1 showing all the constructions used to obtain its shapes.

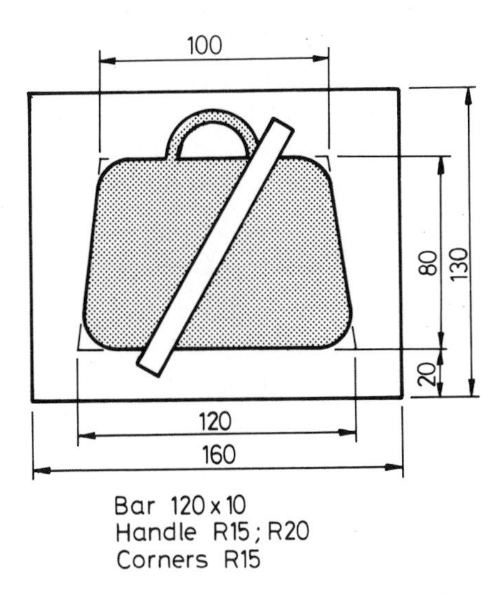

8. A scale drawing of the board for an electrical guitar is illustrated. Draw, to the given dimensions, a geometrically correct copy.

Further geometrical exercises will be found on page 171.

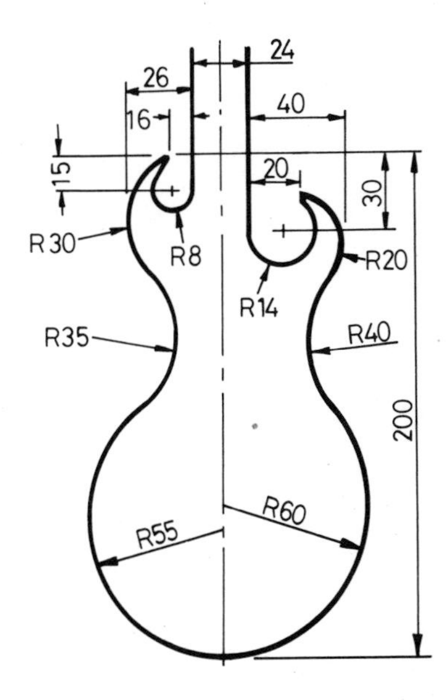

Orthographic projection

First angle orthographic projection

In First Angle projection the object which is to be drawn is imagined as being placed in a space formed by flat planes. These are horizontal plane (HP) and two vertical planes (VPs). All three planes are at right angles to each other (orthogonal to each other). In the example given in drawing 1 the object to be drawn is a plastic shelf clip. The clip is viewed from its front and what is seen is projected onto the vertical plane placed behind the clip. The clip is then viewed from above and what is seen projected onto the horizontal plane. The clip is then viewed from one end and what is seen is projected onto the second vertical plane. The view projected onto the plane behind the clip is its *front view*. The view projected onto the horizontal plane is its *plan*. That projected onto the second vertical plane is its *end view*.

The horizontal plane with its plan and the vertical plane with its end view are now imagined as being swung back, as if hinged, along the plane joint lines, to form a flat sheet with the vertical plane which carries the front view. The resulting three views in First Angle projection then appear as if on a flat drawing sheet as shown by drawing 2. The British Standards symbol for a drawing in First Angle orthographic projection is given in drawing 3. The symbol is a two-view drawing of part of a cone, itself in First Angle projection.

Principles of First Angle projection

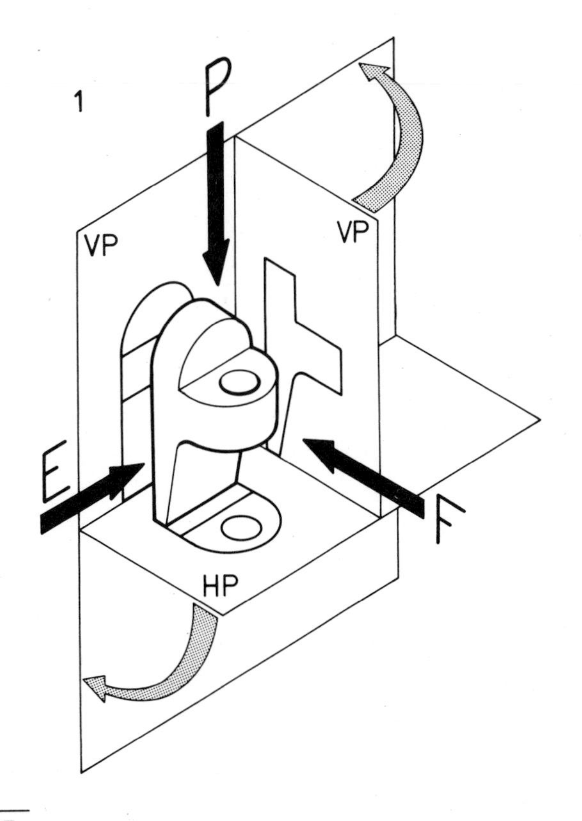

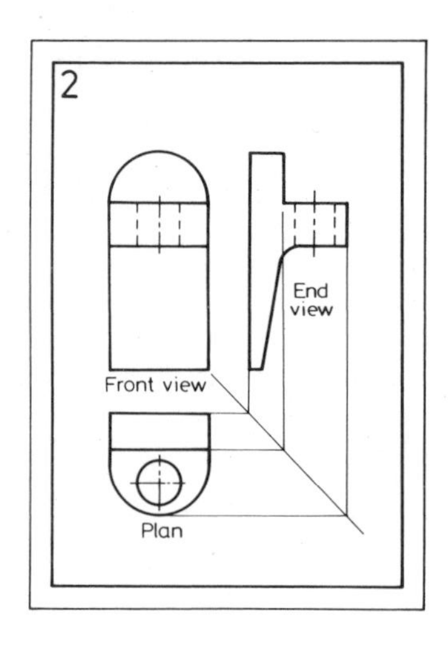

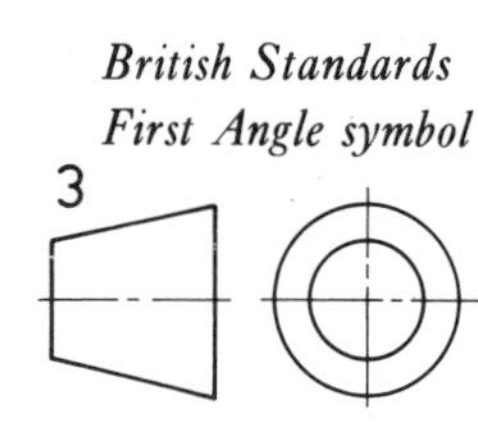
British Standards First Angle symbol

First Angle projection of a shelf clip

See drawing 4. It is a First Angle projection. Note the following details.
1. There is a border to the drawing – 10 mm to 15 mm from the drawing sheet edges.
2. The three views are fully dimensioned.
3. The drawing has a *title block* containing a title, the scale, the dimensioning units and the angle of projection shown by a BS symbol.

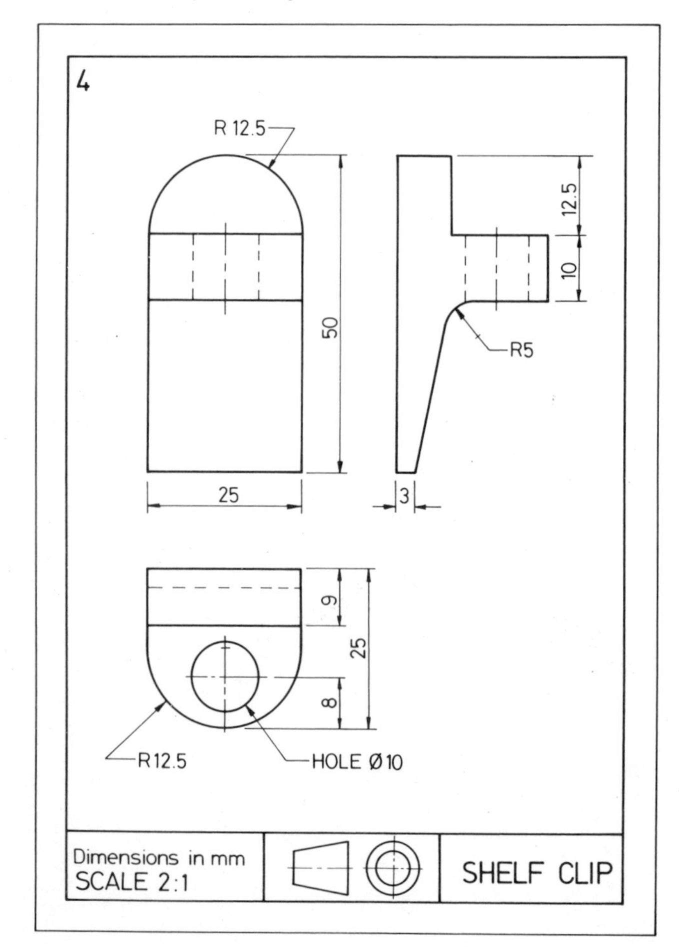

4. The end view is on the right of the front view and in line with it. If the vertical plane onto which the end view is projected had been placed to the left of the clip, then the end view would be placed on the left of the front view.
5. The plan is *below* the front view and in line with it.
6. Both plan and end view face *outwards* from the front view.
7. The views are *not* labelled 'front view' etc.

Third angle orthographic projection

In Third Angle projection, the object which is to be drawn is imagined as being placed in a space behind and beneath vertical and horizontal planes. Drawing 5 shows the shelf clip placed in a Third Angle position.

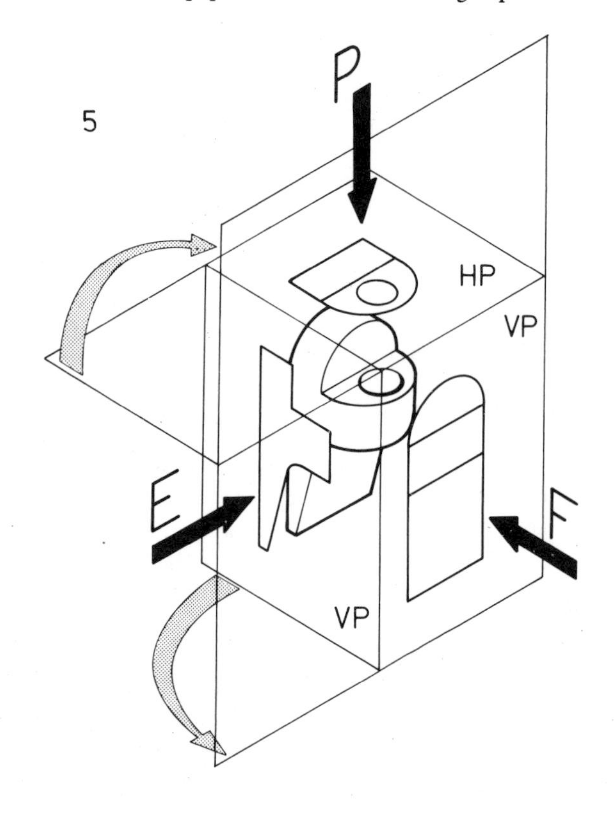

Principles of Third Angle projection

Front and end views are obtained by viewing through the vertical planes in front of and to the end of the clip and then drawing what is seen onto the planes. A plan is obtained by viewing through the horizontal plane from above and drawing the plan view on the HP. The VP and the HP containing the end view and plan are then imagined as being rotated, as shown by the shaded arrows in drawing 5 so as to lie level with the VP containing the front view. The resulting three-view

Third Angle projection of the shelf clip is shown in drawing 6. The British Standards symbol for Third Angle projection is shown in drawing 7. The symbol is a two-view drawing of part of a cone in Third Angle projection.

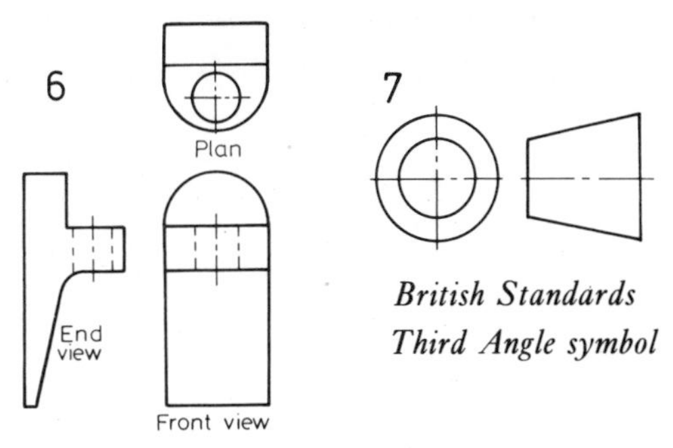

British Standards Third Angle symbol

Third Angle projection of a shelf clip

Drawing 8 is a Third Angle projection of the shelf clip.

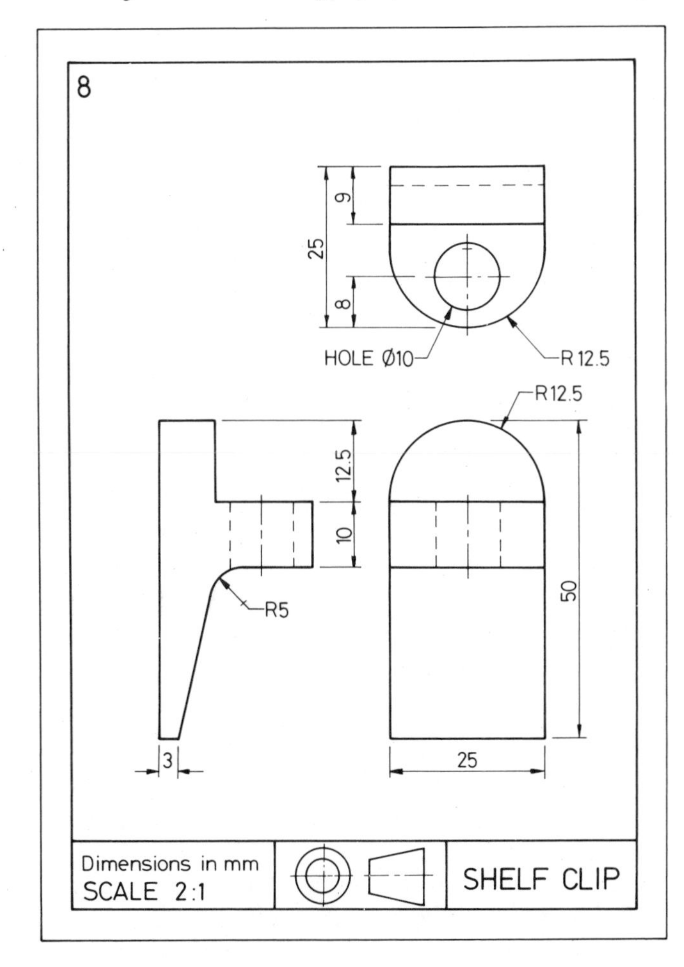

Note the following details.

1. It has a margin and title block and dimensions as for the First Angle example.
2. The end view is on the left of the front view and in line with it. If the vertical plane through which the end view is seen had been placed on the right of the clip, then the end view would be placed to the right of the front view.
3. The plan is *above* the front view and in line with it.
4. Both the plan and the end view face *inwards* towards the front view.

EXERCISES

Dimensions not given may be estimated. All dimensions are in millimetres.

1. Draw a three-view First Angle orthographic projection of the paper towel rack. Work to a scale of 1:2.

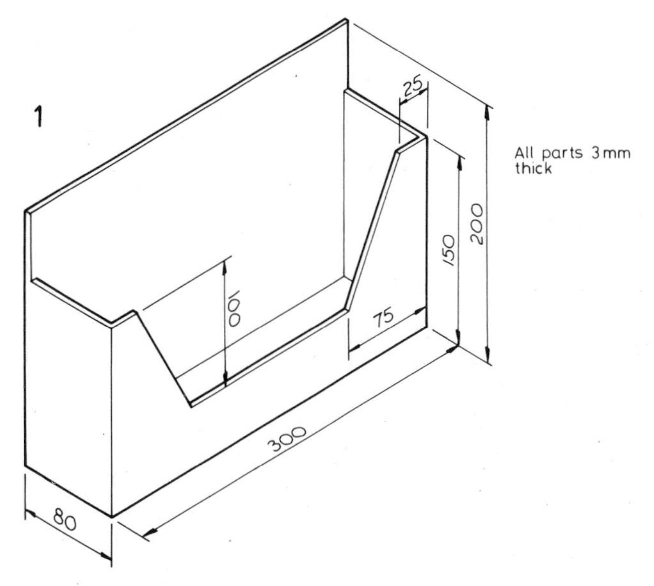

Paper towel rack

2. Draw, scale 1:1, a three-view First Angle orthographic projection of the bird box.

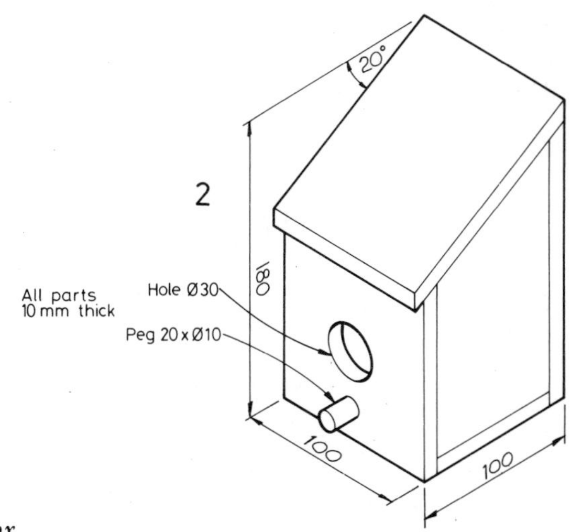

Bird box

3. Working to a scale of 1:1 make a Third Angle orthographic projection of the hand micro-recorder in three views.

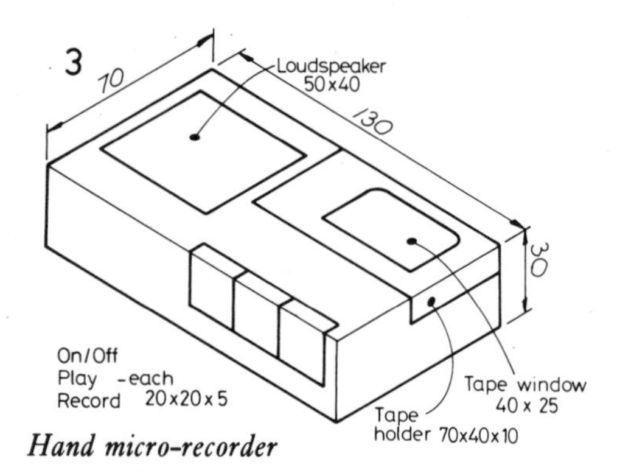

Hand micro-recorder

4. Draw, scale 1:5, in First Angle orthographic projection, a front view, an end view and a plan of the magazine rack.

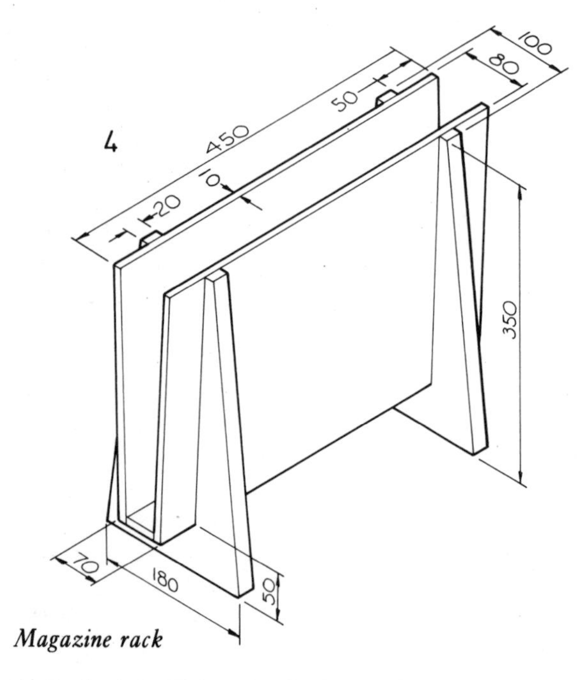

Magazine rack

5. Scale 1:2. Third Angle. Draw front view, end view and plan of the plants rack.

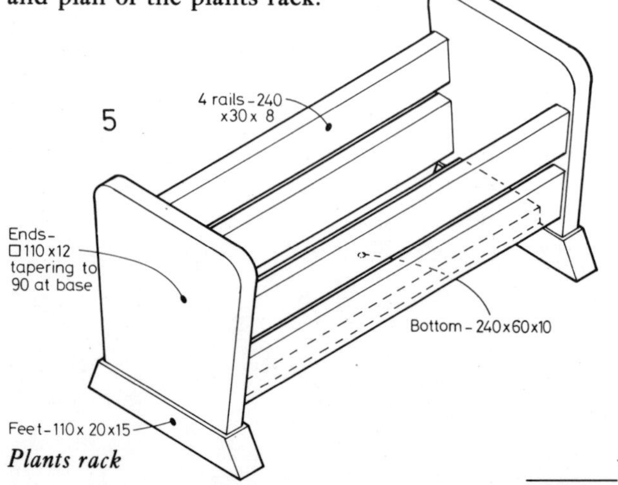

Plants rack

Projecting curves in orthographic drawings

The methods of projecting curves from front and end views into plans are shown in two drawings – a First Angle example and a Third Angle example. The method is the same whichever angle of projection is used.

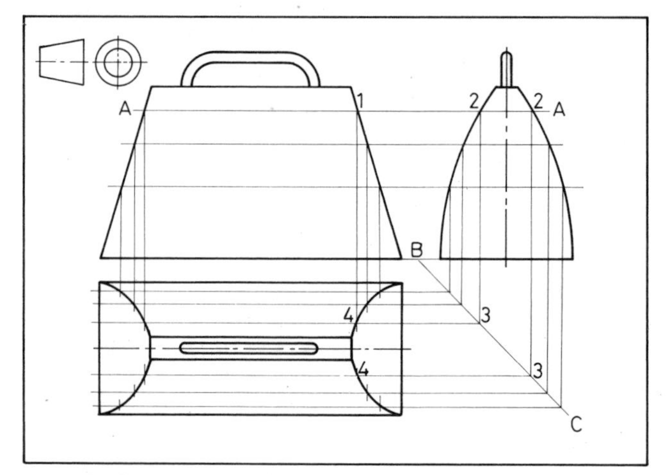

First Angle projection of a dish cover showing curves projected into plan

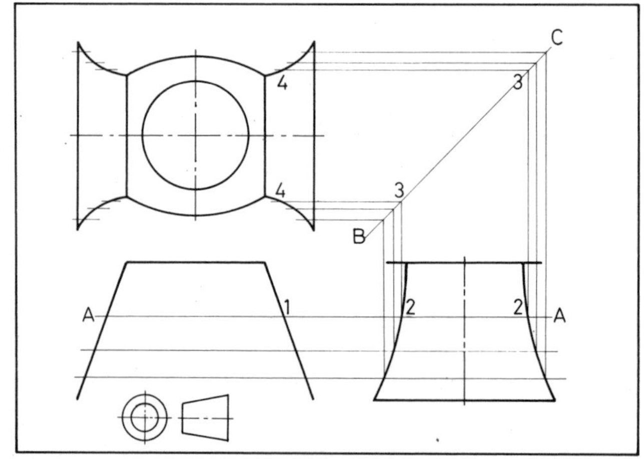

Third Angle projection of an egg stand showing curves projected into plan

1. Draw front and end views.
2. Draw the line BC at 45° to front and end views.
3. Draw a number of lines across front and end views. One such line is AA.
4. The line AA cuts the outline of the front view at 1 and the outline of the end view at the points 2.
5. Project from the points 2 onto BC to give the points 3.
6. Project from the points 3 across the plan.
7. Project from 1 onto the lines from points 3 to give points 4.
8. Points 4 are on the required curves in the plan.
9. Draw fair curves through all such points in the plan.

Sectional views in orthographic projection

An important part of orthographic projection is the drawing of sectional views. If the internal shape of an object cannot be clearly shown in orthographic views when it is seen from the outside, then the object is imagined as being cut by sectional planes. The shapes of the cut surfaces are then drawn as *sectional views* or as *sections*. Using this method, the internal shapes of an object can be clearly shown.

The drawings show the general method of obtaining a sectional view.

Drawing 1 – The outside appearance of an aluminium slide.

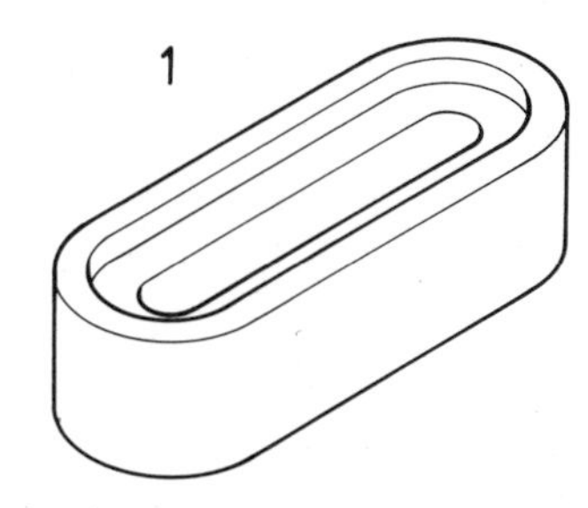

Drawing 2 – The slide is imagined as being cut by a section plane (shaded).

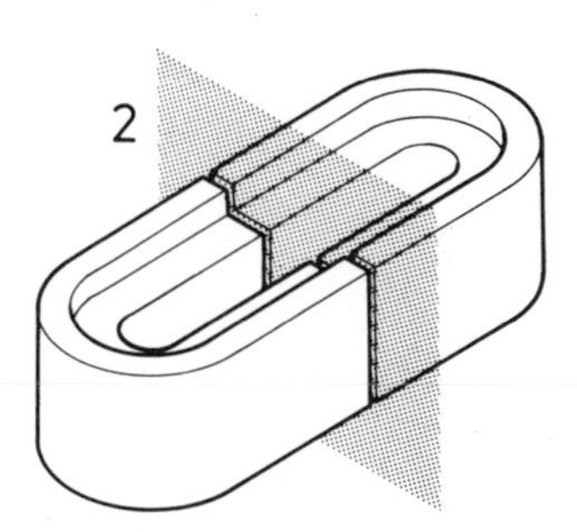

Drawing 3 – The front half of the slide has been removed and the cut surface (shaded) viewed from A.

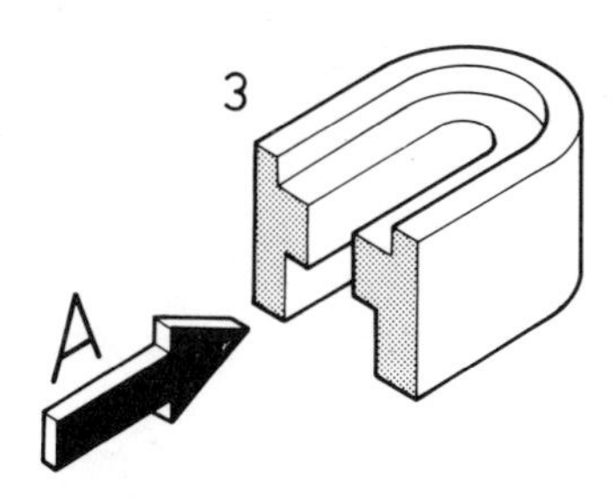

Drawing 4 – The resulting sectional view as seen from A.

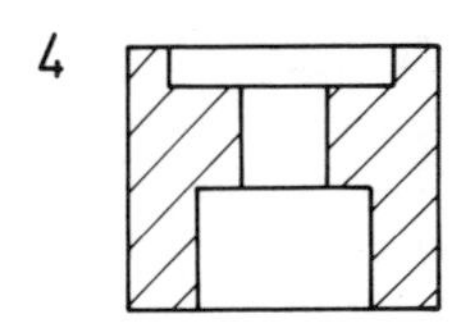

Drawing 5 – A First Angle orthographic projection which includes the sectional view.

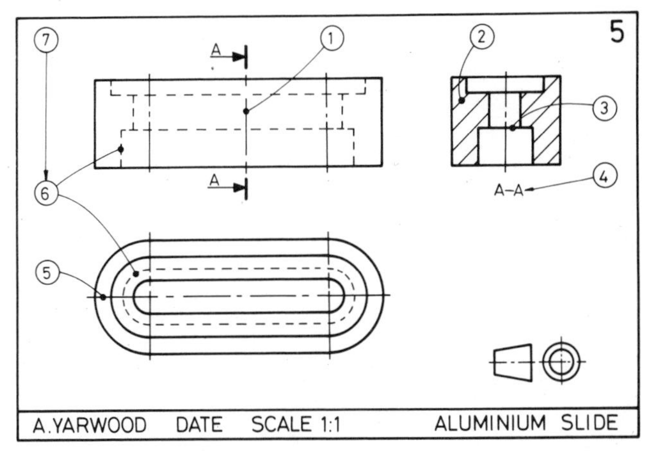

Numbers 1–7 explained below

General rules on sections

See drawing 5.

1. The edge of the section plane is shown by a thin chain line which ends in two short thick lines. The direction in which the surface cut by the section plane is viewed is shown by arrows touching the thick lines.
2. All the surfaces cut by the section plane are 'hatched' with thin lines about 4 mm apart drawn at 45°.
3. All surfaces behind the cut surface are included in the sectional view.
4. The sectional view is labelled A–A.
5. Centre lines are drawn at right angles across each of the arcs of the views.
6. Hidden detail lines show detail of parts which cannot be seen externally.
7. Attention is drawn to the above details by 'leaders' ending in 'balloons'. The balloons are circles around each number. The leaders are thin lines ending in round dots or arrows touching the part to which the numbers refer.

Sections which include outside views

When a section plane cuts through a part such as a bolt, an *outside* view of the bolt is drawn in the sectional view. See drawings 6, 7 and 8. This rule also applies when parts such as screws, nuts, washers, keys, webs

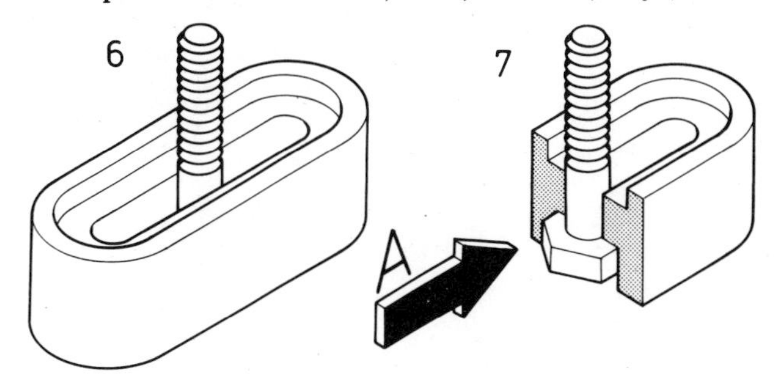

and ribs are to be shown in sectional views. Note the following in drawing 8.

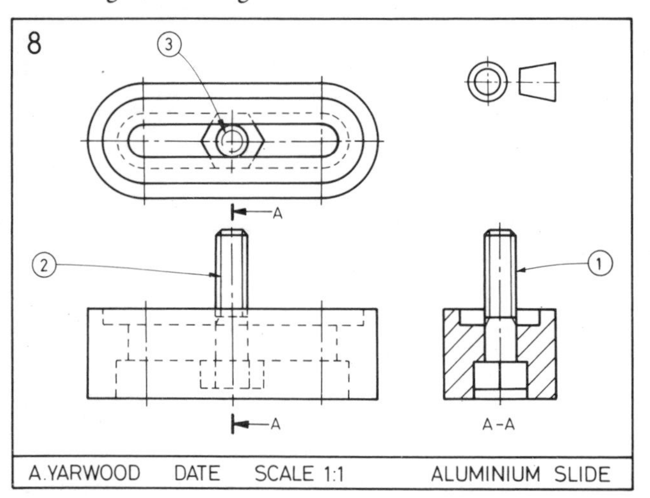

A Third Angle projection showing a sectional view which includes an outside view. Numbers 1–3 explained below

1. The bolt is shown in the section A–A as an outside view.
2. The screw thread of the bolt is shown by thin lines.
3. The screw thread is shown by a thin line in the plan.

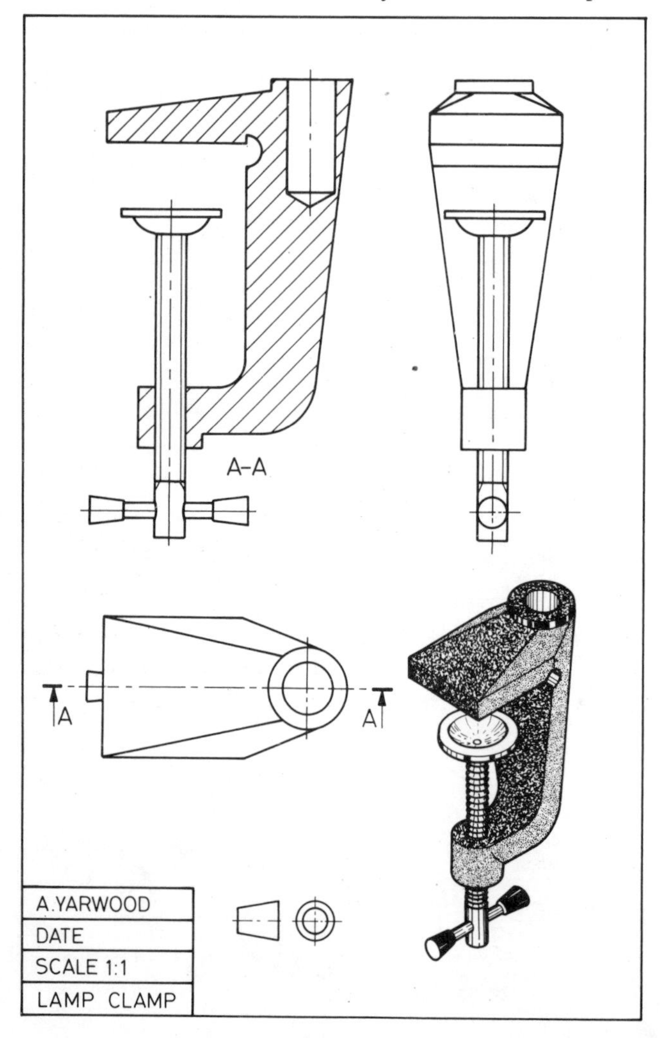

Orthographic projection of a lamp clamp

A First Angle projection of a lamp clamp is given. The projection includes a section. A pictorial drawing of the lamp clamp is included to help the student in his or her understanding of the orthographic views.

EXERCISES

1. State the angle of projection in which the given front view and plan have been drawn. Which of the views, A, B, C or D, are drawn to the same angle of projection? Draw freehand the BS symbol for the angle of projection you have stated. What do the letters BS stand for?

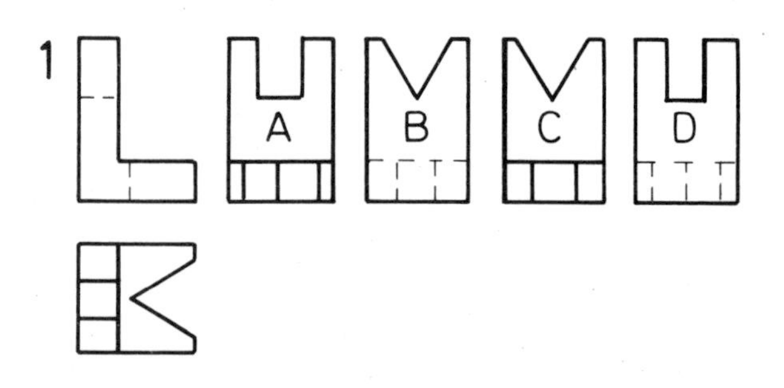

2. The given front and end views are in orthographic projection. Two of the four plans A, B, C or D could be said to be in the same angle of projection. State which two and make notes explaining why both the plans you have chosen are in orthographic projection with the given front and end views.

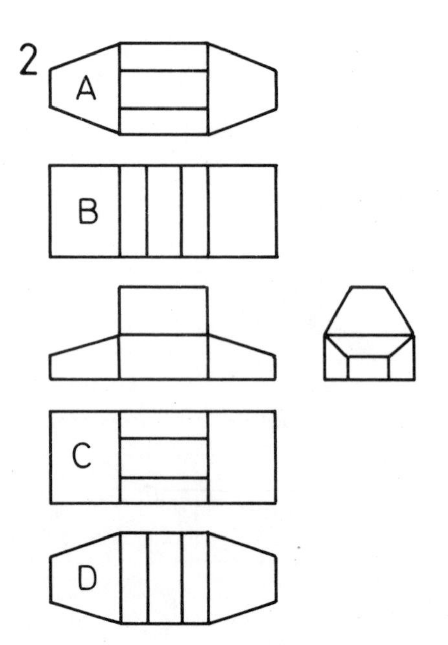

3. In Third Angle orthographic projection all six end views A, B, C, D, E and F could be correct for the given front view. Draw six separate plans in either of the two given positions for each of the six end views. Draw, with instruments, the BS symbol of projection for Third Angle.

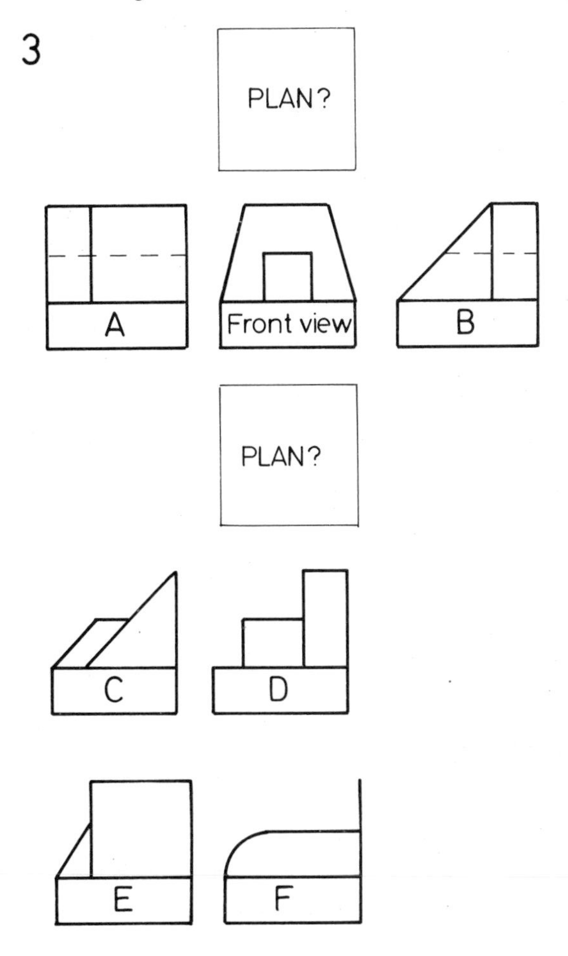

4. A pictorial drawing of a bracket is given. Draw freehand, either on square grid paper or on plain paper, three views of the brackets in First Angle orthographic projection. Choose any suitable sizes for your drawing. Add all hidden detail.

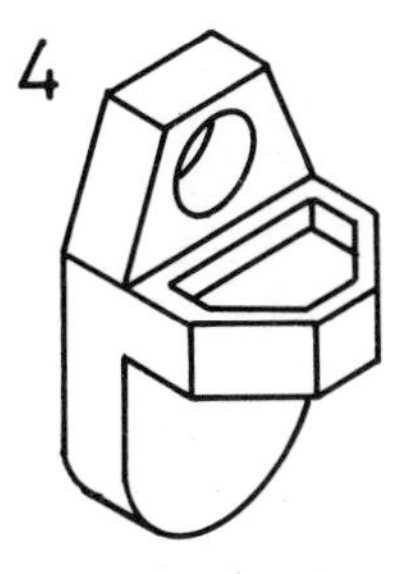

5. Draw in Third Angle orthographic projection, with the aid of instruments, the three views – F (front view), E (end view) and P (plan) as indicated by the arrows.

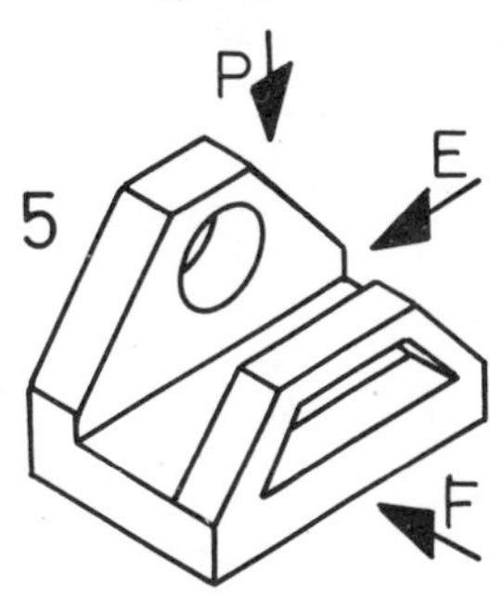

6. Assume that the given isometric grid is one in which each triangle side is 10 mm long. Use the grid lines to determine approximate sizes for each of the nine geometrical solids drawn on the grid. For each of the nine solids draw a front view, an end view and a plan in either First or Third Angle orthographic projection. Name each of the nine solids.

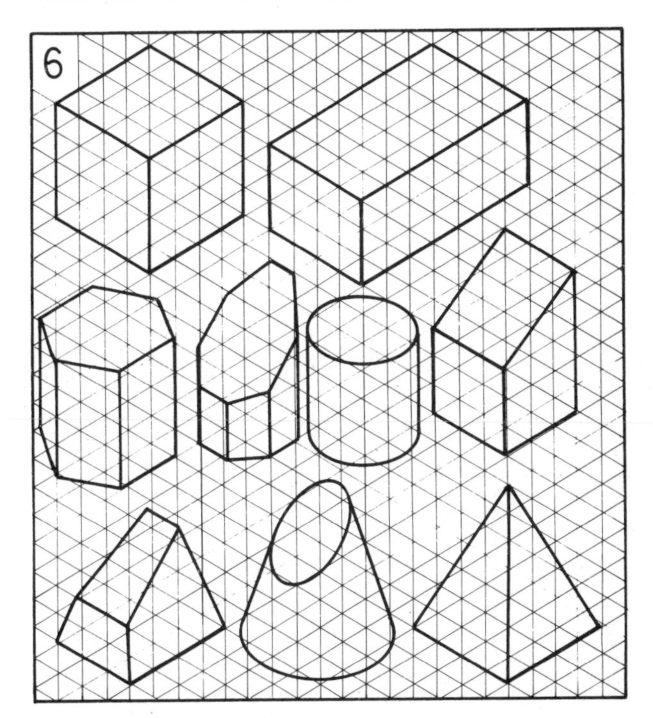

7. Working to dimensions of your own choice draw the given front view, add the sectional end view A–A and a plan. The two given views are in Third Angle projection.

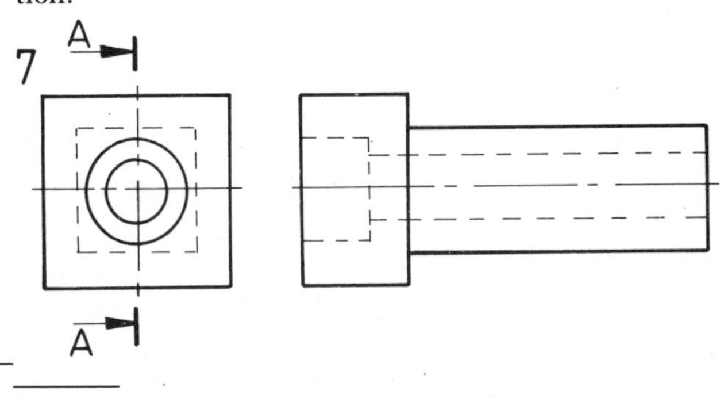

8. Working to dimensions of your own choice draw the given end view and add the sectional front view S–S and add a plan. The two given views are in First Angle projection. Note that a spindle A passes through the part B.

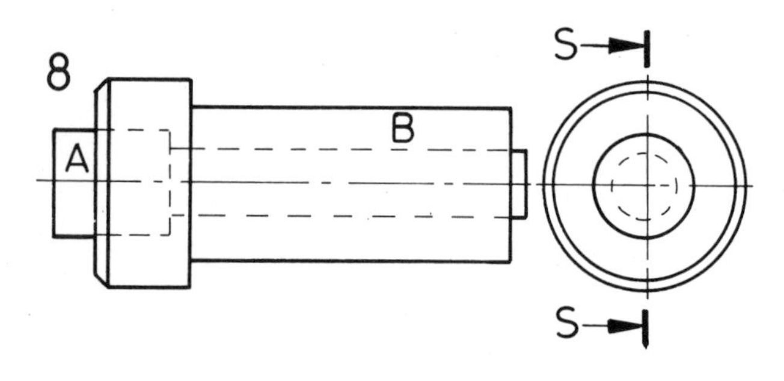

9. A pictorial drawing of a simple support which includes a supporting web is given. Draw the three views of the support in First Angle orthographic projection as indicated on the given drawing. Dimensions not given are left to your own judgement.

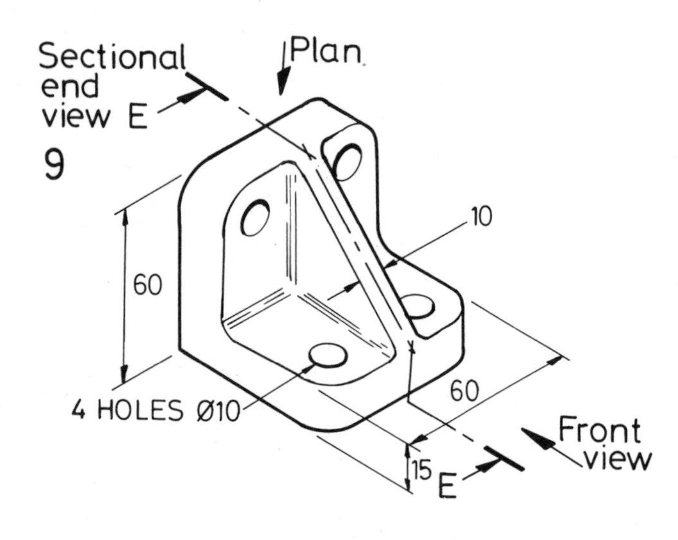

10. Drawing 10 is an exploded pictorial drawing of a steel block from which two dowel pins have been extracted. Draw a sectional front view F–F of the block with the two pins fully within their holes. When fully inserted the pins stand 10 mm above the surface of the block. Add a plan. Work in Third Angle projection.

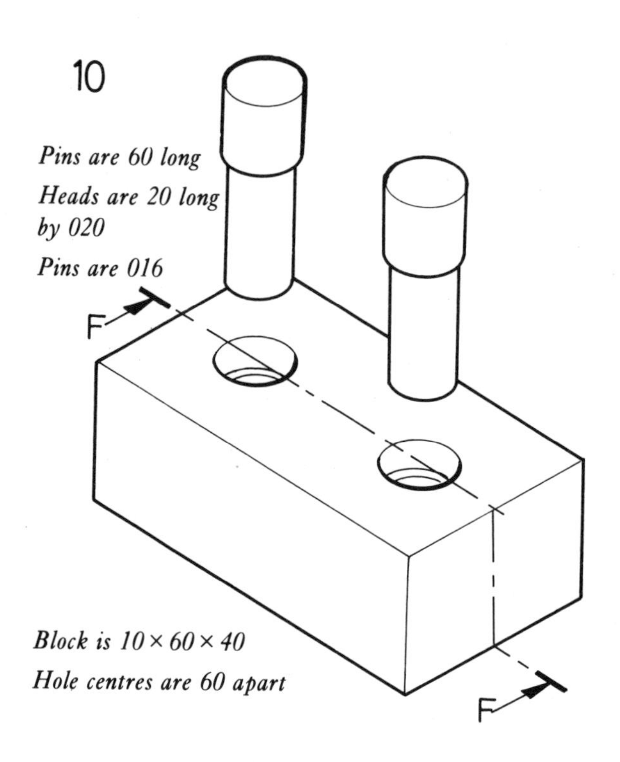

Reading lamp project

The sheet of drawings on page 55 is an example of the inclusion of orthographic projections in a design project. This sheet of drawings is one of a number from a folio in which the student has analysed the design of a reading lamp as a basis for her own design for such a lamp.

Four of the drawings – the main front view, the enlarged part detail B, the enlarged part plan D and the enlarged section A–A – are orthographic projections. In this sheet of drawings end views and plans projected from the main front view have not been thought necessary. This is because pictorial drawings of the base and the part C have been included, together with freehand drawings of alternative shades.

In addition to this sheet, the folio included:

1. drawings showing the balance of the mechanism;
2. other methods of fixing the shade to the stand;
3. other forms of mechanism;
4. photographs of other types of reading lamp;
5. methods of attaching the lamp to the edge of a desk and to the edge of a shelf;
6. other forms of base;
7. a photograph of a model of the link mechanism (see page 120);
8. an evaluation of the designs;
9. drawings and models showing the student's own designs based on this analysis.

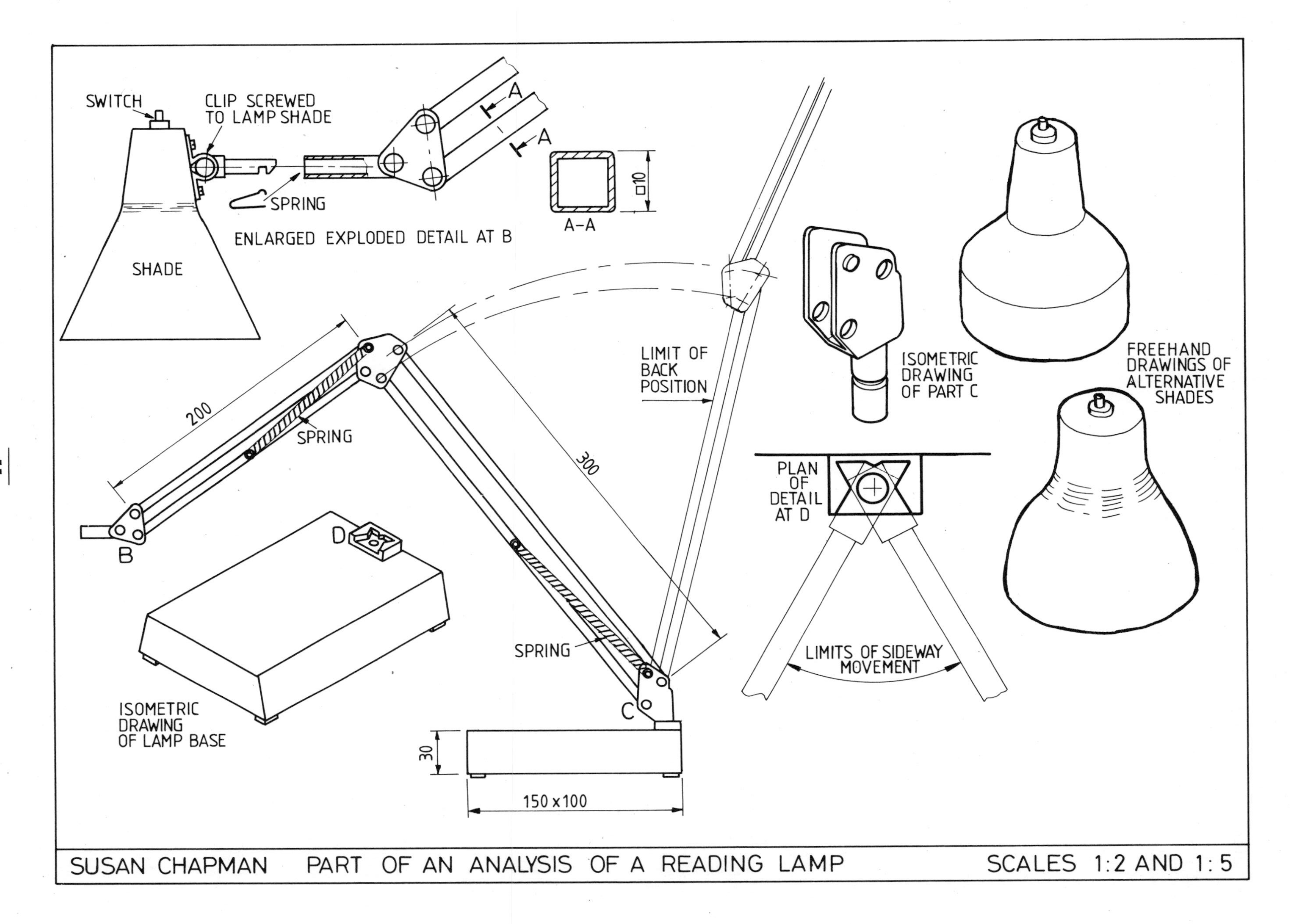
SWITCH
CLIP SCREWED TO LAMP SHADE
A
A
SPRING
A–A
□10
ENLARGED EXPLODED DETAIL AT B
SHADE
200
SPRING
300
LIMIT OF BACK POSITION
ISOMETRIC DRAWING OF PART C
FREEHAND DRAWINGS OF ALTERNATIVE SHADES
PLAN OF DETAIL AT D
D
B
SPRING
LIMITS OF SIDEWAY MOVEMENT
ISOMETRIC DRAWING OF LAMP BASE
C
30
150 x 100
SUSAN CHAPMAN
PART OF AN ANALYSIS OF A READING LAMP
SCALES 1:2 AND 1:5

Pictorial drawing with instruments

Isometric drawing

Isometric drawing is one of the easiest and quickest methods of producing pictorial drawings. A series of four drawings shows the basic method of isometric drawing.

Drawing 1 – With the aid of a 30°, 60° set square draw the outline of the box. The lengths L, W and H must only be measured along the 30° lines and the vertical lines.

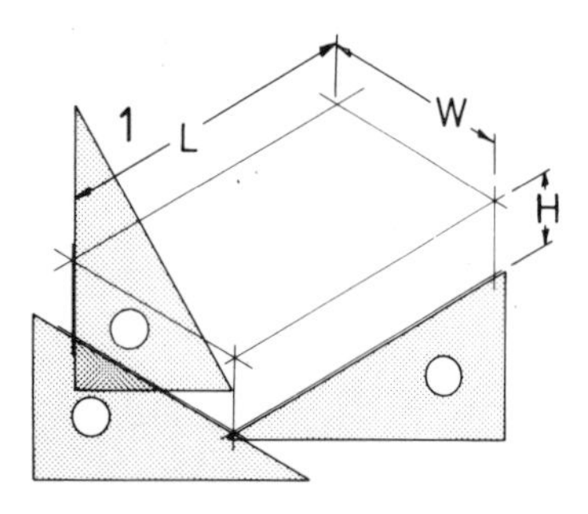

Drawing 2 – Fill in details of side and end thicknesses. Add the lines of the bottom of the box.

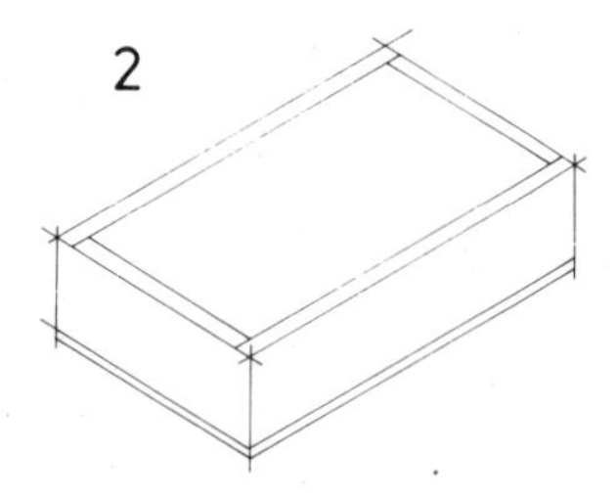

Drawing 3 – Complete all drawing details by adding the lines of the interior of the box.

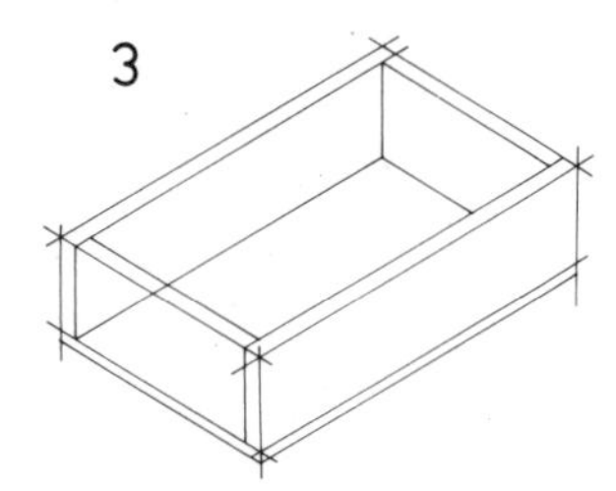

Drawing 4 – Line in to obtain the completed drawing. Erase unwanted lines.

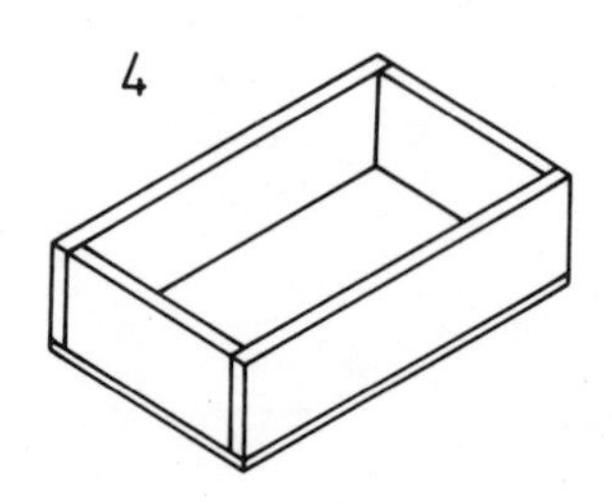

Sloping lines in isometric drawings

A firm rule in isometric drawing is that measurements can only be made along either the 30° or the vertical lines. Measurements must never be taken along any other lines. To obtain the sloping face of the perpetual calendar shown, the lengths A and B must first be measured along the 30° and vertical lines. Only then can the sloping line of the face be drawn between the points so measured. Once this sloping line has been drawn, other lines along the slope can be drawn parallel to it.

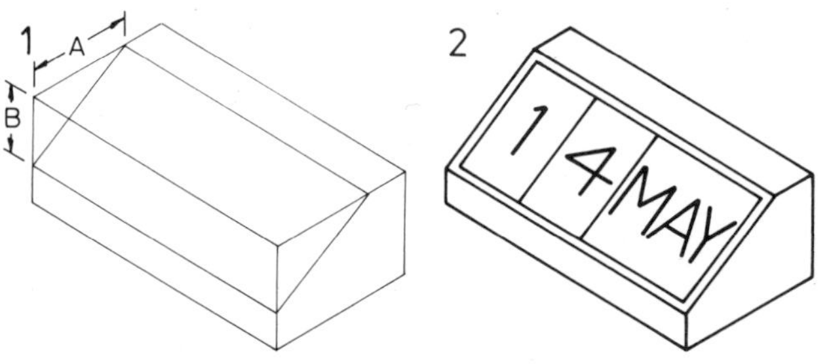

Sloping lines in isometric drawings – a drawing of a perpetual calendar

A second example of drawing lines in an isometric drawing which are not parallel to the isometric axes is given.

Drawing 1 – Measure the lengths A, B, C and D along the isometric axes to obtain the sloping lines on the face of the object.

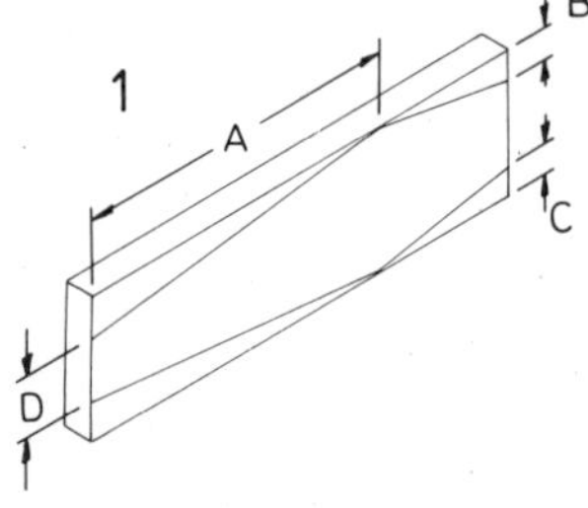

Drawing 2 – The length E can now be measured, again along a 30° axis and lines from F and G drawn to the points so obtained.

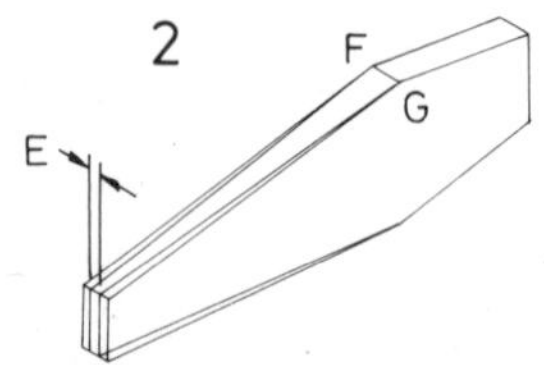

Drawing 3 – The drawing can now be lined in and unwanted lines erased.

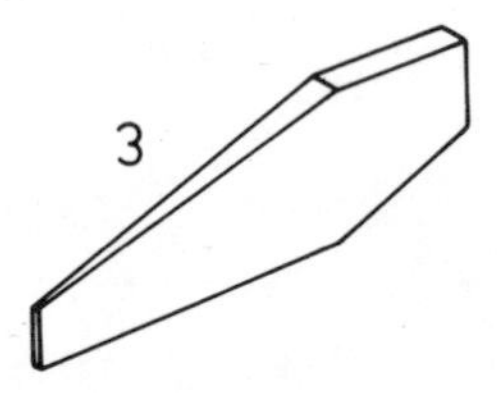

Drawing circular parts in an isometric drawing

The rule that measurements must only be taken along the isometric axes of 30° and 90° is again applied when drawing curved lines in isometric drawings.
Drawing 1 – Draw a front view of the curve – in this example, a circle of radius c. Draw the two diameters at right angles to each other. Mark a number of vertical lines at any convenient position – in this example at distances a and b from the vertical diameter. These lines are known as *ordinates*.

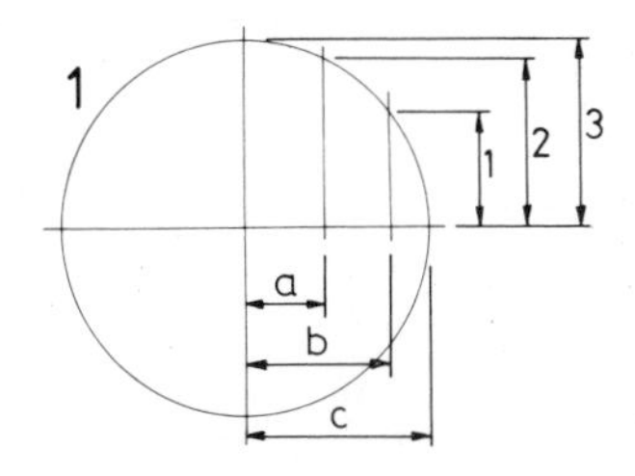

Drawing 2 – Draw the two 30° axes of the isometric drawing. Measure the distances a, b and c along one of the 30° axes. Draw 30° lines through these points. Going back to the front view, measure the lengths of the lines at a, b and c to give 1, 2 and 3. Mark these lengths in turn along the isometric lines. Draw a neat freehand curve through the twelve points.

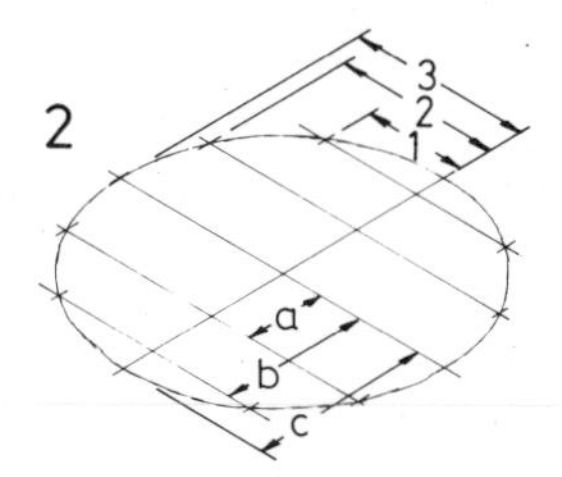

Drawing 3 – Shows the completed drawing with construction lines rubbed out.

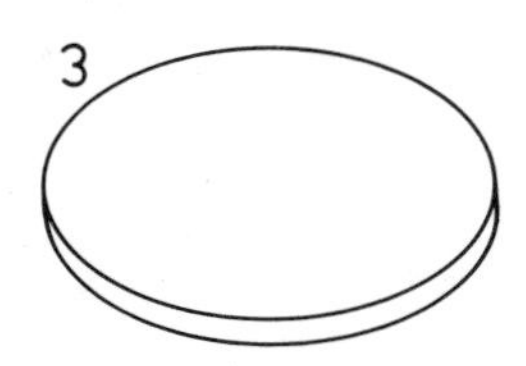

An example of an isometric drawing containing circular parts is shown.
Drawing 1 – Make an accurate front view drawing and mark convenient ordinates on the front view.

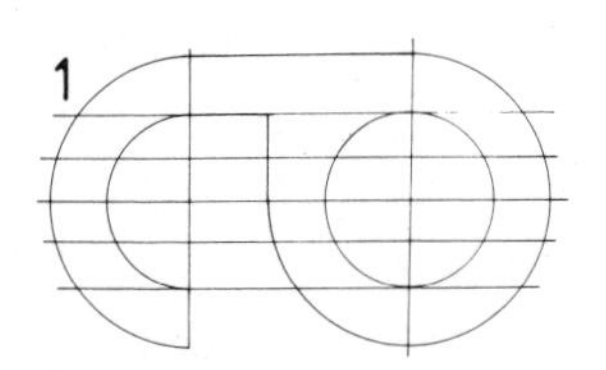

Drawing 2 – Transfer the ordinates to the isometric axes and draw a fair curve through the points so obtained. Erase unwanted details.

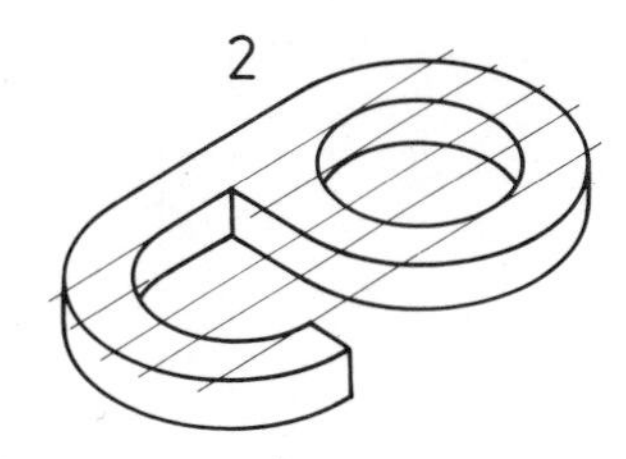

This method of transferring measurements from ordinates from a front view onto an isometric drawing is not restricted to the drawing of arcs of circles. An example of the method applied to an isometric drawing of a handle which is shaped on two half ellipses is given.

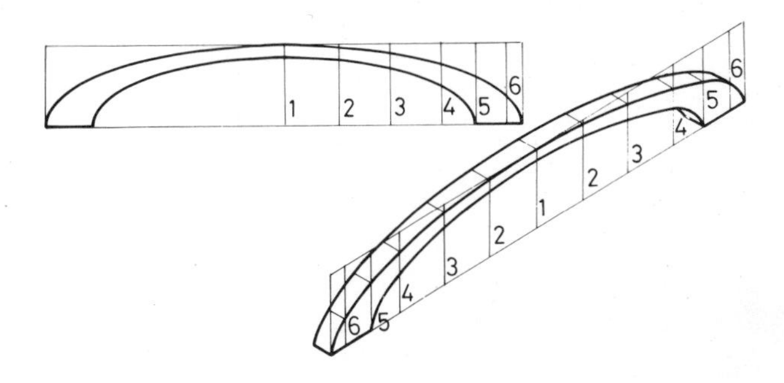

Exploded isometric drawings

Exploded drawings are made to show how the parts of an object fit together. Examples of exploded isometric drawings are given.
Drawing 1 – A simple halving joint between two pieces of wood.

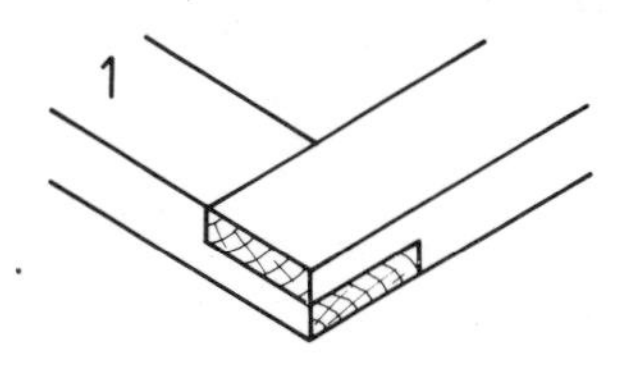

Drawing 2 – An exploded isometric drawing of the halving joint.

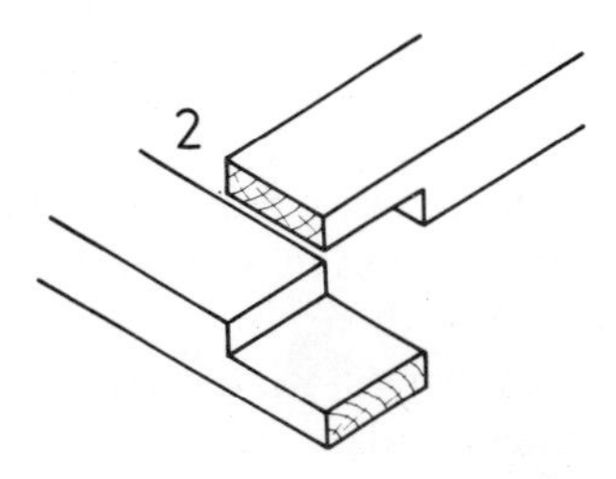

Drawing 3 – A dowelled mortise and tenon joint.

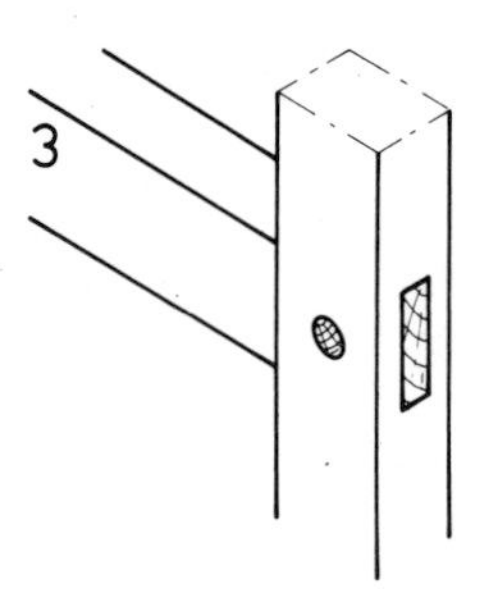

Drawing 4 – An exploded isometric drawing showing the three parts of the joint.

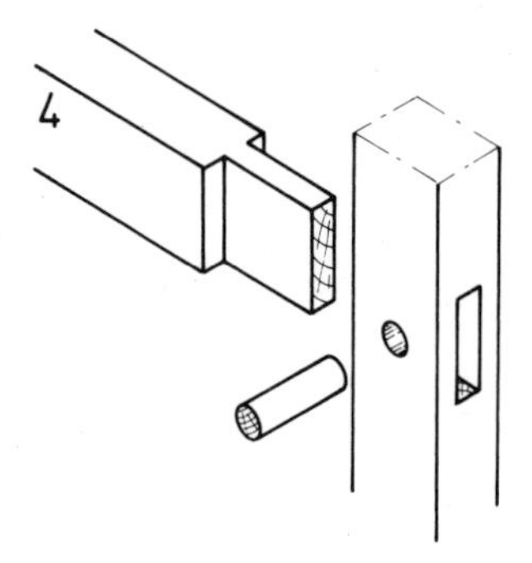

Drawing 5 – An exploded isometric drawing showing how the parts of a sharpening guide fit into each other.

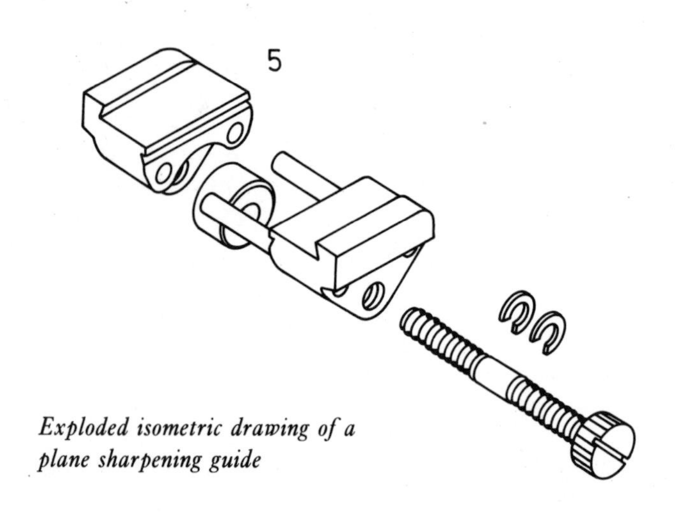

Exploded isometric drawing of a plane sharpening guide

Note: The 'explosion' always takes place with the parts of the object in line with each other in the direction in which the parts fit together.

The ellipses of drawing 5 were drawn with the aid of an isometric ellipse template.

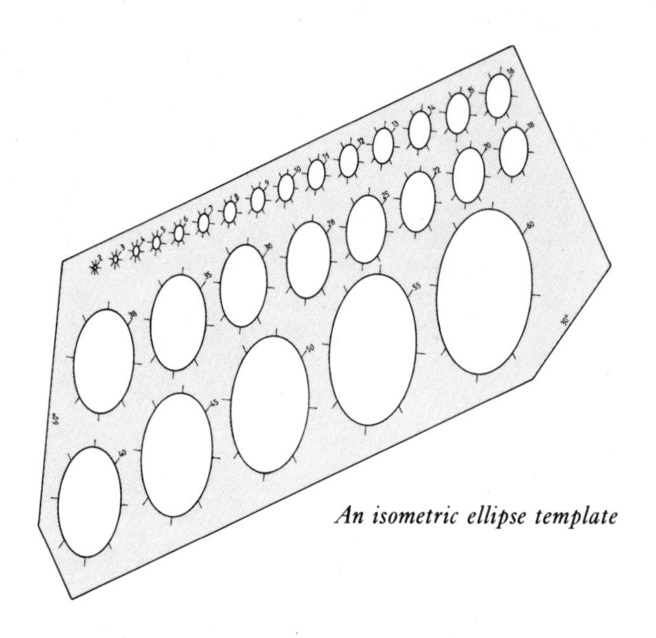

An isometric ellipse template

Isometric grid papers

Two examples of isometric drawings made on grid papers are given. That on a 5 mm grid is of a small table. The one on a 10 mm grid shows an exploded isometric drawing.

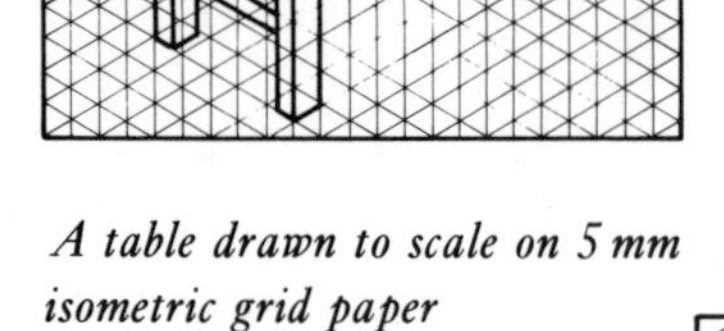

A table drawn to scale on 5 mm isometric grid paper

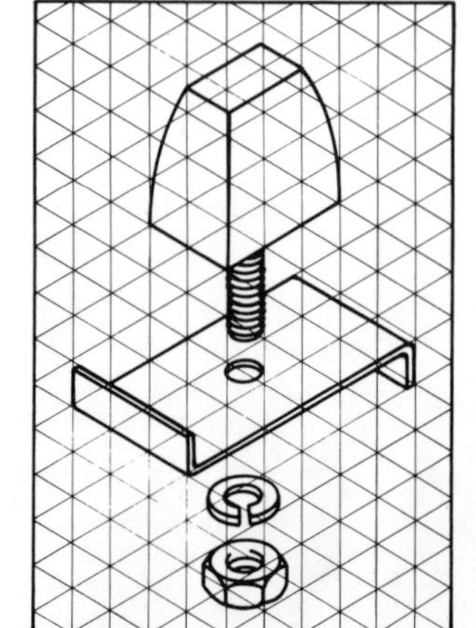

A car shock absorbing rubber mount on 10 mm grid paper

Oblique projection

Commence with a front view of the article. Take sloping lines from this front view at a suitable angle. Measure the distances of depth along the sloping lines. Then draw in the rear of the article. Three oblique projections are shown. Drawing 1 is with the slope lines at 60°. Drawing 2 is with the slope lines at 45°. In drawing 3 the slope lines are at 30°. Oblique drawings of this type are known as *cavalier* projections.

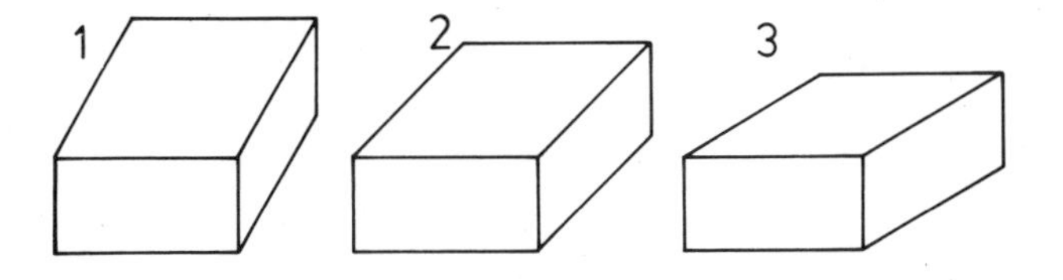

Cabinet projection

This is the form of oblique projection most commonly used. All slope lines are drawn with the aid of a 45° set square. Measurements along the slope lines are made at half size – to a scale of 1:2. Drawing 1 shows the construction lines for a simple box. Drawing 2 shows the lined-in completed cabinet projection of the box.

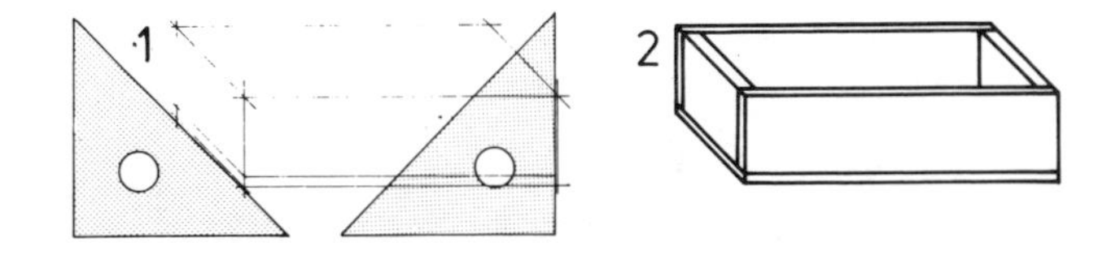

Two examples of cabinet drawings are given. In each case, drawing 1 shows the construction and 2 the completed projection. In the drawing of the thermostat control all circles and arcs were drawn with the aid of a compass. In the drawing of the wall switch curves occur both in the front view and along the slope lines. To construct the sloping curves each measurement along the 45° lines must be halved. Ordinates a, b and c had to be reduced to $\frac{1}{2}$a, $\frac{1}{2}$b and $\frac{1}{2}$c before being measured along the 45° lines.

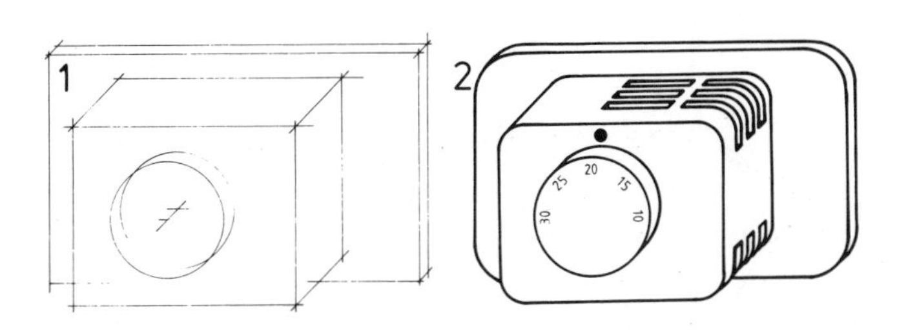

Cabinet projection of a thermostat control

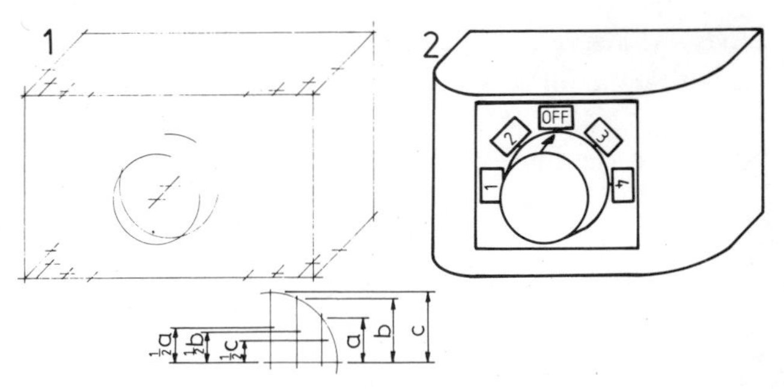

Planometric drawing

When making planometric drawings, the plan is drawn first. The plan may be drawn either with 30° and 60° sloping lines or with 45° sloping lines. Then verticals are taken at appropriate corners. When using 30°, 60° slope lines use the same scale for vertical lines as for the plan. When using 45° slope lines for the plan, it is better to scale verticals to about $\frac{3}{4}$ size. Otherwise your drawings may look too high.

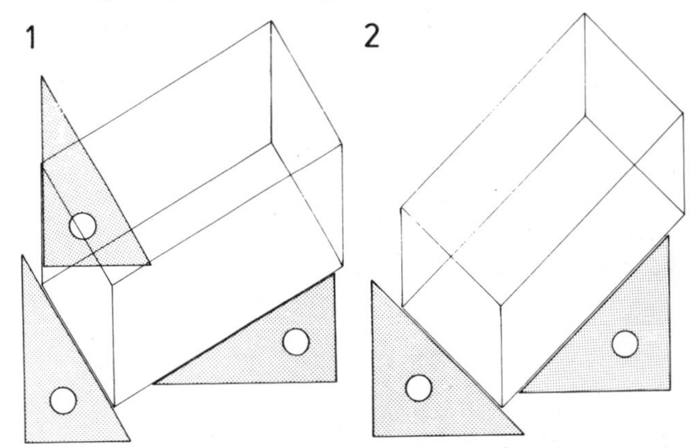

Methods of planometric projection

Two examples of planometric drawings are given. In that of the room corner fitment drawn with 30° and 60° plan lines, the construction is shown in drawing 1 and the completed planometric in drawing 2. The plano-

A room corner fitment drawn in 30°, 60° planometric projection

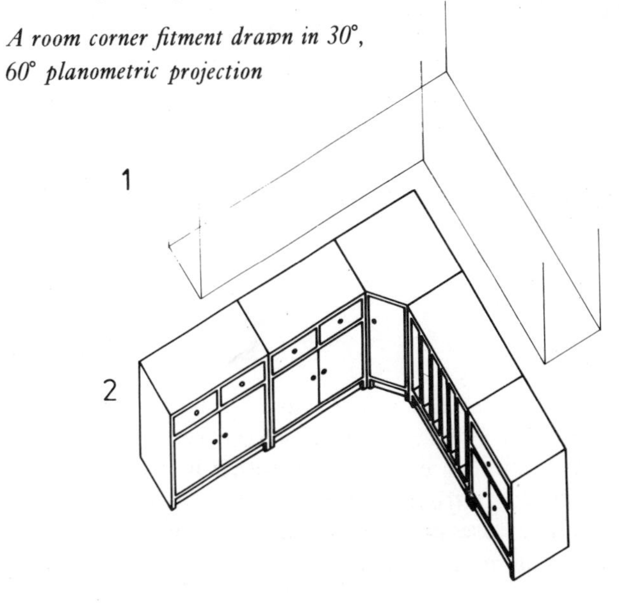

metric drawing of the two floors of a house was drawn with the aid of a 45° set square and is an exploded drawing. Note the shading on the walls as if light is falling on the house from the left and from behind.

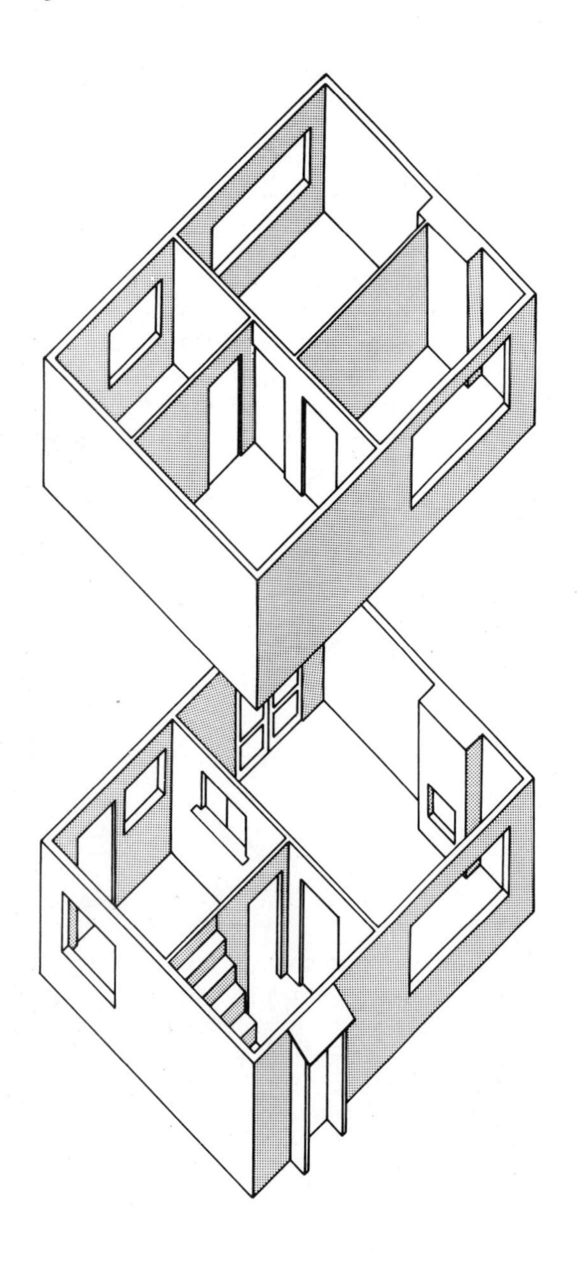

An exploded 45° planometric drawing

EXERCISES

1. The three drawings of a cassette, 1a, 1b and 1c, have been drawn using three different types of pictorial drawing. Name each type of drawing.

Select one from the three which you consider is the best for showing the cassette and copy it freehand. Shade or colour your drawing.

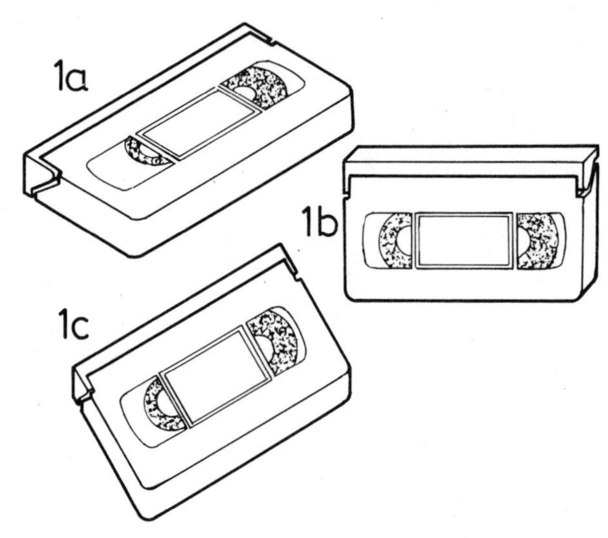

2. The dimensioned orthographic projection shows the plastic box in which a video cassette is stored. Draw with the aid of instruments a full size isometric drawing of the box. Shade or colour your drawing.

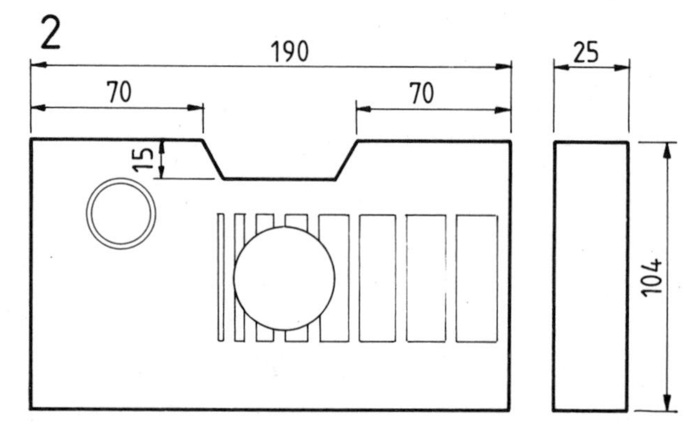

3. The given drawing shows a cube of 100 mm edge lengths. Imagine that all eight corners of the cube are cut away on the lines drawn on the surfaces of the cube. Drawing 3a shows one corner cut off in this fashion.
(a) In planometric drawing, draw the given complete cube, showing the cutting lines.
(b) In isometric drawing, draw the shape remaining after ALL eight corners have been cut off.
(c) In cabinet drawing, draw the corner b after it has been cut from the cube.

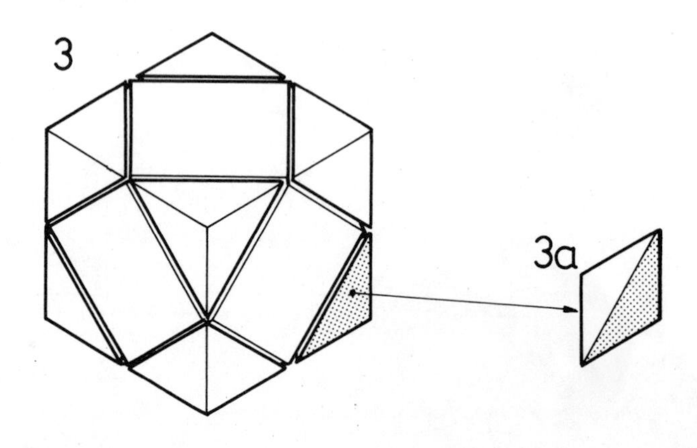

4. The two drawings 4a and 4b show two joints used for joining pieces of wood together.
(a) Name the drawing methods by which the two drawings have been made.
(b) Make a full size isometric drawing of 4a with its two parts together.
(c) Make a full size exploded isometric drawing of joint 4a.
(d) Make a full size cabinet drawing of joint 4b with its two parts assembled.
(e) Make a full size exploded cabinet drawing of joint 4b.
(f) Examine your finished drawings and write an evaluation of each. Include comments on the following.
  (i) Which of the two methods of drawing shows such joints to best advantage?
  (ii) Have you selected the best viewing position from which to 'see' the joints?
  (iii) Could you have placed the joints in a better position for viewing?
  (iv) Would shading or colouring improve the value of your drawings?

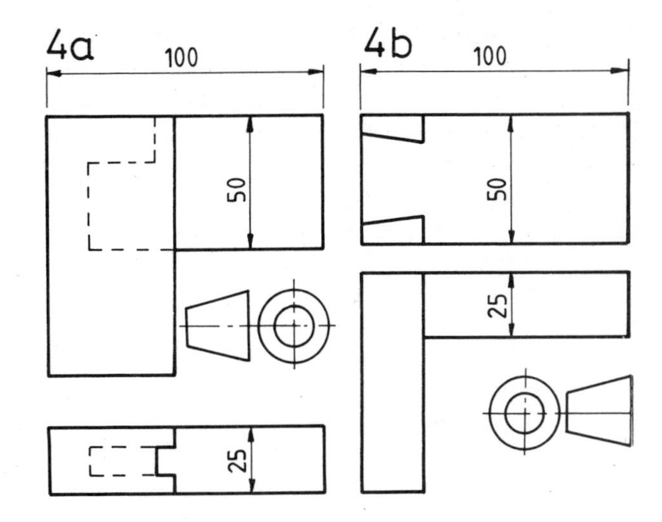

Perspective drawing

When any object is looked at in such a way that its sides are sloping away from you, lines which are parallel in the object no longer appear to be parallel. They appear to converge on one another as if they would meet at a distant point. Perspective drawing takes account of this convergence of parallel lines. In perspective drawing the points at which parallels apparently meet are called *vanishing points* (VPs). A number of geometrical methods of perspective drawing can be used. In this book only one of them will be described – *two-point perspective*. In schools and colleges a more commonly used method of perspective drawing is *estimated perspective*. This will be dealt with later (see pages 63 to 65).

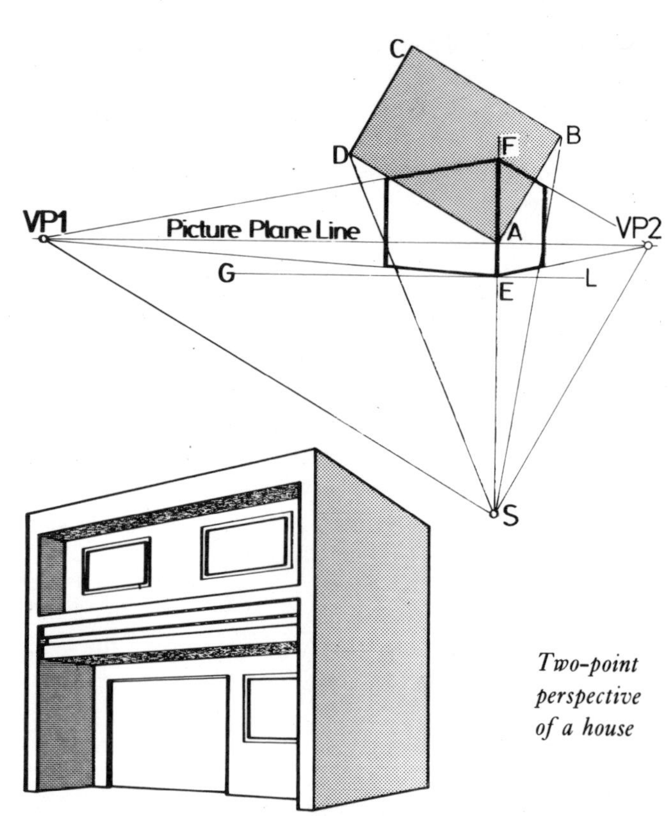

Two-point perspective of a house

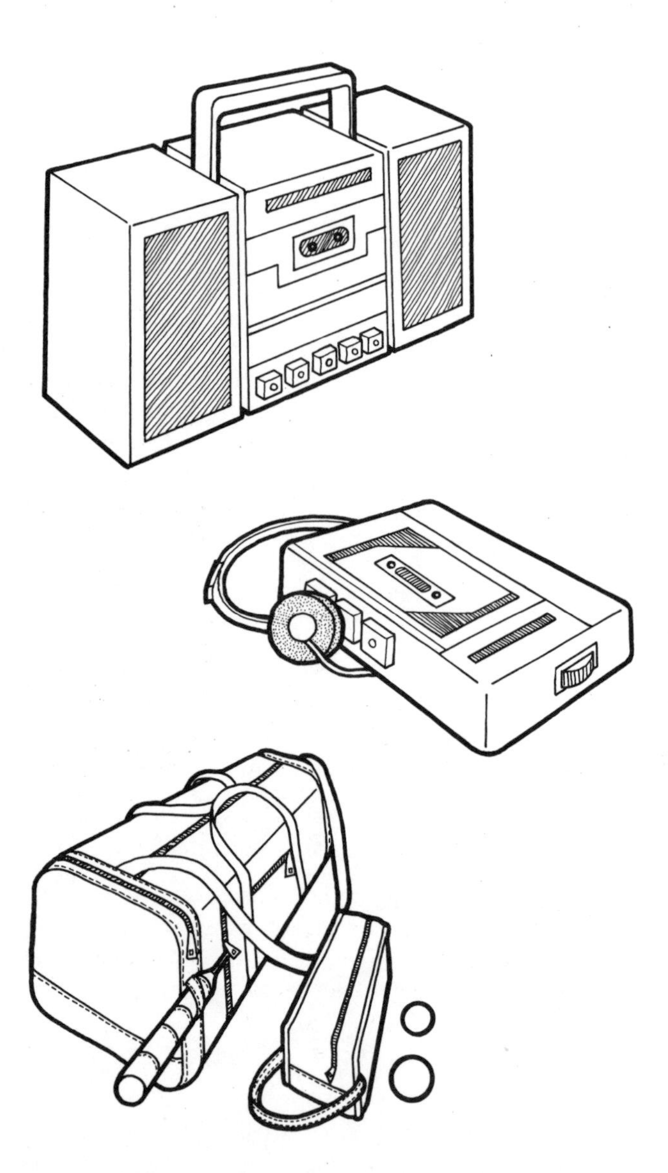

Two-point perspective drawing

A two-point perspective drawing of a house is shown, together with the method by which the drawing was made.

The method of two-point perspective

1. Draw a plan, ABCD, of the building to a convenient scale.
2. Mark the position S of the person looking at the house – this gives the *Station Point*. Draw SA.
3. Draw the *Picture Plane Line* (PPL) – in the example this has been drawn through A.
4. Draw lines from S to PPL parallel to AB and CD of the plan. This gives the two VPs – VP1 and VP2.
5. Draw a *Ground Line* GL at a scaled 1.7 metres below the PPL. 1.7 m is taken to be the average eye level height. GL crosses SA at E.
6. Join E to the VPs. This gives the base lines of the house.
7. Mark F above E so that EF is the scaled height of corner A of the house.
8. Join F to the two VPs to give the roof lines of the house.
9. Join S to B and D in the plan.
10. Where SB and SD cross the Picture Plane Line draw verticals. The outline of the perspective drawing of the house can now be drawn.

Note: To complete all the details of the perspective drawing of the house, each of the vertical dimensions must be measured along the vertical EF. Each horizontal dimension must be measured on the plan. The actual points of details in the perspective drawing are then found at the intersections of lines from the plan S with lines from the vertical EF to the two VPs.

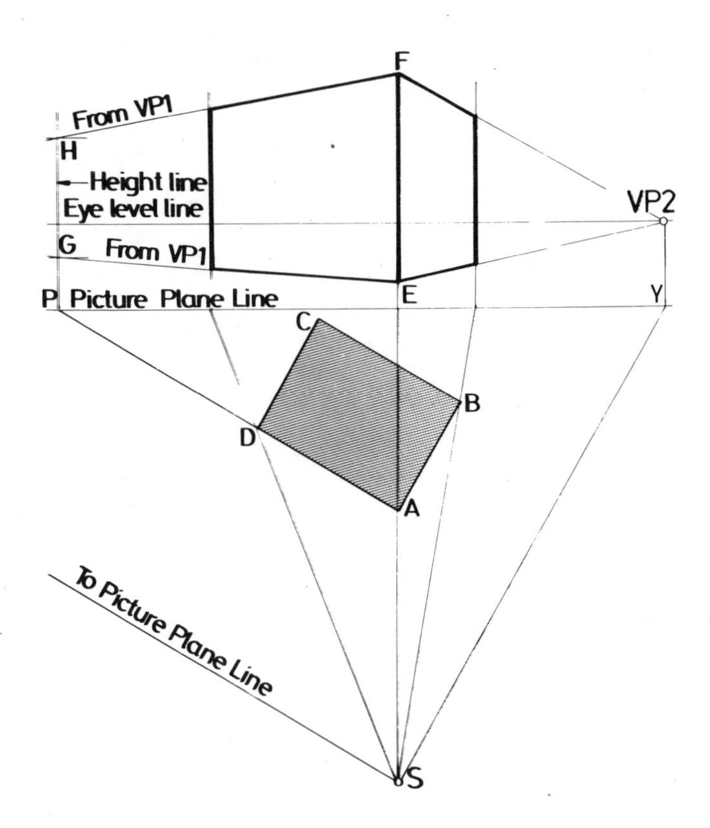

The method of two-point perspective

You may find that a more convenient size drawing can be obtained by placing the Picture Plane Line above the plan of the house.

1. Draw the plan ABCD.
2. Mark the Station Point S.
3. Draw a Picture Plane Line above the plan.
4. Draw lines from S parallel to AB and AD onto the Picture Plane Line to give X and Y. Because of limitations of page space X is not shown.
5. Draw an Eye Level Line at any convenient point above the Picture Plane Line.
6. Draw verticals from X and Y onto the Eye Level Line to give VP1 and VP2. VP1 is off the page.
7. Project AD onto the Picture Plane Line to give P.
8. At P draw a perpendicular. This is the *Height Line* along which all vertical heights must be measured.
9. Along the Height Line measure the scaled ground level point and the house height – G and H. The scale is that to which the plan has been drawn.
10. From VP1 draw lines through G and H on to SA produced to give the front edge of the building – EF.
11. From E and F draw lines to VP1 and to VP2.
12. From S draw lines through B and D onto the Picture Plane Line.
13. From the points so obtained draw verticals to give the outline of the house.

Two-point perspective of a digital clock

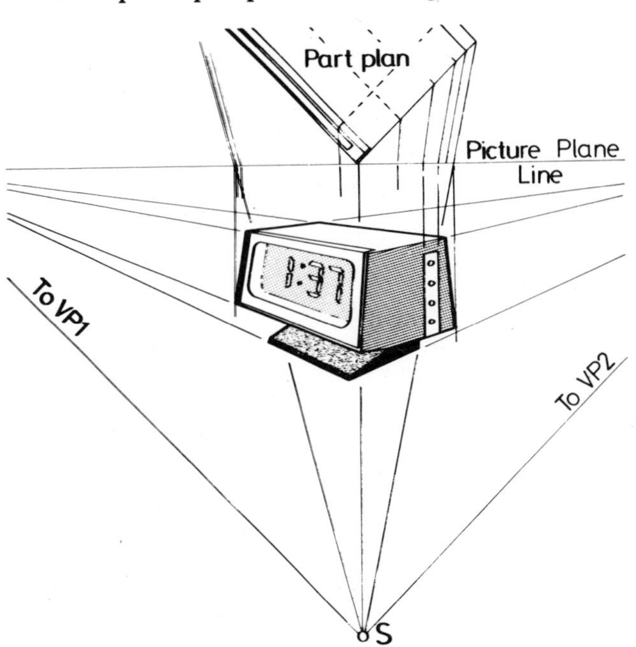

This drawing shows the methods of two-point perspective applied to the drawing of a small article. The whole of the drawing is below the Picture Plane Line.
Note: All vertical measurements have been made along the vertical line between S and the front corner of the plan. All horizontal dimensions are found by drawing lines from S to the relevant point on the plan. Where these lines cross the Picture Plane Line, verticals are taken down onto the perspective drawing.

Estimated perspective drawing

A method of perspective drawing which is commonly used is to *estimate* the positions of VPs by placing them at any position suitable to the size of the drawing sheet or drawing board. Two methods are used – *single-point* and *two-point estimated perspective*.

The principles of estimated perspective

Drawings 1 and 2 – In single-point estimated perspective the single VP can be placed either to one side or above the drawing, its position being estimated in relation to the size of the object being drawn.

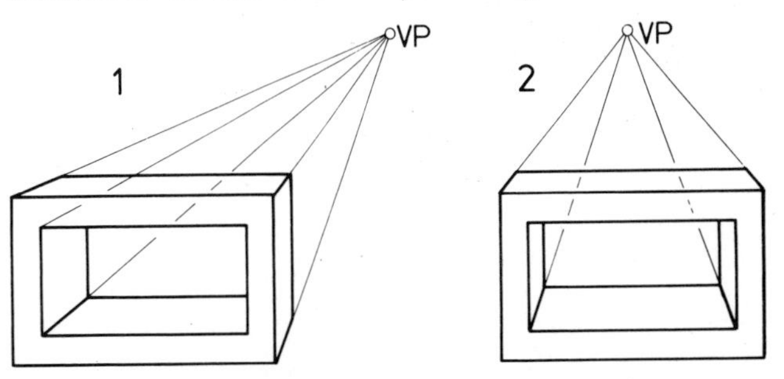

The principles of one-point estimated perspective drawing

Drawing 3 – If the drawing is of a large object such as the interior of a room, the single VP may have to be placed within the boundaries of the drawing at about eye level.

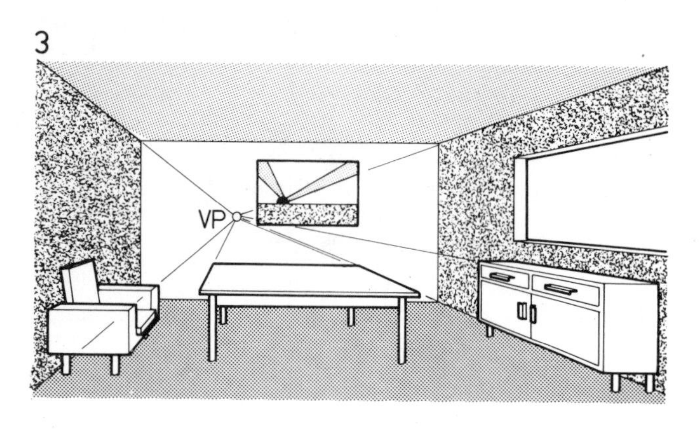

A one-point estimated perspective drawing

Drawing 4 – In two-point estimated perspective, the two VPs can be placed at any convenient position, the only rule being that they must be in line with each other horizontally.

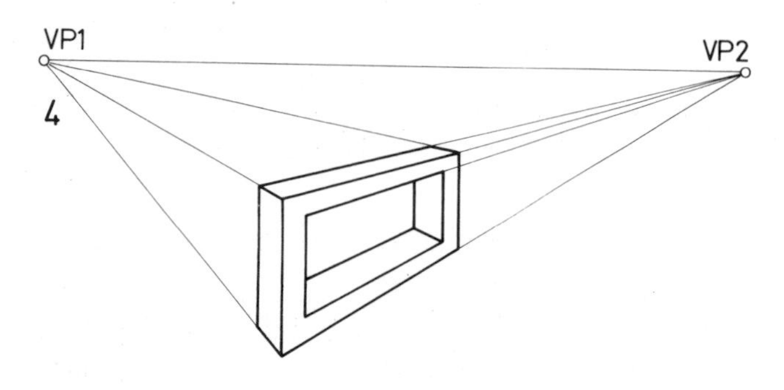

The principles of two-point estimated perspective drawing

Drawing 5 – A two-point estimated perspective drawing of a set of filing drawers. Have the positions of the VPs been well chosen in this example?

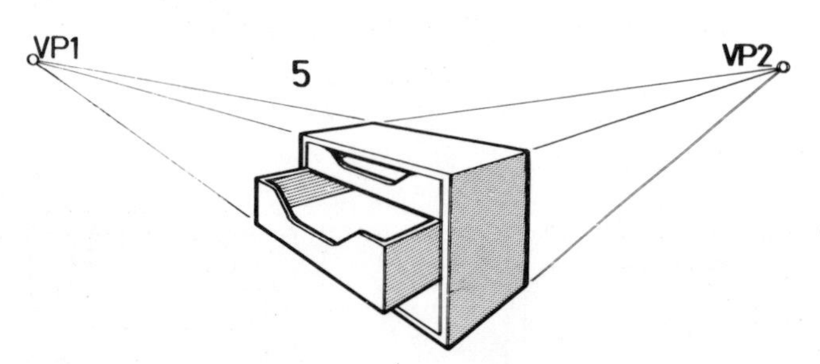

An estimated two-point perspective drawing of a filing cabinet

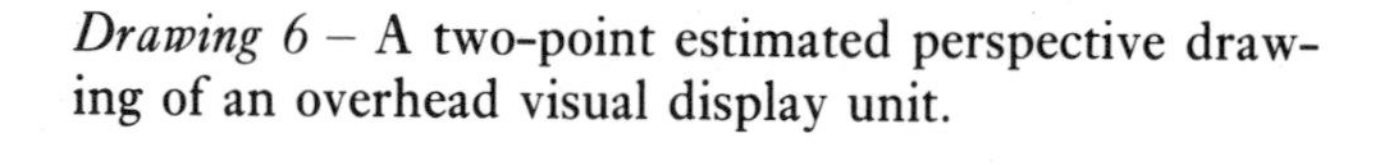

Drawing 6 – A two-point estimated perspective drawing of an overhead visual display unit.

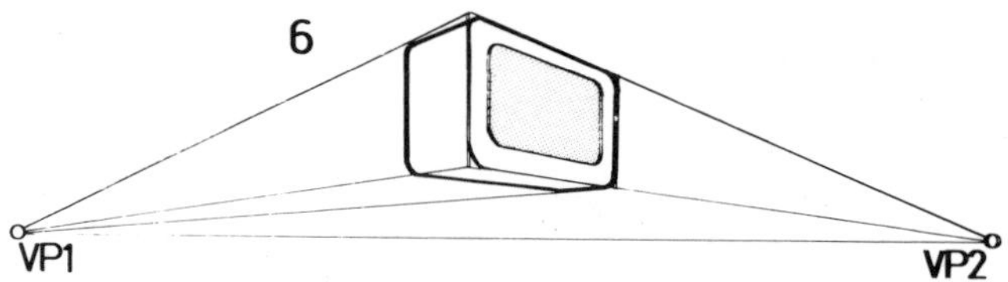

An estimated two-point perspective drawing of an overhead visual display unit

To find the position of equally spaced verticals in perspective drawing

Equally spaced verticals along the sides of an object being drawn in perspective will appear to come closer and closer together the further the verticals are from the front edge of the object. Three methods of locating the positions of such equally spaced verticals in a perspective drawing are shown.

Drawing (a) – ABCD is part of a perspective drawing with the first vertical at 1. Draw B1 and parallel lines to B1 from the bottom of vertical 1 to find vertical 2. Repeat from 2 to give vertical 3.

Drawing (b) – EFGH is part of a perspective drawing. Divide EF into four equal parts. Draw lines from the points so obtained to the VP. Draw the diagonal FH. Where FH crosses the lines to the VP, draw the required verticals.

Drawing (c) – JKLM is part of a perspective drawing. Draw the diagonals JL and KM. Draw vertical 2 at the intersection of the diagonals. Repeat in the smaller areas to give verticals 1 and 3.

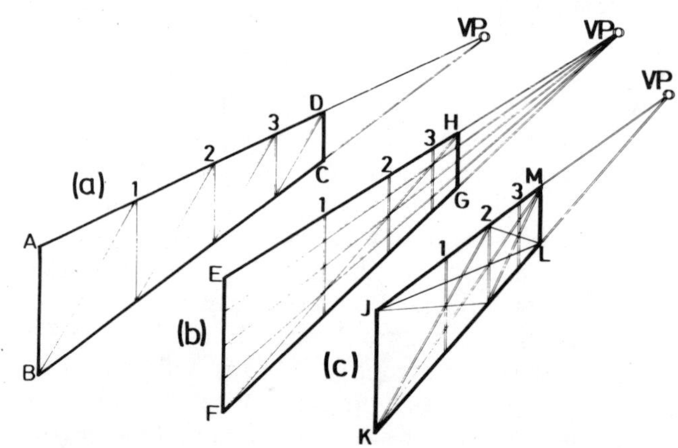

Methods of constructing equally spaced verticals in a perspective drawing

Perspective grids

Perspective grids can be drawn following the methods of two-point estimated perspective and of spacing of verticals. By placing tracing paper over the grid and constructing the required perspective drawing with the aid of the grid lines, the grid can be used over and over again. An enlarged exploded perspective drawing on a grid is shown. The drawing shows the parts of a wall paper clip.

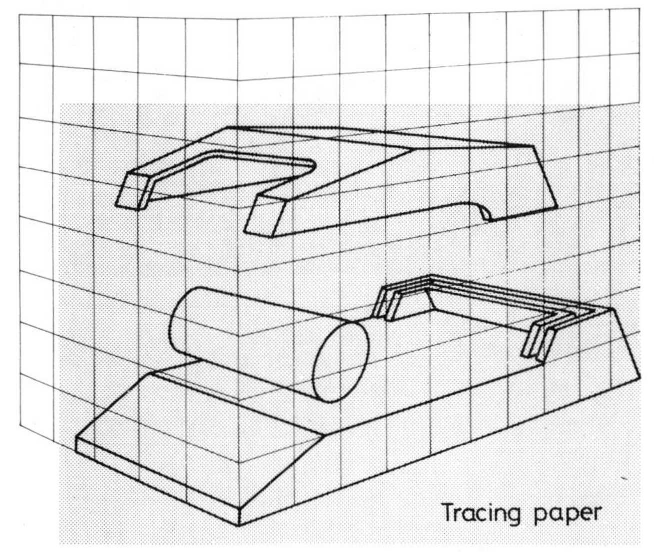

Two-point estimated perspective drawing of a wall paper clip drawn on a grid

EXERCISES

1. In order to achieve a good standard of perspective drawing you need to practice. Commence by drawing in both single-point and two-point estimated perspective with the aid of a ruler, drawings of simple household items such as:
   a match box; a book; a food packet; a spectacles case.

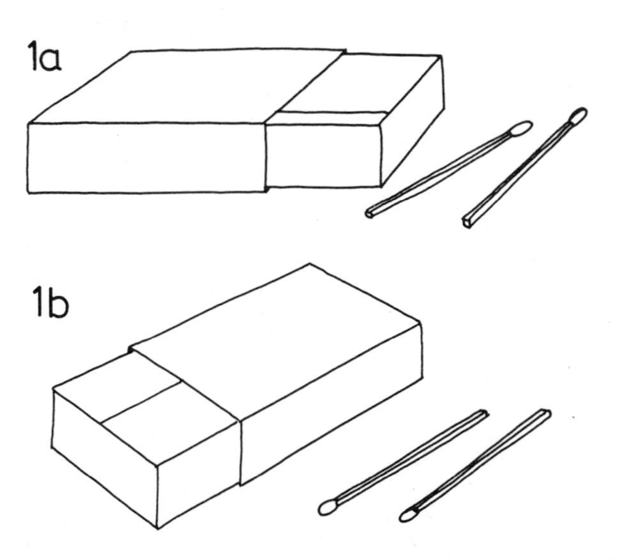

2. Proceed to perspective drawings of items which contain circles, such as:
   a wastepaper basket; a saucepan; a frying pan; drinking glasses; simple childrens' toys (see drawing 2).
Now attempt more difficult items such as:
   a television set; a transistor radio; a music system.

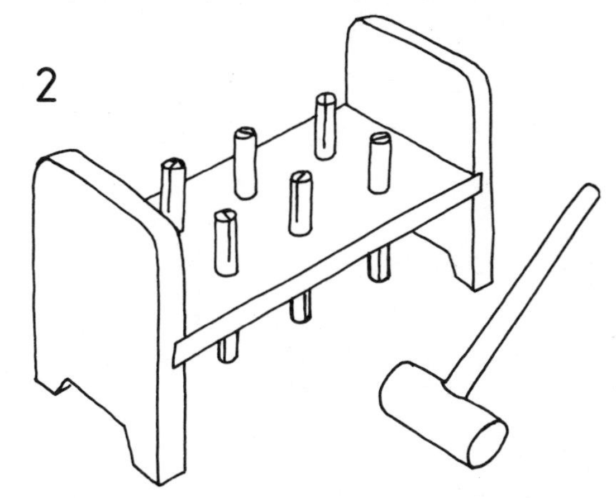

3. Make two-point estimated perspective drawings of some tools such as:
   a mallet (drawing 3); a cutting knife; a chisel; a pair of pliers; a spanner; a saw.

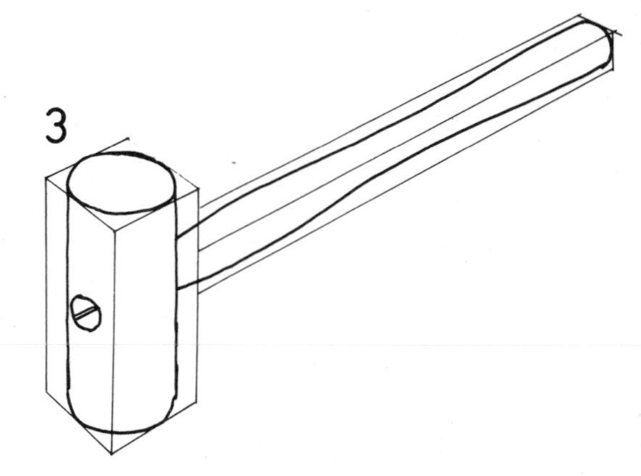

4. Drawing 4 is an outline two-point estimated perspective drawing of a packet which is to hold first-aid finger plasters. Copy the given drawing with the aid of a ruler and design suitable shapes for labels which indicate the contents of the packet.

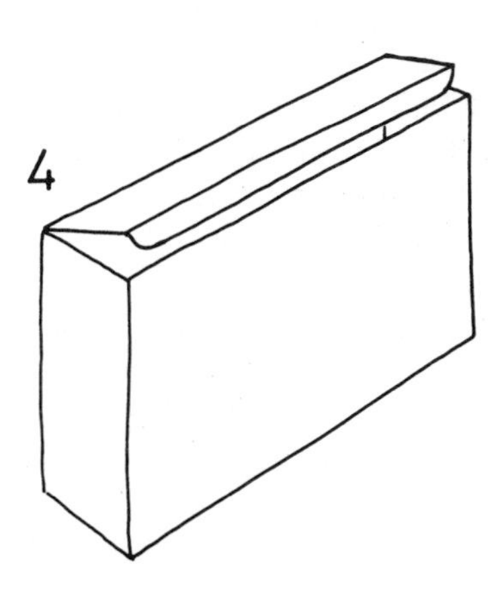

5. Drawing 5 is a First Angle orthographic projection of a toy castle. Make a two-point perspective drawing of the toy. Add suitable shading or colour to give your drawing a 3D appearance.

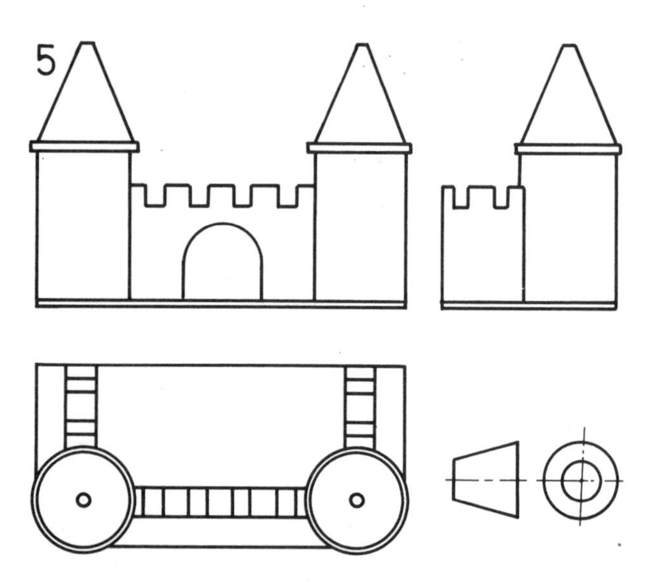

Freehand drawing

The methods of orthographic projection, isometric drawing, cabinet drawing, planometric drawing and perspective drawing can be applied to drawings made freehand either without the aid of instruments or with minimal aid from them.

If pencils are used for freehand drawing, grades B or HB are suitable. Technical pens or ballpoint pens are quite suitable for freehand work, although drawings made in ink are probably best if first lightly drawn in with pencil, then inked in over the pencil lines.

Freehand drawings in orthographic projection

On square grid paper

Square grid papers are very suitable for freehand drawings of orthographic views. The lines of the squares give good assistance when drawing horizontal and vertical lines. The side lengths of the squares are a good guide to scaling of the sizes needed for the views.

Two orthographic projections drawn freehand on square grid papers are shown. Such drawings can be made as a preparation before commencing a finished orthographic projection which is to be made with instruments. Errors of detail or errors of projection can be seen in such preparatory freehand drawings before commencing work on a final drawing. Note that the sprayer handle is shown again on page 104 as a coloured isometric drawing.

6. A photograph of a simple clothes peg box is given. Draw several estimated perspective drawings showing other designs for a clothes peg box.

7. Make single-point or two-point perspective drawings for designs for items of jewellery such as:
a brooch; a neck pendant; a ring; an earring.

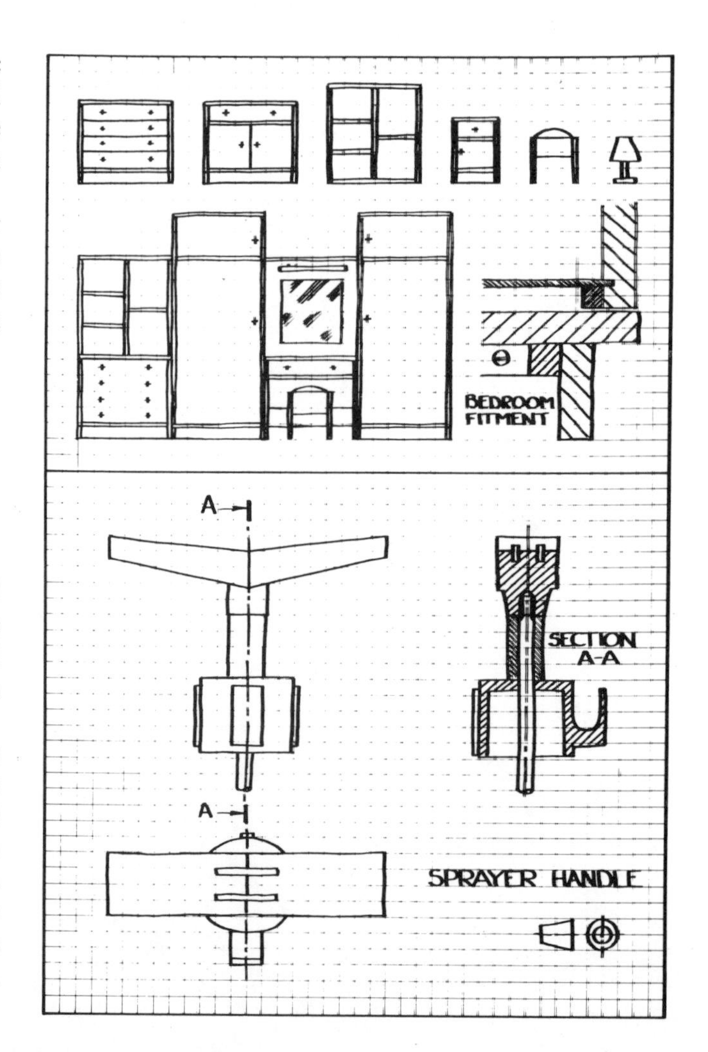

On plain paper

If you are making freehand orthographic projections on plain paper, a few guide lines can first be drawn with the aid of a ruler. Once the guide lines have been drawn, the ruler can be set aside and details of the drawing can be filled in freehand. Two examples are shown. The first shows a suggestion for double glazing a window. Such a drawing might have been made to allow a discussion on how double glazing can be constructed. The second drawing shows a picnic table drawn in Third Angle projection.

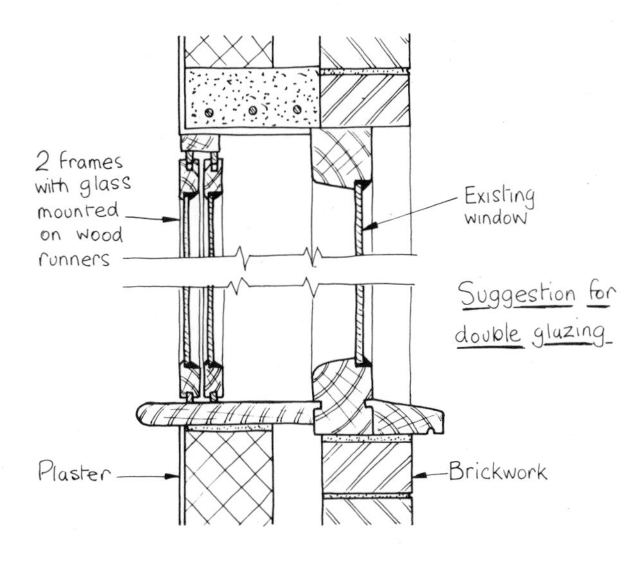

Sectional view drawn freehand on plain paper

Freehand three-view orthographic projection on plain paper

Freehand isometric drawings

On isometric grid paper

The lines of the grid give an accurate guide to the sloping and vertical lines of the drawing at correct isometric angles. The lengths of the sides of the triangles of the grid (usually either 5 mm or 10 mm) assist in obtaining good proportions between parts of the drawing. Drawing 1 shows a series of suggestions for allowing parts to slide into each other; drawing 2 shows a clip which is to fit on a cylindrical rod; drawing 3 shows a case for a fan heater; drawing 4 is an aluminium hat and coat peg; drawing 5 shows a garden cold frame.

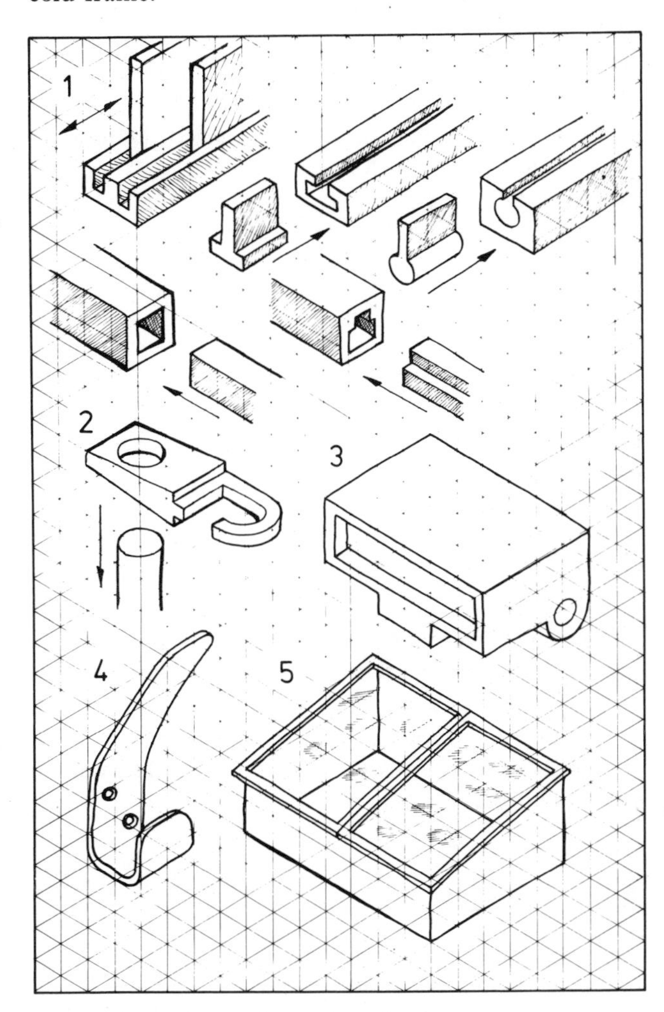

Freehand drawings on isometric grid paper

On plain papers

When making freehand isometric drawings on plain paper, an isometric 'box' of overall length, width and height may first be drawn in faint pencil lines with the aid of a 30°, 60° set square, before filling in details freehand within the 'box'. The two stages of making such a drawing are shown in the freehand isometric of an infant's walking trolley.

Drawing 1 – The outlines of the box, handle supports and wheels have been drawn on approximate 30° and 90° lines with faint pencil lines.

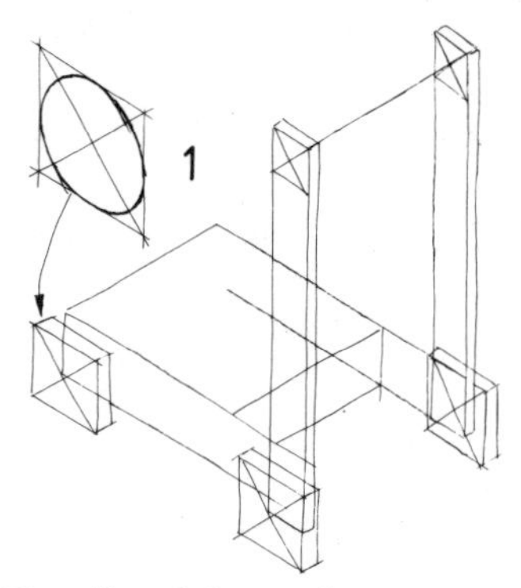

Drawing 2 – Details of the various parts of the trolley have then been lightly drawn within the isometric outlines of the 'boxes' and, when satisfied that reasonable shapes have been constructed, the completed sketch is lined-in and construction lines rubbed out.

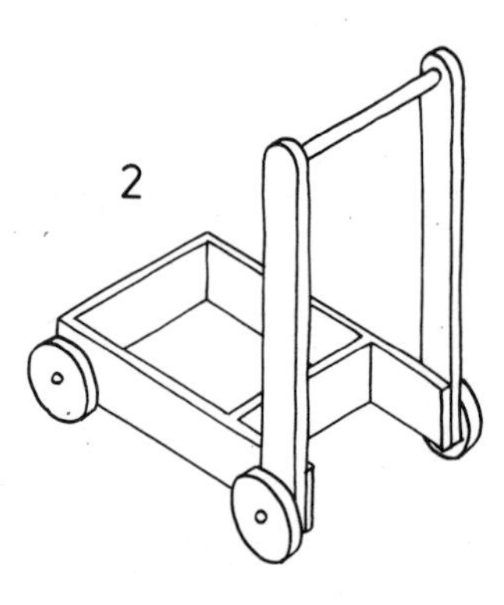

A second freehand isometric drawing shows an exploded drawing of the parts of a wall-light fitting. 'Box' outlines of each of the five parts of the fitting were first drawn with faint pencil lines. The finished drawing was then lined-in within the 'boxes' and unwanted constructions erased.

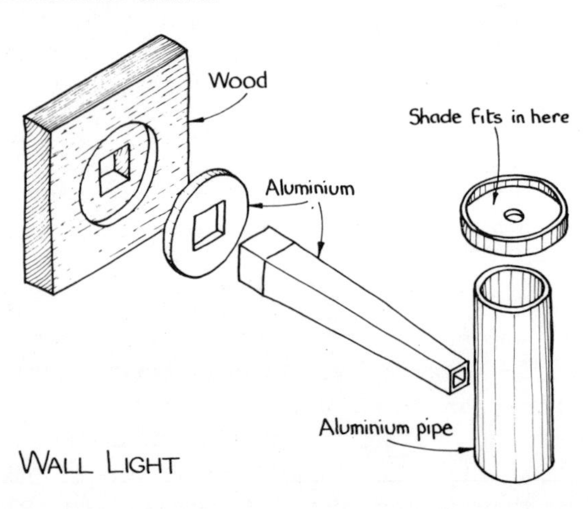

Freehand exploded isometric drawing of a wall-light fitment

Freehand cabinet projection drawing

Cabinet projection provides an easy and quick method of drawing a pictorial view of an object, but the resulting illustration can look somewhat distorted. Two examples of freehand cabinet drawings are shown. Start with a 'box' outline and then fill in details within the 'box' before lining-in and erasing unwanted construction lines.

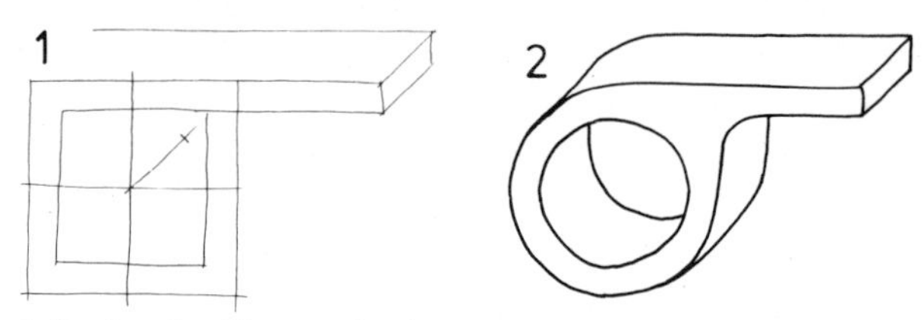

A freehand cabinet projection

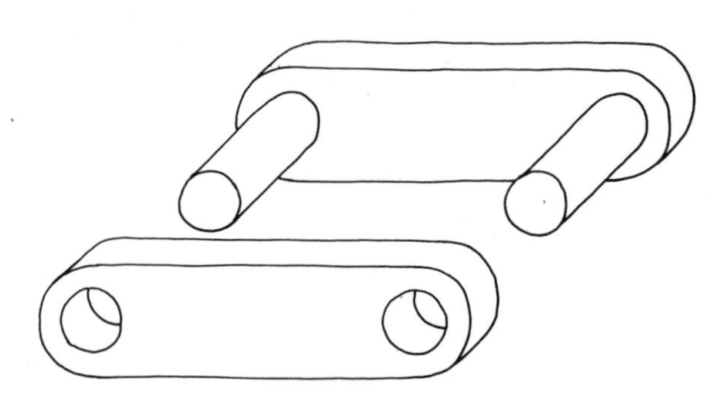

A freehand exploded cabinet projection

Freehand planometric drawing

The given example shows a design suggestion for the positioning of display stands and a small office in a shop. A plan of the floor was first drawn on approximate 30° and 60° axes. Verticals taken at the corners of each rectangle give the third dimension of the drawing. Although a ruler could be used to assist in making the drawing, the finished lines are drawn freehand.

A freehand 30°, 60° planometric drawing

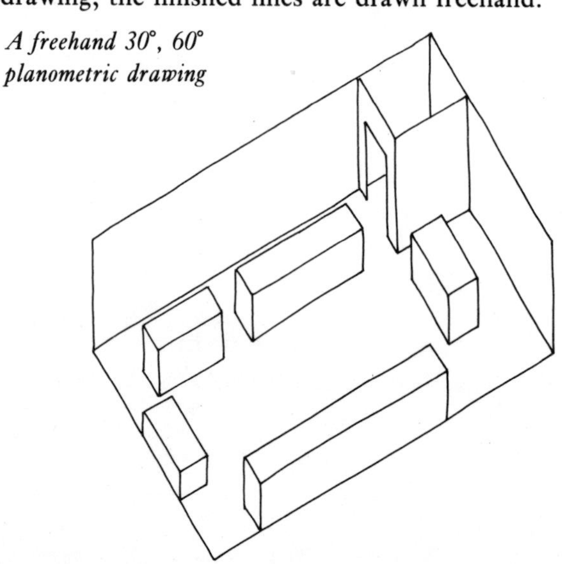

Freehand perspective drawing

The methods of perspective are those most commonly used for making freehand pictorial drawings in graphic design work. The general method of producing a simple freehand perspective drawing is shown in stages as follows. The drawing is of a transistor radio.

Drawing 1 – Construct a 'box' outline into which the completed drawing will fit. Use faint pencil lines. Most people find a sideways movement of the pencil from left to right will produce better straight lines than attempts at an up and down movement. Because of this, it is easier to move the drawing paper around to enable lines to be drawn sideways. Drawing 1 describes the movement. A clutch pencil with an HB grade 'lead' is shown. Note that the sloping lines of the 'box' converge together towards vanishing points imagined as being in the distance to the left and right of the front vertical line of the 'box'.

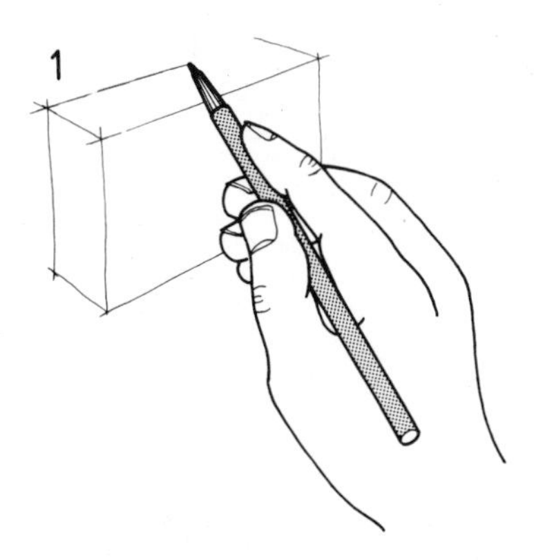

Drawing 2 – Within the 'box' outline draw the perspective squares and rectangles outlining parts of the required illustration.

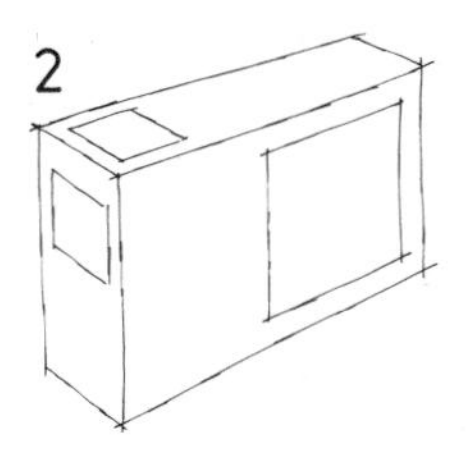

Drawing 3 – This shows how a method of drawing perspective circles can be applied. Draw a circle. Draw its diagonals. Where the diagonals cross the circle draw lines parallel to the sides of the square. It will be seen that each of the resulting small squares has sides which are about $\frac{1}{6}$ of the length of the square outlining the circle.

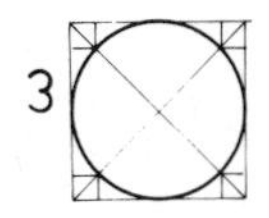

Drawing 4 – Using the method of drawing 3, draw a perspective 'square'. Draw its diagonals. Draw smaller squares with sides of approximately $\frac{1}{6}$ of the sides of the main square. Where the small squares meet the diagonals are points through which the perspective circle can be drawn.

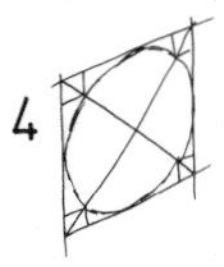

Drawing 5 – Two further examples of perspective 'circles' using the methods shown in 3 and 4.

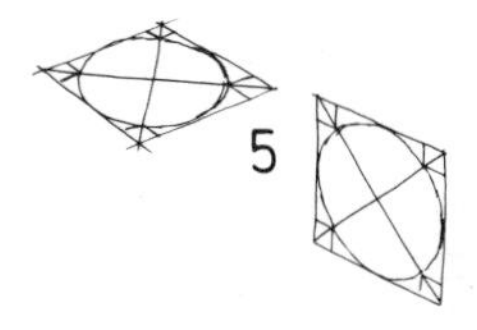

Drawing 6 – The three perspective circles of the controls and loudspeaker of the radio are constructed as shown in 3, 4 and 5.

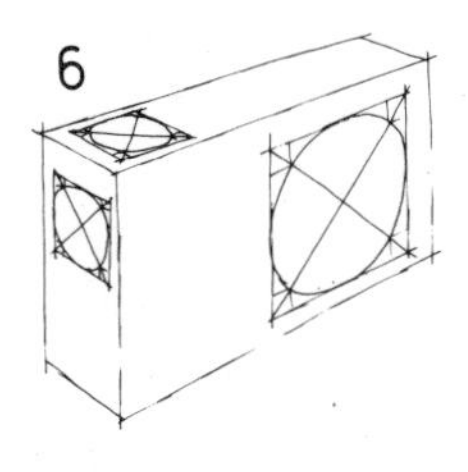

Drawing 7 – Further details of the required pictorial drawing are then constructed and the finished lines lined-in. Unwanted construction lines can now be erased.

Dimensioning a freehand drawing

Dimension lines are taken parallel to the outline of the drawing and the dimensions are placed above the lines at an angle comparable to the lines of the drawing.

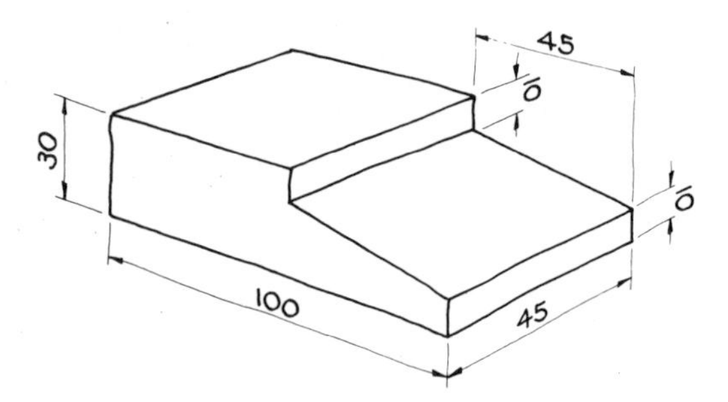

Method of dimensioning a pictorial drawing

Examples of freehand perspective drawings

Three examples are shown – a small dressing-table stool, a door handle and a daylight photographic transparency slide viewer. In each example the stages in producing the 'boxes' and the finished drawings are shown. A fourth drawing shows part of a design sequence of drawings in which the construction of a bookcase is given in a sheet of freehand drawings – all except one being perspective drawings.

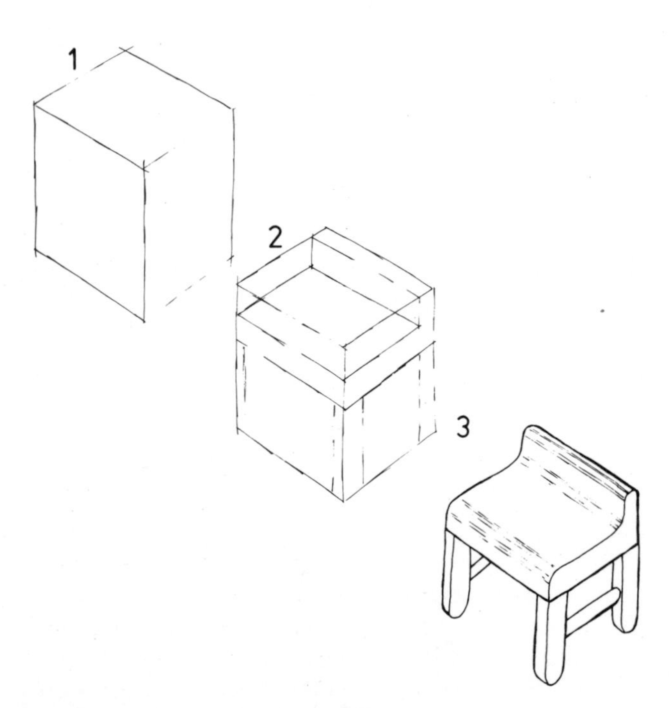

Stages in making a freehand perspective drawing

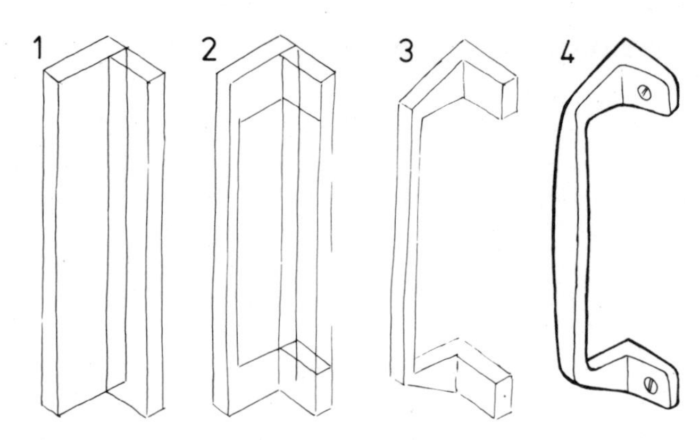

A second example of the stages in drawing a freehand perspective drawing

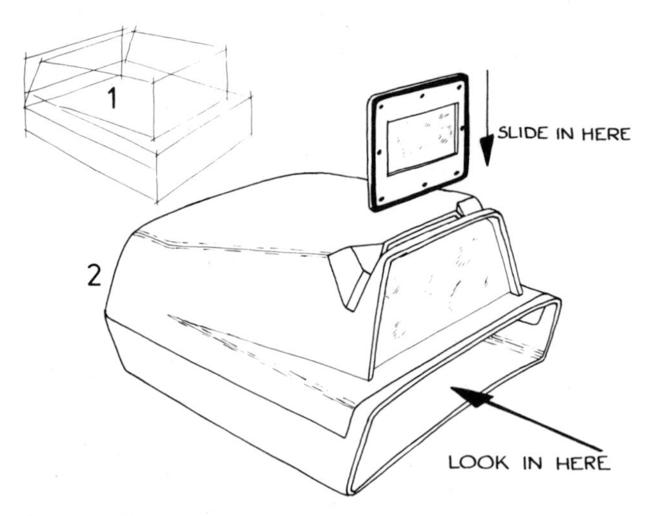

An example of a freehand perspective drawing of a daylight transparency viewer

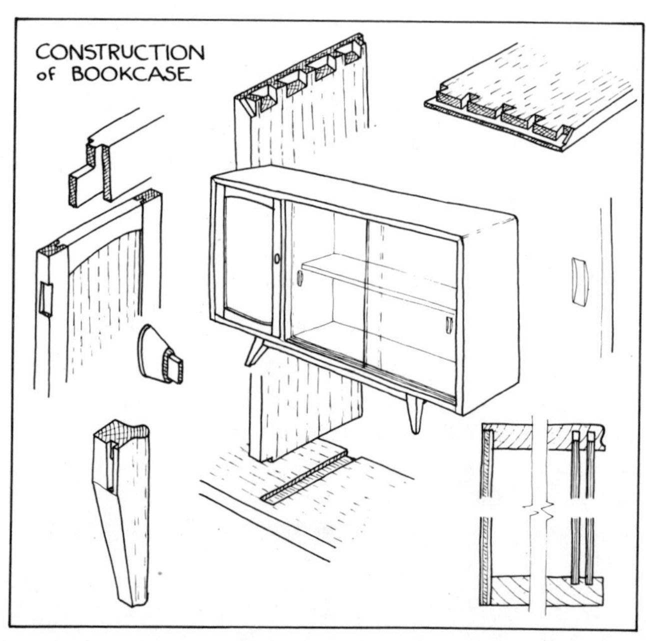

A design consisting of freehand perspective drawings showing the construction of a bookcase

Graphs and charts

Graphs and charts play an important role in communicating by means of graphics. A very large range of different types of graphs and charts is used. They should make a speedy impact by showing in an easily understood visual form a clear message. Graphs and charts may show comparisons or trends, movements, increases or decreases, sequences, a series of operations or statistical information in a visual form. They may include symbols designed by the person who draws them or symbols such as those given in BS 4058. A chart involving symbols from BS 4058 is given on page 34.

Line graphs

Line graphs are commonly used to indicate mathematical and statistical trends. Two line graphs are included here. The first shows the results of the series of censuses of population taken in the United Kingdom at ten-yearly intervals between 1801 and 1981. There was no census in 1941 and the position for that year was *interpolated*. The second graph gives three line graphs on the same drawing.

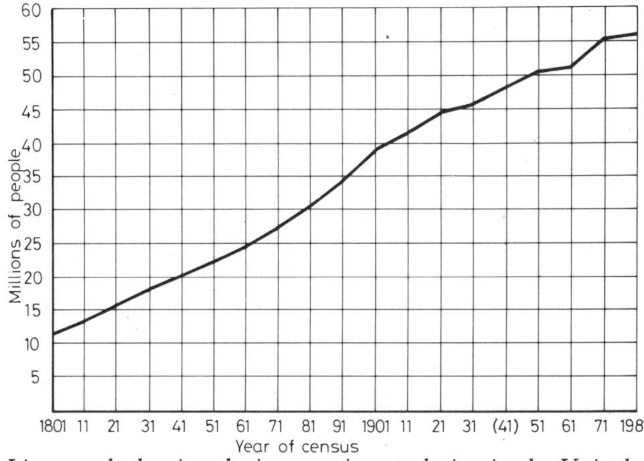

Line graph showing the increase in population in the United Kingdom

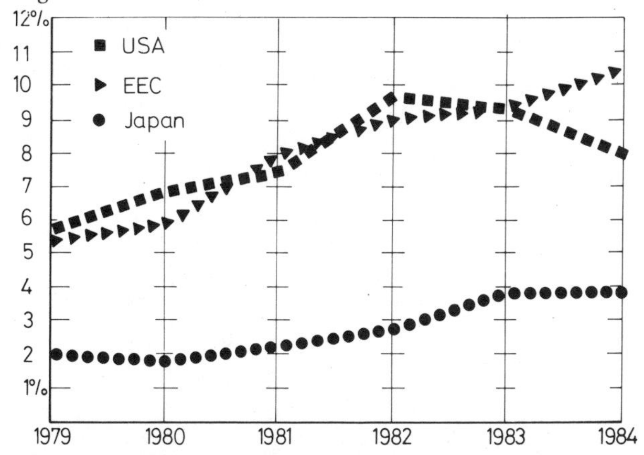

Line graphs showing the percentages of unemployed people in the USA, the EEC and Japan

Note the similarities between these two graphs.

1. The base line (abscissa) in each is divided into equal spaces, each space representing an interval of time.
2. The verticals (ordinates) are also divided into equal spaces, each representing millions of people or units of percentages respectively.
3. Lines are drawn through plot points so that the graph lines are more or less continuous. A rising line denotes an increase in numbers, a line falling downwards denotes a decrease in numbers.

In the second of these two line graphs the lines are in the form of symbols, each of which denotes a country or group of countries. If this type of symbol is used on a graph, a key to the identity of each symbol must be included in the drawing.

Bar graphs

In bar graphs the length of each bar is drawn in proportion to the size or number of the item it represents. The bars may be drawn vertically or horizontally. Vertical bar graphs are usually called *histograms*.

Three histograms are shown here. The first compares the relative incomes per head of population of the ten countries of the European Economic Community in 1981. Each bar is the same width. The average income per head of the whole population of the EEC is shown as 100. Each bar is labelled with the name of the country it represents. There is no need for a vertical scale because this graph is only intended to represent comparisons.

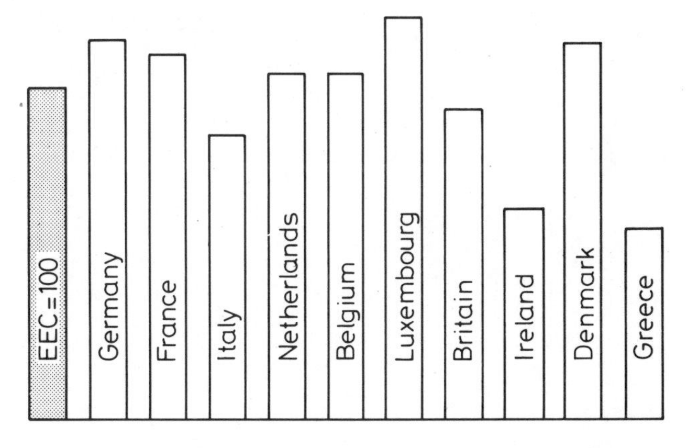

Histogram showing relative incomes of members of the EEC in 1981

The second of the three histograms consists of twelve double bars. This graph shows the average monthly temperatures in Delhi. The base (abscissa) is divided into twelve parts, each representing a month of the year. The vertical ordinate is divided into degrees Celsius. This form of graph depends upon temperature readings taken over a period of years and the resulting figures can only be regarded as averages.

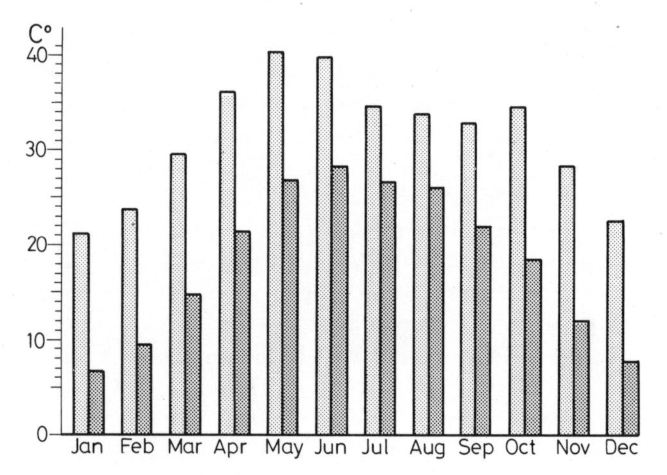

Histogram showing monthly average maximum and minimum temperatures in Delhi, India

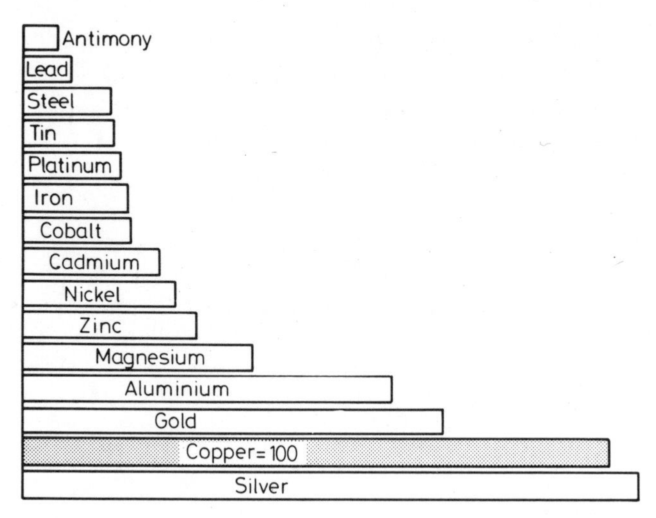

Bar graph showing comparative electrical conductivity in metals with copper taken as 100

The third of the three bar graphs illustrates the same census figures as those given in the line graph on page 71. In this graph, however, instead of taking the whole population of the UK for each census, the separate populations of England and Wales, Scotland and Northern Ireland have been illustrated in graph form. The graph, therefore, gives comparisons between the growth of population in the three parts of the UK. Different depth of shading indicate each of the three populations.

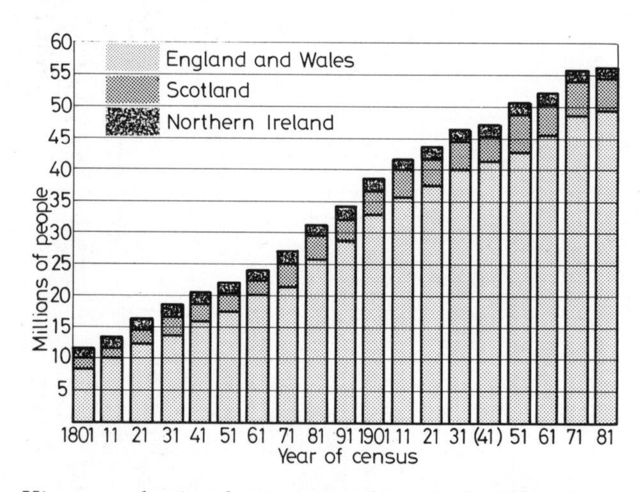

Histogram showing the rise in population in the UK split into England and Wales, Scotland and Northern Ireland

Horizontal bar graphs

A typical example of a horizontal bar graph is given. This compares the relative electrical conductivity of fifteen metals with that of one of them – copper. Note that the graph does not show actual values, but only comparisons with copper. The graph is arranged in ascending order of conductivity, with the poorest conductor at the top and the best at the bottom.

Pie graphs

Pie graphs (often called *pie charts*) consist of circles divided into sectors, each of a size proportional to the item the sector represents. Because each sector size is based upon the angle its arms makes to the circle centre, before a pie graph can be drawn the angles of its sectors must be calculated.

In the pie graph shown, the proportional amounts of money spent on the items making up a family's total budget are illustrated. The calculations required to calculate the angles of the sectors are:

Housekeeping	$25\% = \frac{25}{100} \times 360° =$	90°
Mortgage	$23\% = \frac{23}{100} \times 360° =$	82.8°
Transport	$12\% = \frac{12}{100} \times 360° =$	43.2°
Furniture etc.	$10\% = \frac{10}{100} \times 360° =$	36°
Clothing	$8\% = \frac{8}{100} \times 360° =$	28.8°
Savings	$7\% = \frac{7}{100} \times 360° =$	25.2°
Energy	$5\% = \frac{5}{100} \times 360° =$	18°
Rates	$5\% = \frac{5}{100} \times 360° =$	18°
Leisure	$5\% = \frac{5}{100} \times 360° =$	18°
Total =	100%	Total = 360°

In a pie graph there is no real need to work to a decimal place of a degree so, constructing the chart with the aid of a protractor, the angles used were 90°; 83°; 43°; 36°; 29°; 25°; 18°; 18°.

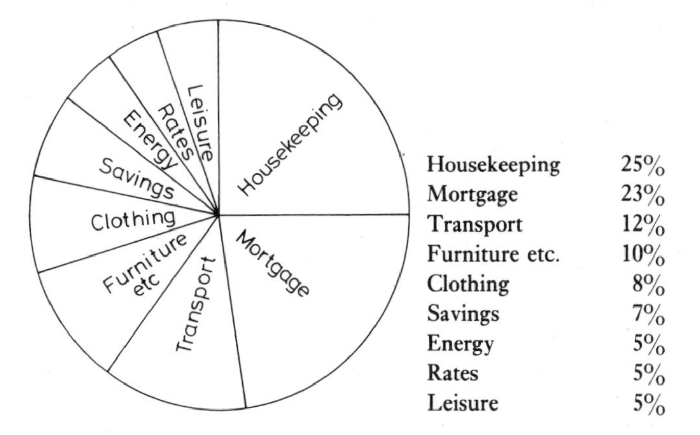

Pie graph showing the distribution of expenses within a household budget

Pictorial pie graph

An example of a pie graph drawn in a pictorial form is given. The graph is an isometric drawing showing a disc as if it were to be cut into a number of different size wedge-shaped portions. The angles of the sectors were first worked out on a circle and the sizes so found were then transferred to the isometric drawing. An electronic calculator was used in the calculations and the angles, taken to the nearest degree, were:

Salaries	$£1\,000\,000 = \frac{8}{21} \times 360°$	$= 137°$
New Plant	$£450\,000 = \frac{6}{35} \times 360°$	$= 62°$
Materials	$£400\,000 = \frac{16}{105} \times 360°$	$= 55°$
Power	$£335\,000 = \frac{67}{525} \times 360°$	$= 46°$
Transport	$£300\,000 = \frac{4}{35} \times 360°$	$= 41°$
Depreciation	$£140\,000 = \frac{28}{525} \times 360°$	$= 19°$
Total	£2 625 000	360°

The fractions used in the calculations are the fractions of the total cost. As an example:

$$\text{Power at } £335\,000 = \frac{335\,000}{2\,625\,000} = \frac{67}{525}$$

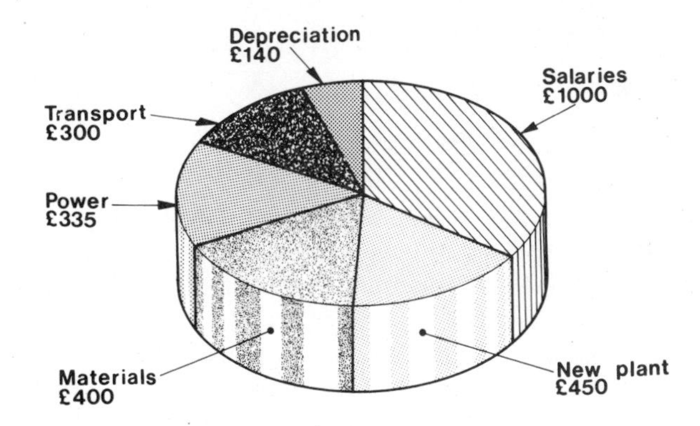

Pictorial pie graph showing expenses (in £1000s) of a corporation undertaking

Other pictorial graphs

Two other pictorial graphs are shown. The first shows the growth in numbers of employees in a factory over a period of five years. The second shows the increase in billions of barrels of proven oil reserves discovered throughout the world between 1976 and 1981. Each one billion barrels is represented by a pictorial drawing of a barrel.

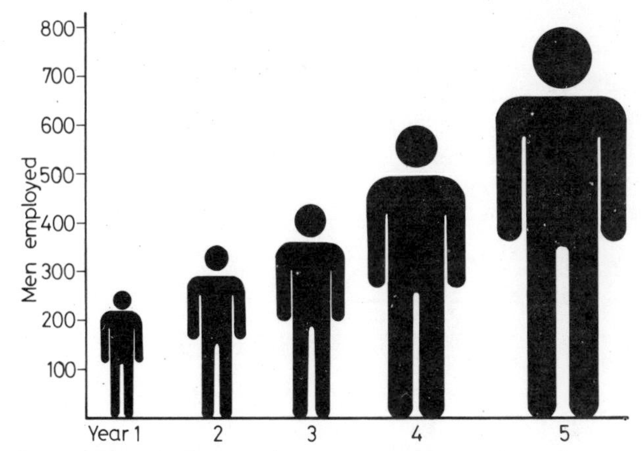

Pictorial graph showing the increase in the numbers of employees in a factory over a period of 5 years

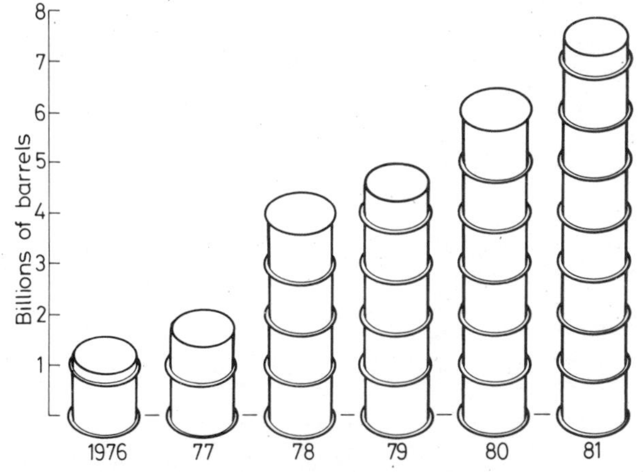

Pictorial graph showing the increase in world-wide oil reserves discovered between 1976 and 1981

Flow charts for planning and control of projects

A discussion on flow charts for project planning is far beyond the scope of this book, but the three flow charts here describing the construction of a wooden garden fence will give an indication of the types of graphics associated with the planning and control of a project.

The following table, on which all three charts are based, describes the stages in erecting the fence, together with the order of procedures and the dependency of each stage in the process on the others. The table also gives estimates of the time necessary for the completion of each stage.

Stage	Activity	Dependent upon	Estimated hours required
1	Mark out post holes	—	1
2	Bore post holes	Stage 1	6
3	Test hole depths	Stage 2	1
4	Mix Concrete	—	1
5	Concrete posts	Stages 1, 2, 3 & 4	3
6	Brace posts	Stage 5	1
7	Collect all wood	—	2
8	Fit rails	Stages 6 & 7	2
9	Saw boards to length	Stage 8	2
10	Fit boards	Stages 8 & 9	4
11	Paint with preservative	Finish	2

Simple flow chart

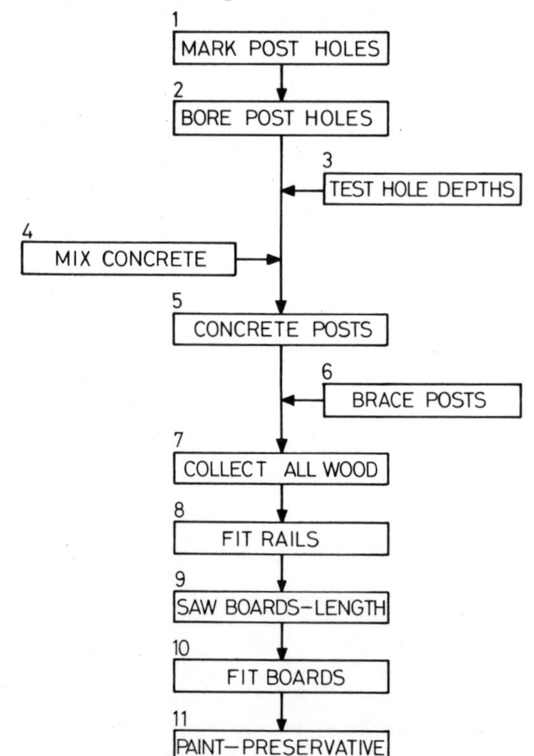

The first of the three charts shows the various activities arranged in sequence, each being printed in a 'box'. Arrows on the lines indicate the order in which the activities occur and each activity is numbered. Activities 3, 4 and 6 are placed to one side because, although essential to the construction of the fence, they are activities subsidiary to the construction.

Activity-on-node chart

The second of the three charts shows nodes as a series of 'boxes'. The activities are printed within the boxes and arrows point to the next activity in the sequence.

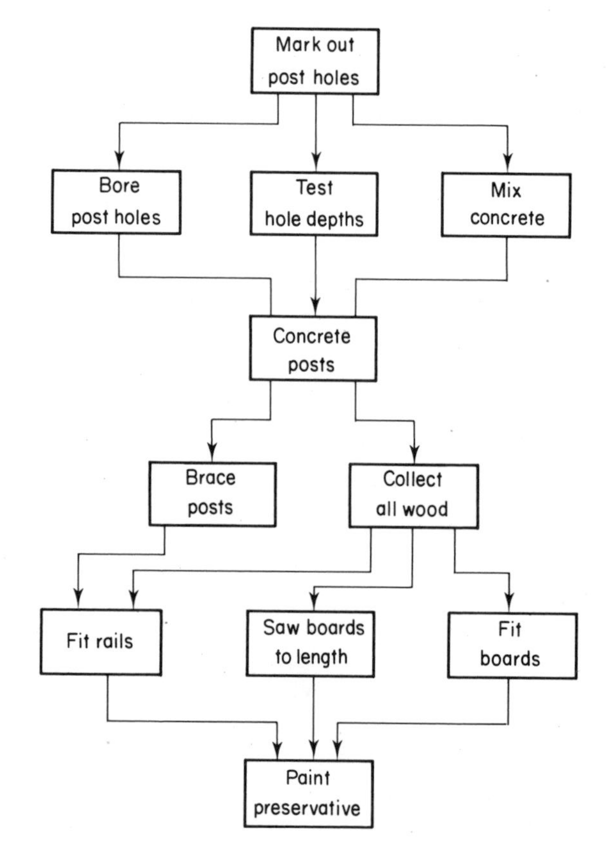

An activity on node sequential chart

Activity-on-arrow charts

The third of the three charts is an activity-on-arrow chart. Each activity in the procedure is printed above the arrowed line, the figures within the circles denoting the completion of the activity. Broken lines in the chart indicate dependent activities which do not necessarily take up time. The times taken by each activity are shown by numbers at the end of each arrow.

Activity flow charts such as those given are sometimes referred to as *sequential charts*.

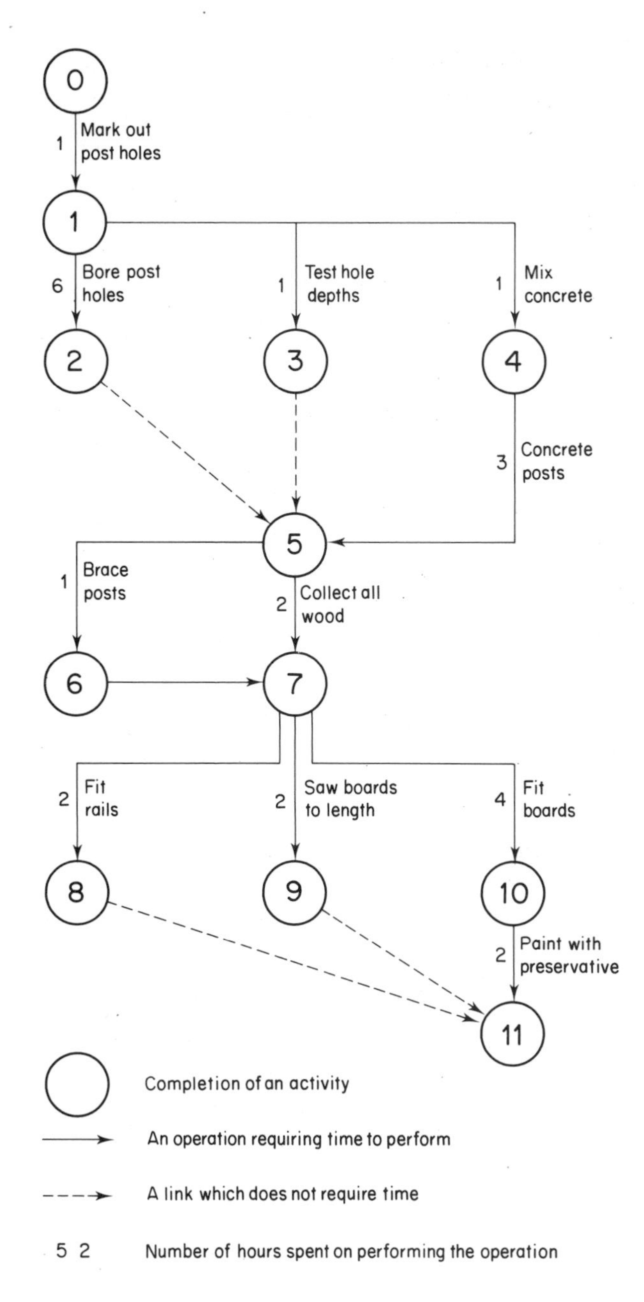

An activity-on-arrow sequential chart

Pictorial sequence charts

Sequence charts include a number of illustrations describing the sequence of the processes required for making a construction or assembling the parts of a construction. The sequence chart given here shows how to solder the joints of a sheet metal box. This chart describes, in six freehand drawings, the sequence of operations. Each illustration has been drawn from sketches made when watching another person working through the processes involved. Parts of each drawing have been shaded. The shading could be line shading, shading with dry transfer tints, with coloured crayons, with marker pens or with water-paint colours.

Because the drawings were developed from sketches made at the place where the work took place, the six drawings are in perspective. Each drawing includes only those tools and items necessary to show clearly the processes taking place; any other tools or equipment actually present when the sketches were made have been ignored and not included in the finished drawings. A title in quite bold stencil lettering has been included with each drawing, and labels given to details such as tools and materials. The labels are in smaller stencil lettering to avoid confusion with the drawing titles. Each drawing is also labelled with a large stencilled figure within a 'balloon' to indicate clearly the order in which the various processes should occur. When attempting a sequence drawing of this type, it is essential that the processes involved are clearly understood by the person making the drawings. If he or she is not fully conversant with the methods by which the processes are performed, then they should question the operator to make quite certain that each drawing conveys correct information. It is also as well to ask the operator to look at the completed drawings to check that they are technically correct.

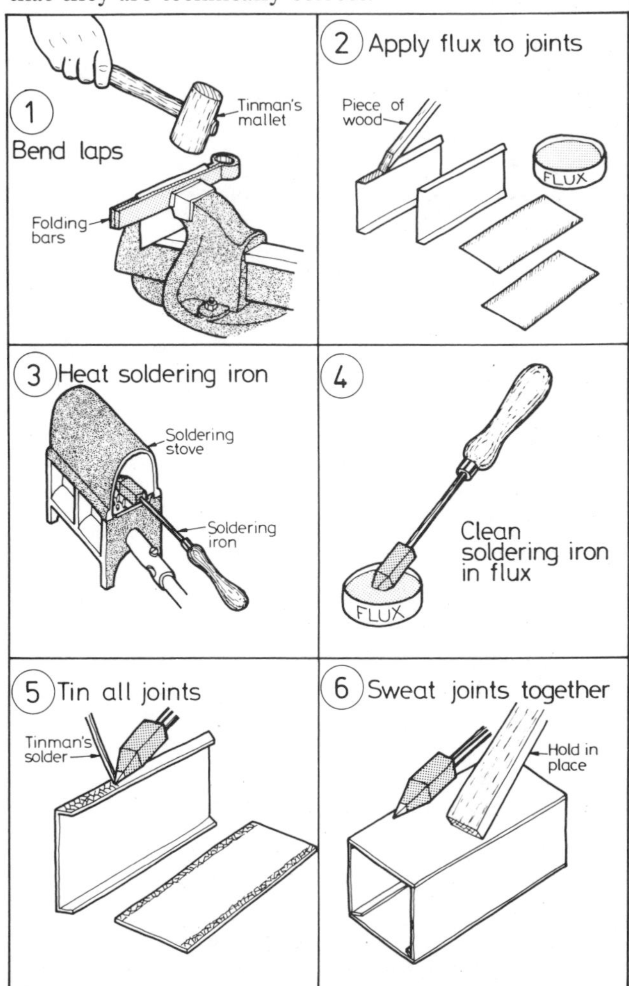

A pictorial diagram showing the sequence of operations involved in soldering the joints of a sheet metal box

SECTION 3

Models and model making

It is best to regard the making of models as an important part of the design process. During the investigation stage and when you consider that a good solution has been found, the making of a model may be necessary. Models allow one to examine the appearance, the shape and form, the colours and the proportions of a suggested design before it is finally made. Models allow tests for strength, for correct working and, in general, for whether a design is fit for the purpose for which it is intended. The making and testing of models will take time and may appear to be unnecessarily expensive in materials and equipment. However, if a model shows up faults in a design, time and money will eventually be saved. This is because there is a better chance of the design being successful after faults have been 'ironed out' in a model.

Purposes of model making

Models are made:
1. to satisfy oneself that a design solution looks as if it is a successful answer to a design brief;
2. to assess which materials are most suitable for constructing the design;
3. to test the efficient working of electrical, electronics, pneumatic or hydraulic circuits which have been designed on paper;
4. to test the efficient working of mechanical systems which have been designed on paper;
5. to test the strength of structures which have been designed on paper;
6. to suggest methods by which a design may be constructed;
7. as mock-ups to show others what a design looks like and how it works; to allow others to suggest possible modifications before a design is realised.

Types of models

Scale models – Made in sizes which are smaller than the final design.
Full size models – When a design is itself small, e.g. an item of jewellery.
Larger than full size – When a design is itself tiny.
Electrical and electronics circuits – Made up from modules or kits on which components are mounted and between which conductors can be temporarily fitted to form a circuit.
Pneumatics and hydraulics modules – To allow circuits to be built up for testing.
Mechanical models – To test correct functioning of a design or part of a design.
Structural models – Mainly to test strength.
Mock-ups – Usually full size, not necessarily made to exact shapes, to allow features such as ergonomics to be tested.
Working models – To explain how mechanisms function.
Models made from paper or cardboard – To assist in learning what geometrical solids such as cubes, cylinders, cones, pyramids look like or how they can be assembled together to create new forms.
The making of model aeroplanes – Model road vehicles, model trains and the like as a hobby.
Human figure models – To check on ergonomic problems.
Stage models – To show how stage scenery for a play will appear.

Materials for model making

Sheet materials
Sheet materials such as paper, cardboards, thin plywood, hardboard, plastics sheet, aluminium foil, tinplate, sheet aluminium and sheet steels can be used to represent surfaces. They can also be made into 3D forms by constructing surface developments of the forms and joining the surfaces of the developments together (see pages 79 to 86).

Solid materials
Solid 3D forms can be modelled from blocks of wood – in particular balsa and jelutong – , blocks of expanded polystyrene foam, old plastics container bottles, empty food packets etc.

Strip materials
Strips of wood from off-cuts from sawing wood to size are of particular value. Dowels, strips of wire, pins, Meccano, string and cotton can also be used.

Glues
PVA wood glue is possibly one of the best general glues to use when making models. It forms strong joints and is clean to use. Pastes are suitable for paper and card.

Other joining materials
Self-adhesive tapes such as clear Sellotape, masking tapes and double sided self-adhesive tapes are of particular value. Nails and screws, pins, staples fixed with the aid of stapling machines, paper clips and 'bulldog' clips are good for holding small parts together while glue is setting. 'Blu-tack', soft iron wire and fuse wire are also useful.

Pliable materials
Materials such as Plasticine, clays, plaster of Paris and cement can be used.

Modular components
When building model circuits for electrics, electronics, hydraulics or pneumatics, purpose-made modular components which can be easily and quickly linked together to form test circuits are excellent. A number of firms make this form of modular equipment. Among others can be mentioned LEGO®, Meccano, Fischertechnik, the Danum Trent kits, Hybridex kits and other kits as supplied by firms such as Economics.
Note: It is only rarely that models are made from actual materials from which the final design will be made, but the surfaces of some models must be painted or drawn on to represent the materials to be used in the finished design.

Models for graphics

Mock-ups showing how a sheet of graphics will appear can be regarded as models. Instead of finished drawings, areas can be represented by blocks of lines, by coloured papers, by shading or by colouring in order to show the general effect or appearance of a sheet of drawings. When satisfied that a mock-up shows what is required, the finished item of graphics can then be drawn. Some examples of this form of modelling are given on pages 24 to 25.

Models of buildings

Models of buildings, either made from painted blocks of solid materials, or made up as developments from paper or cardboard, are frequently used to show how a building wall will be set in its gardens, how a street of houses will appear when a whole street has been built, how an area in a town will appear after it has been rebuilt or to place suggestions for town design changes before the public in model form.

Examples of models

A number of photographs of models are given on this page and pages 78 and 79. Further photographs of models are given on page 127. These photographs show examples of the value of models in the design process, covering topics such as:

- models for craft projects;
- models for technology projects;
- models for testing structures;
- models for checking the function of mechanisms;
- models to check correct functioning of circuitry.

Note: The models for checking the correct functioning of circuitry would be of value in the design process as follows.

1. The circuit is designed 'on paper' and drawn using British Standard symbols to represent the components in the circuit.
2. A model of the circuit is constructed using modular kit units.

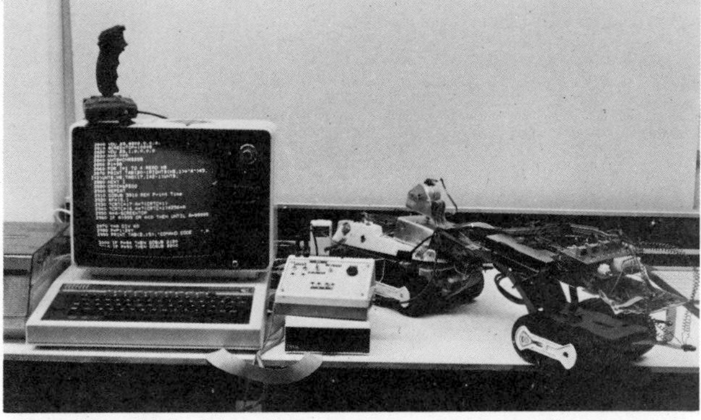

A model vehicle driven by a computer program via an interface

3. The model circuit is tested to check whether it functions properly or not.
4. If the model fails to function, the circuit is re-designed.
5. When tests with the model show correct functioning, the final circuit is constructed as a permanent feature to be incorporated into the design in which it is to function.

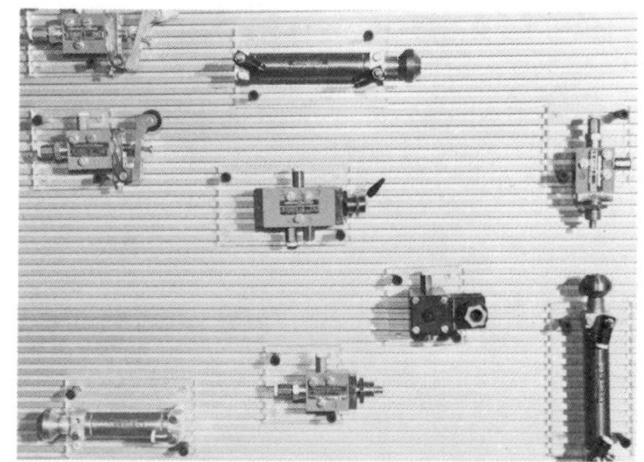

Pneumatics components mounted on 'bread board' with T-bolts

A base unit designed for mounting models

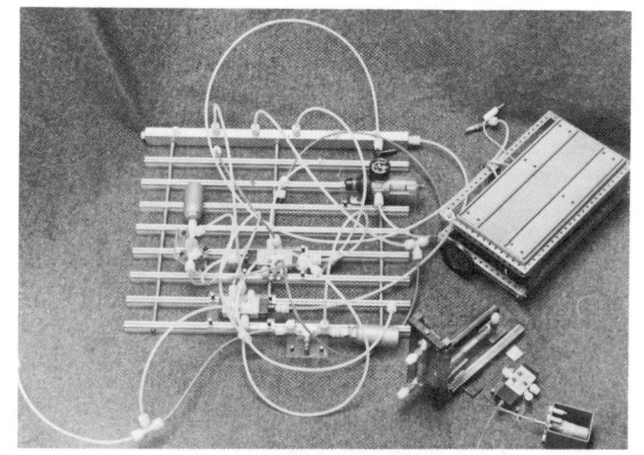

Building a pneumatics circuit model for testing the circuit

Industrial models

Construction kit arranged as a structures model

Models to test the action of pop-up cards

Construction kit designed to show action of a mechanical unit

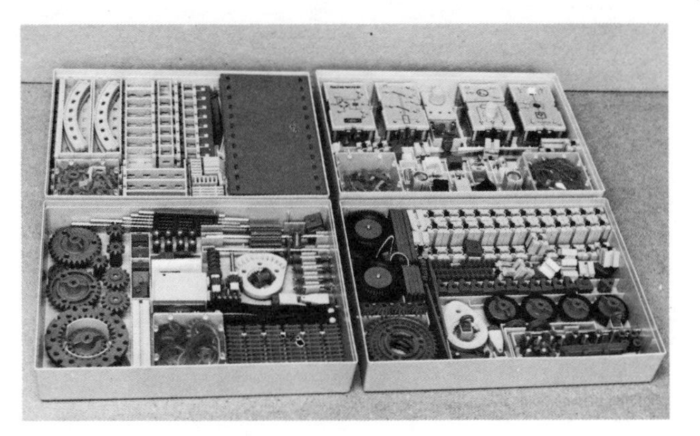

Construction kit packed ready for use

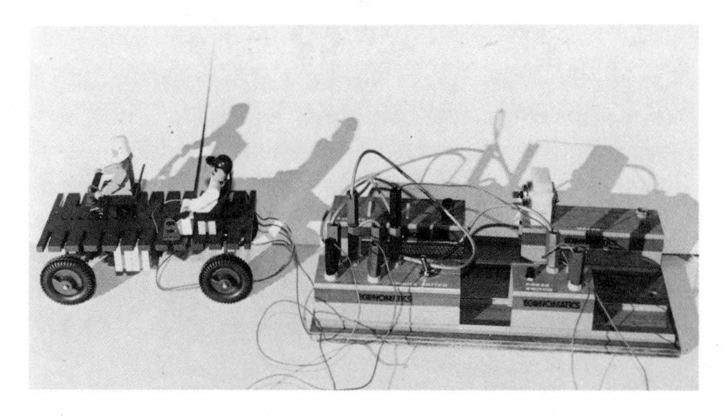

Electronics modules mounted on plastic base section

The photographs on pages 77 and 78 were either supplied by Economatics (Education) Ltd or taken by the author at their showrooms in Handsworth, Sheffield. The firm supplies modules, kits and work benches for the construction of models for electronics, electric, pneumatics and hydraulic circuits, for construction of structural models and mechanical models and for models worked via interfaces from computer programs.

Surface development

When an article is to be made from materials such as papers, cardboards, sheet metals or other sheet materials, a surface 'development' for the article may be needed before it can be made. The methods of development should produce exact shapes of the surfaces of the article.

Articles which are in the following geometric forms are those most likely to require surface developments: cubes and square prisms, prisms, cylinders, cones, pyramids.

Surface development of cubes and square prisms

Each of the six faces of a cube is a square. All six squares are of equal size. The surface development for a cube is, therefore, six equal-sized squares, arranged as shown.

The surface development of a square prism (or cuboid) differs from that of a cube in that its four sides are rectangles. The development of a cuboid therefore consists of four equal-sized rectangles and two equal-sized squares, making up the six faces of the cuboid.

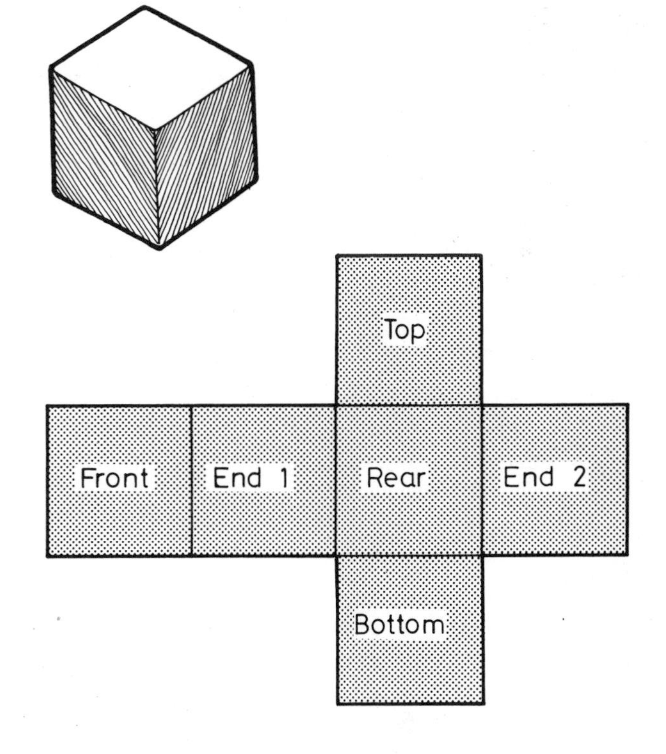

Surface development of a cube

Development of a packet

Many packets are in the form of square prisms. Their surface developments will thus consist of four rectangles and two squares. In order to be able to construct such a packet from its six-sided surface development, gluing tabs and tucking-in flaps are required as shown in the drawings. Some packets are in the form of rectangular prisms. The developments of these will consist of six rectangles.

Development of model of a 'Shuttle' cab

The 'Shuttle' model is square in section. The back and the front slope in opposite directions. When constructing a surface development for the Shuttle, two details must be observed.

1. The developments of front and rear are 'taller' than the sides because they are sloping surfaces.
2. Outlines of windows and doors must be drawn and coloured before the development is cut out and the model made.

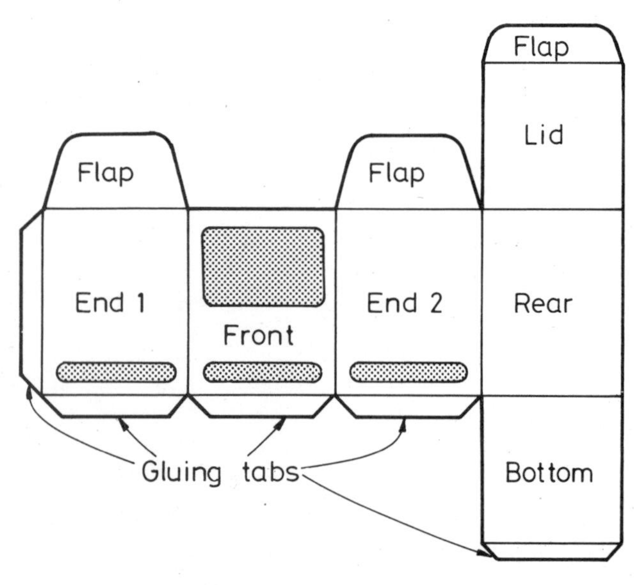

Surface development of a packet

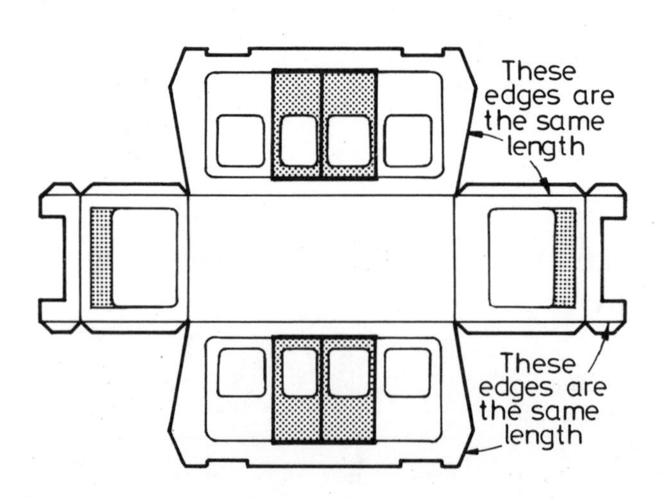

Note: When the surfaces of a developed form have to be decorated, for example with a description of the contents on the sides of a packet, such decoration should be added to the development before it is cut out and assembled.

Development of a 'Shuttle' cab

Surface development of prisms

The surface development of a truncated hexagonal prism shows the general method for developing the surfaces of prisms.

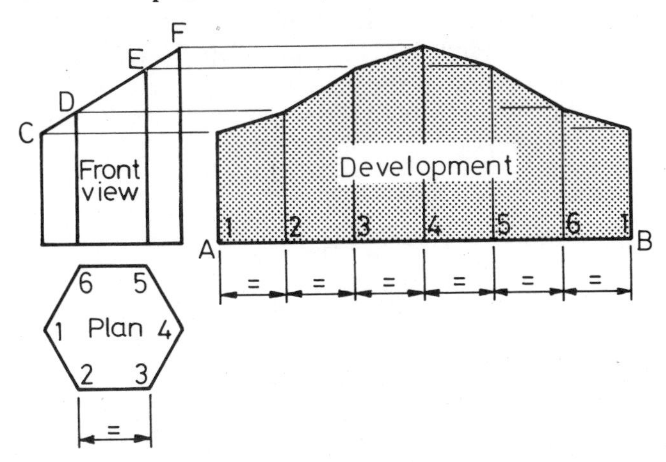

1. Draw the front view and plan of the prism.
2. Draw the straight line AB projected from the base of the front view.
3. Set compasses to the length of one edge in the plan.
4. Step off six equal spaces 1 to 6 along AB with the compasses as set.
5. Draw verticals to AB at the points 1 to 6 to 1.
6. From C, D, E and F draw lines parallel to AB to determine the heights of the lines from 1 to 6 to 1.
7. Complete the surface development as shown.

Development of a hexagonal prism with top and bottom

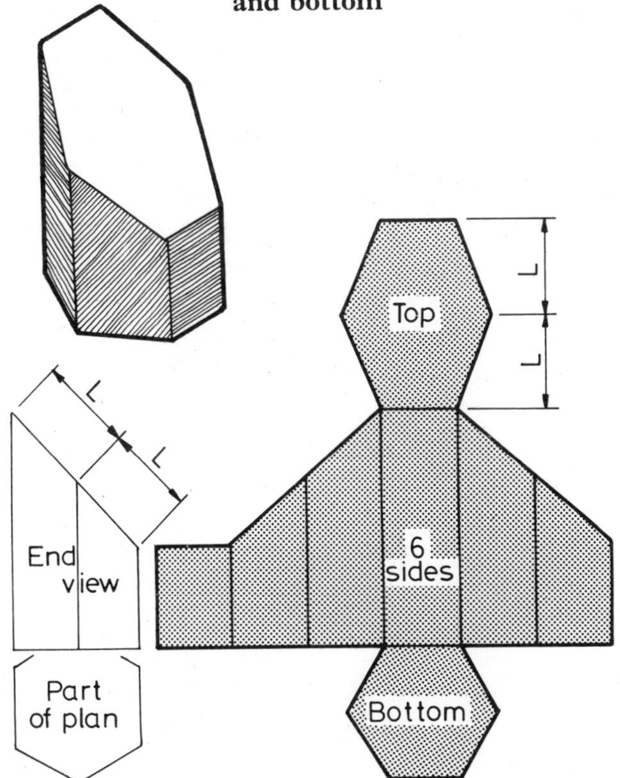

1. Proceed as before for the sides of the prism.
2. The bottom is a regular hexagon.
3. The lengths L for the top must be taken from the lengths L in the end view.

Development of a packet in the form of a pentagonal prism

1. Draw the front view.
2. The development of the five vertical sides of the prism can be obtained in a similar manner as for the hexagonal prisms described above.
3. The bottom is a regular pentagon.
4. Three edges of the top are the same as three sides of the regular pentagonal bottom.
5. To construct the other two edges of the top, the length L must be taken from L in the front view.
6. Note the difference in lengths of the five edges of the top of the packet.
7. Add appropriate gluing tabs.
8. Cut out and assemble the packet.

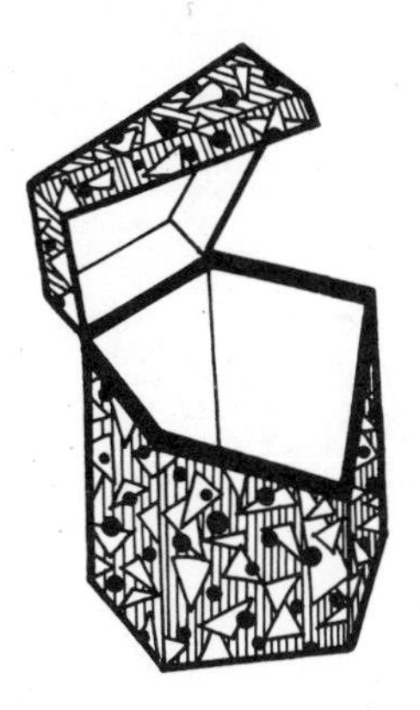

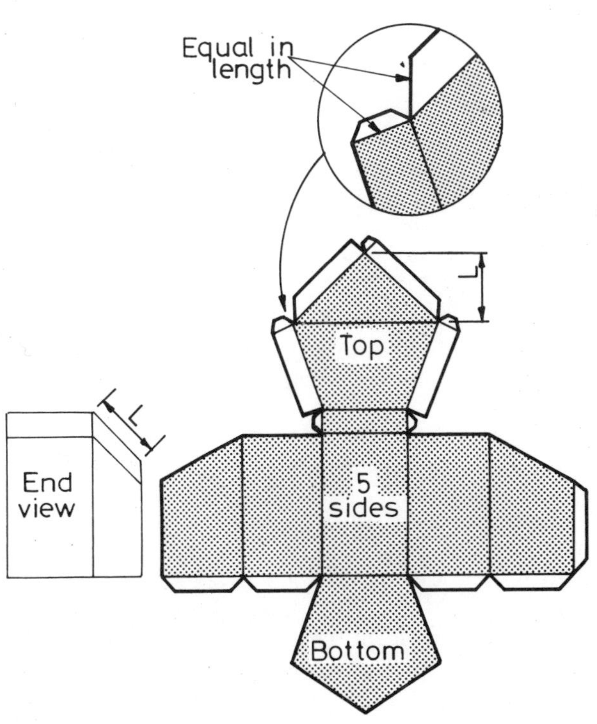

Development of a model for a hexagonal prism carrier

1. Construct a front view and plan.
2. Construct the required development following the principles given in the examples above, noting that the length L in the development must be taken from the length L in the front view and that the joining edges of the development must be equal in length.
3. Cut out the development and assemble the carrier.

Surface development of a cylinder

1. Draw a front view and plan of the cylinder.
2. Draw AB projected from C, the base of the front view.
3. Project from D a line parallel to AB.
4. Find the value of πD and complete the rectangle of the development of the vertical part of the cylinder.
5. Add top and bottom – circles of cylinder diameter.

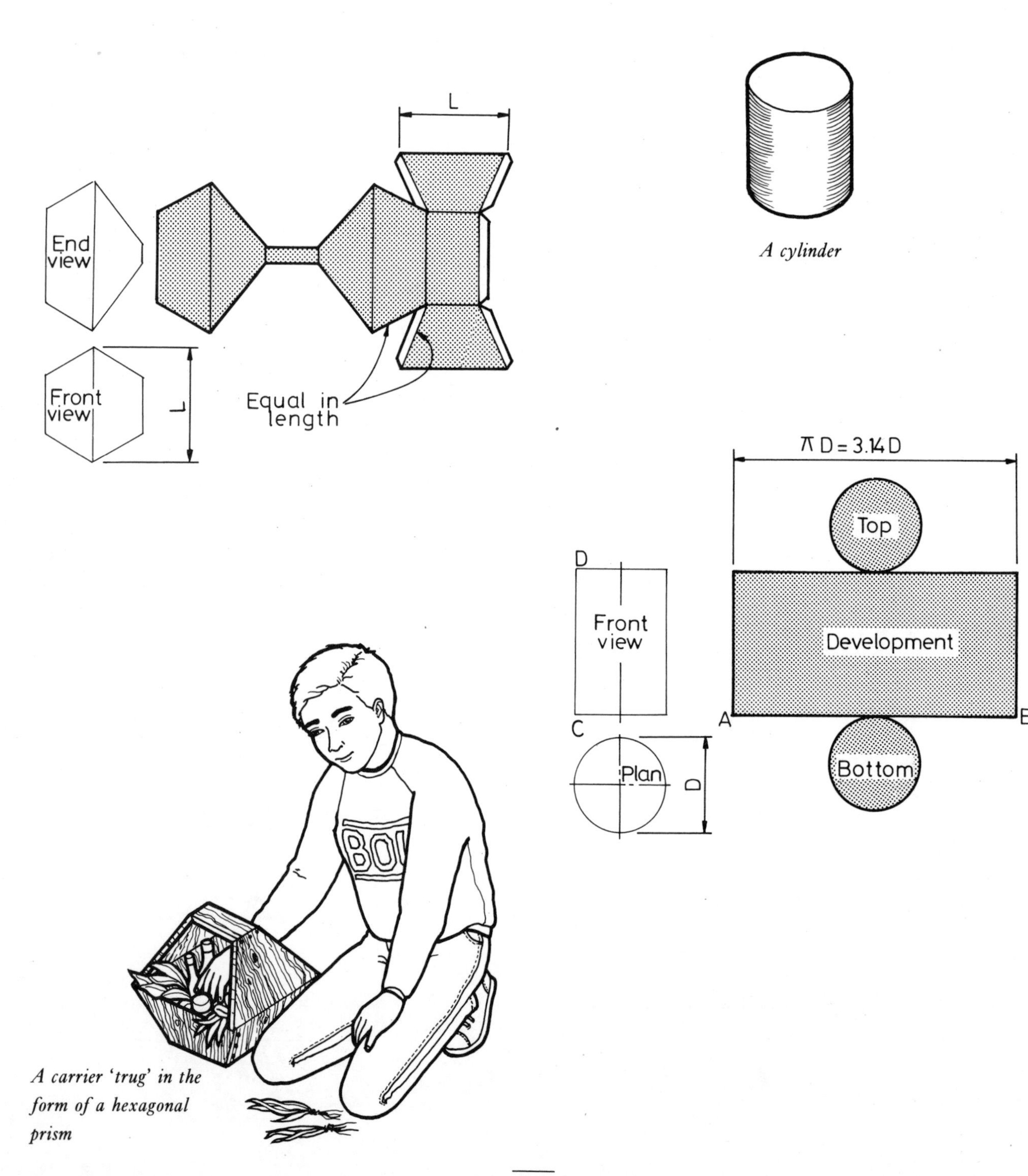

A cylinder

A carrier 'trug' in the form of a hexagonal prism

Development of a cylindrical wall lamp

1. Draw front view and plan.
2. Divide plan into twelve equal parts with the aid of a 30°, 60° set square.
3. Project the points 1 to 12 so obtained into the front view to give A to G on the truncation line of the cylinder.
4. Project a horizontal line from the base of the front view to produce a base line for the development.
5. Set a compass to one of the twelve equal divisions in the plan.
6. Step off twelve of these divisions along the development base line to give points 1 to 12 to 1.
7. Draw verticals at the points 1 to 12 to 1.
8. From A, B, C, D, E, F and G in the front view project lines parallel to the base line of the development onto the respective numbered verticals in the development.
9. Complete the development as shown with a fair freehand curve.
10. To draw the development of the back of the wall lamp transfer the lengths between A, B, C, D, E, F and G along length L to a back development line; transfer the lengths which points 2 to 12 are each side of the plan centre line to each side of the back development centre line; draw a fair freehand curve through the points so obtained; add gluing tabs.

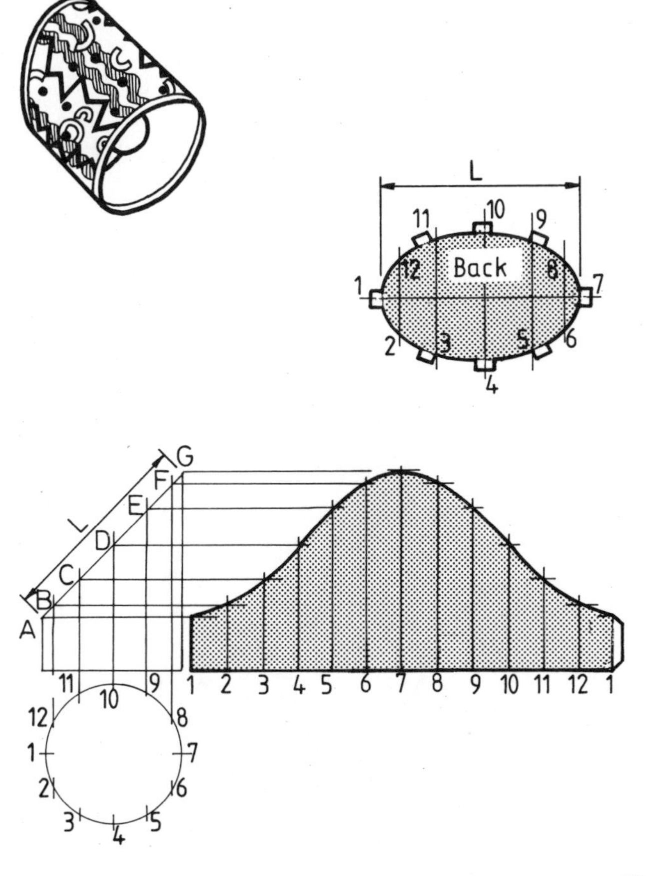

Development of a cylindrical shopping bag

The development for the bag follows the method of developing the truncated cylindrical wall lamp given above. Note the method shown for obtaining the true shape of the two ends of the bag by projecting at right angles to an end.

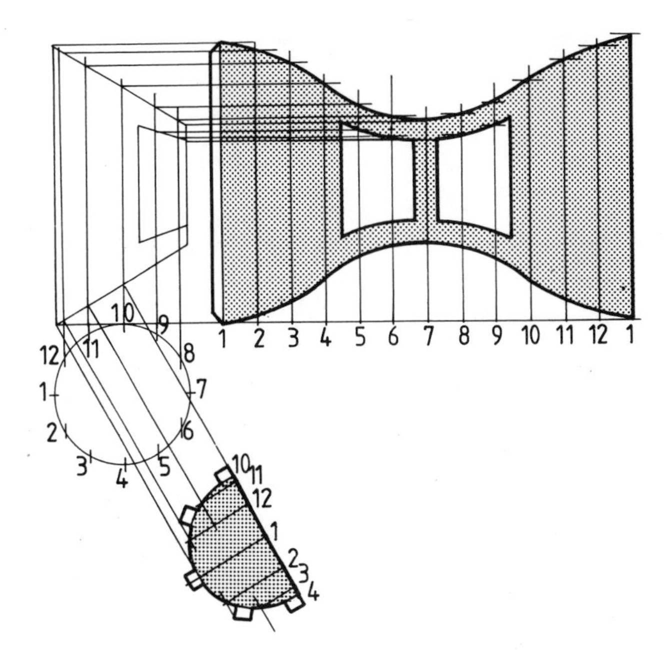

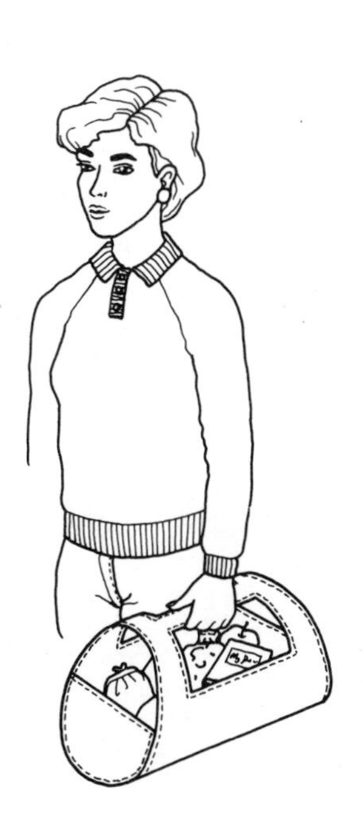

Surface development of a cone

1. Draw front view and plan.
2. Divide the circle of the plan into twelve equal parts with the aid of a 30°, 60° set square.
3. Set a compass to the length of the slope line of the front view of the cone – the length TL.
4. With the compass draw an arc of radius TL.
5. Set a compass to any one of the twelve equal divisions around the plan circle.
6. Step off twelve of these equal spaces along the arc of radius TL.
7. Complete the development as shown.

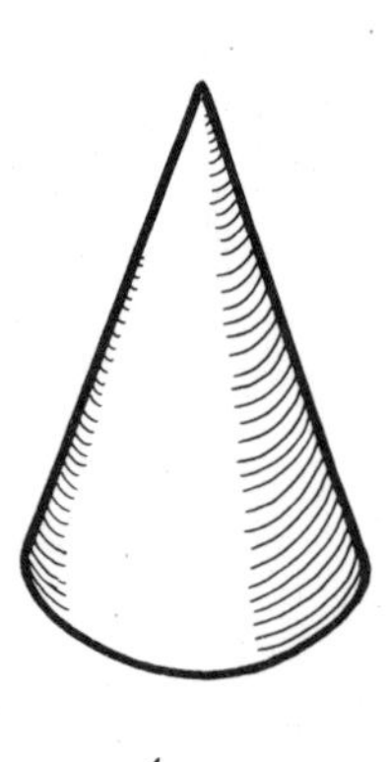

A cone

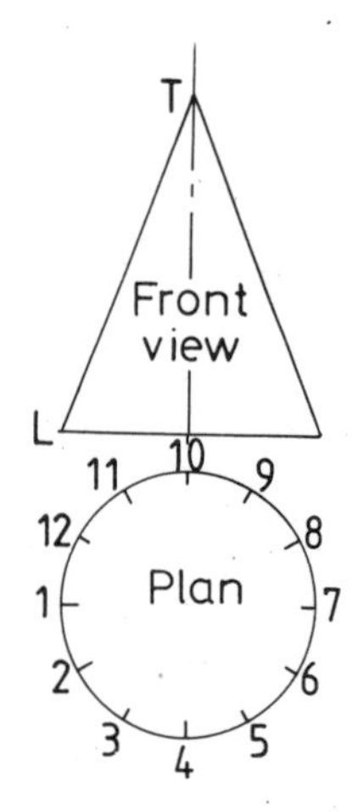

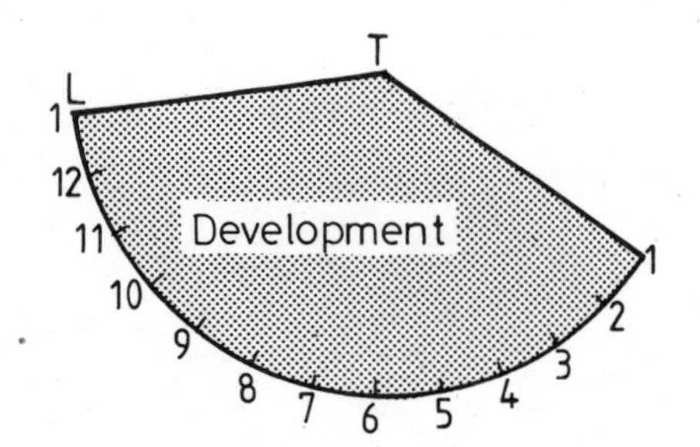

Development of a conical party hat

The hat is in the form of a truncated cone.

1. Draw the front view and plan of the hat.
2. Divide the plan circle into twelve equal divisions.
3. Project the twelve points in the plan onto the base line of the front view.
4. Join the points so found on the base to the apex T of the front view.
5. Set a compass to the slope line of the cone – TL.
6. With the compass set to TL draw an arc.
7. Along the arc TL step off twelve spaces, each equal to one of the divisions in the plan.
8. Where the sloping lines in the front view cross the truncation line draw lines parallel to the base of the cone to give the lengths a, b, c, d, e, f and g.
9. Set a compass to each of the lengths a to g in turn and with it mark off the respective lengths along the lines from T to 1 to 12 to 1 in the development.
10. Complete the development as shown.
11. Add a gluing tab and any necessary decoration to the development before cutting out its shape.

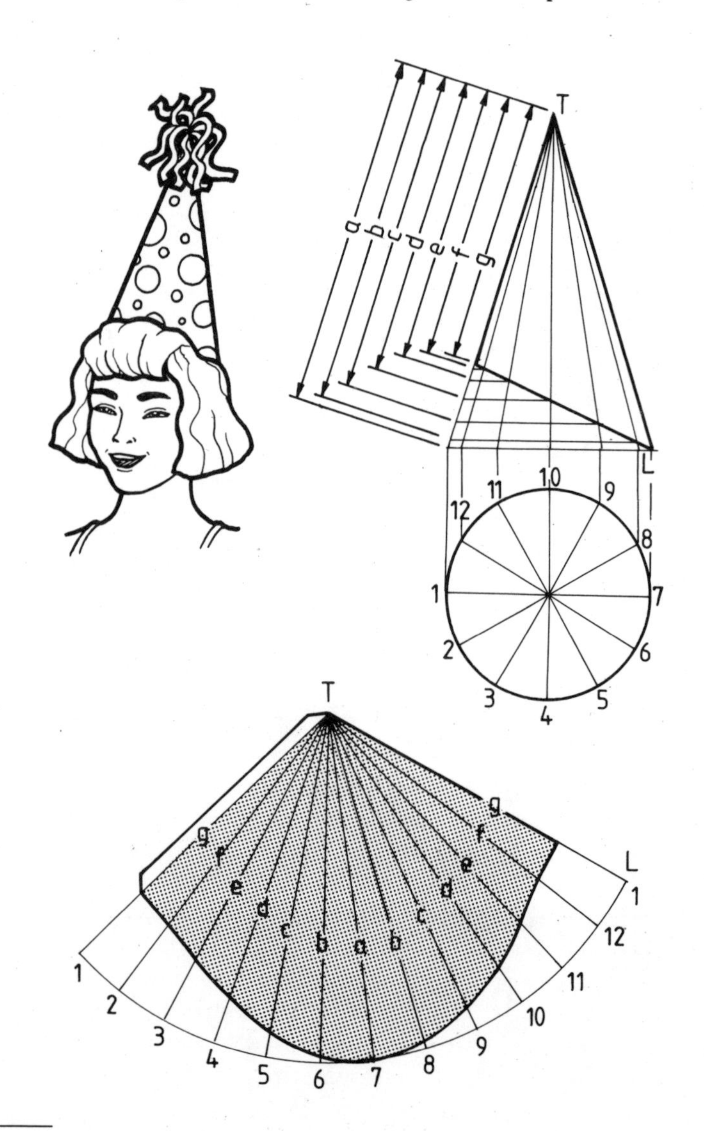

Development of a conical lamp shade

This development follows the same method as that given for constructing the development of the truncated cone of the party hat, except that the cone is truncated at an angle at both top and bottom.

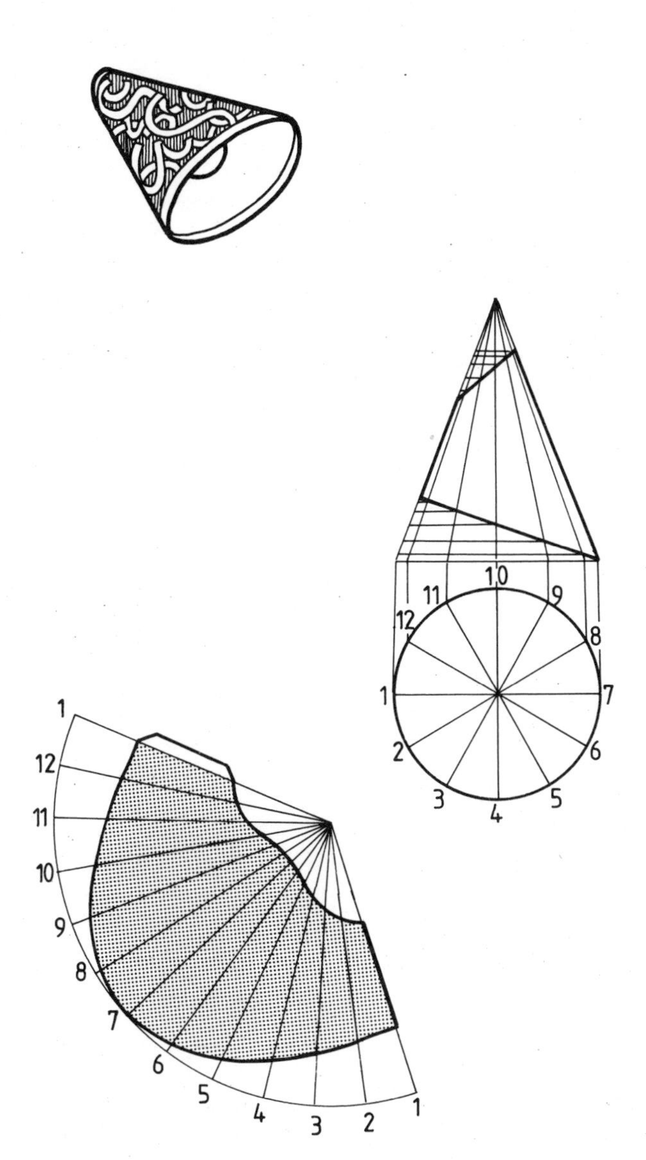

Surface development of pyramids

Note: The most important feature of the methods for the construction of developments is to make quite certain that all *lengths* in a development are *true lengths*. A front view or plan of a face sloping away from or towards an observer is not a *true shape* of that face. Therefore lines on such a front view of a sloping face may not be *true length* lines. When constructing any development, you must always ensure that *all* lines on the development are true lengths. Failure to do so will result in an incorrect development.

Development of a square pyramid

1. Draw the front view and plan. See note above. The sloping faces of this front view and plan result in all sloping edges of the pyramid showing as lines shorter in length than they actually are. Therefore the *true length* of the slope lines of the pyramid must be constructed before starting the development.
2. To find the true length, draw line VP in the plan; set compass to V1 in plan and draw arc to P; project P to base line of front view; join V to the point so found.
3. With a compass set to the true length draw an arc.
4. Set a compass to the length of one edge of the square base of the pyramid.
5. With this compass step off the spaces 1 to 2, 2 to 3 and so on to 1 along the true length arc.
6. Complete the development as shown.
7. Add a bottom – a square – to one edge of the development of the pyramid's sides.

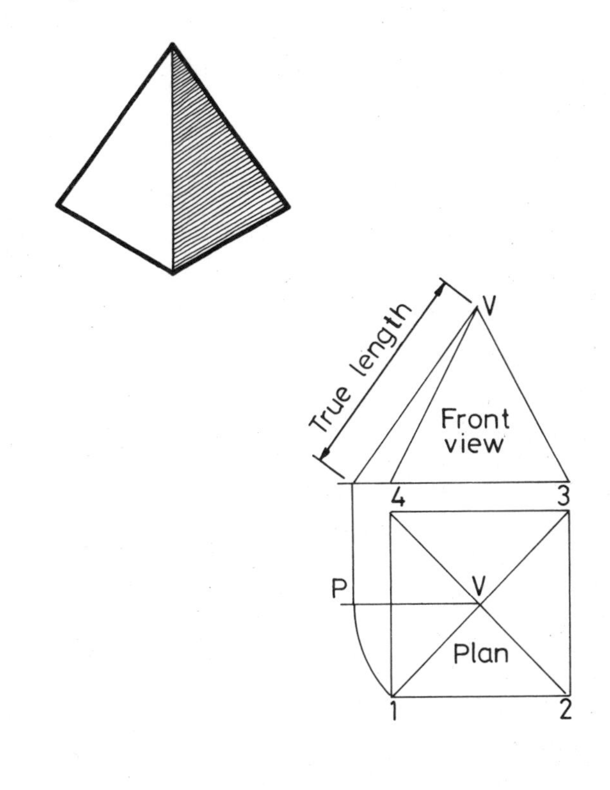

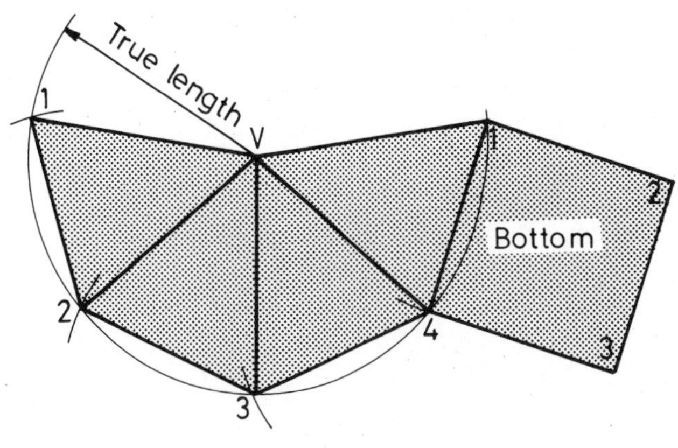

Development of a rectangular pyramid for a candle stand

Note: This example is based upon a rectangular pyramid.

1. Draw front view and plan.
2. Construct and find the true length of one sloping edge of the complete pyramid.
3. Draw an arc of true length radius.
4. Step off along this arc the four lengths of the edges of the rectangular base.
5. Find L – the true length of a sloping edge of the part cut from the top of the pyramid.
6. With a compass set to L draw an arc on the development to determine the ends of the four top edges of the development. Complete the development of the pyramid sides as shown.
7. Add a base, a top, any decoration required on the sides of the stand and gluing tabs.

Development of a model of a wheelbarrow body

The body of the wheelbarrow is in the form of an inverted and truncated rectangular pyramid.

1. Draw front view and plan.
2. Find the true length of a sloping edge of the complete pyramid.
3. Set a compass to this true length and with it draw an arc.
4. Step off along this arc the lengths of the sides of the base of the pyramid to give the points 1 to 4 to 1.
5. Join the points so found to V.
6. On the front view draw lines parallel to the pyramid base to obtain points A and B. These points give the ends of the true lengths of the sloping edges of the truncated pyramid.
7. Set a compass to VA and VB in turn and step off the true lengths from V in the development.
8. Add a base for the wheelbarrow body.
9. Complete the development as shown.

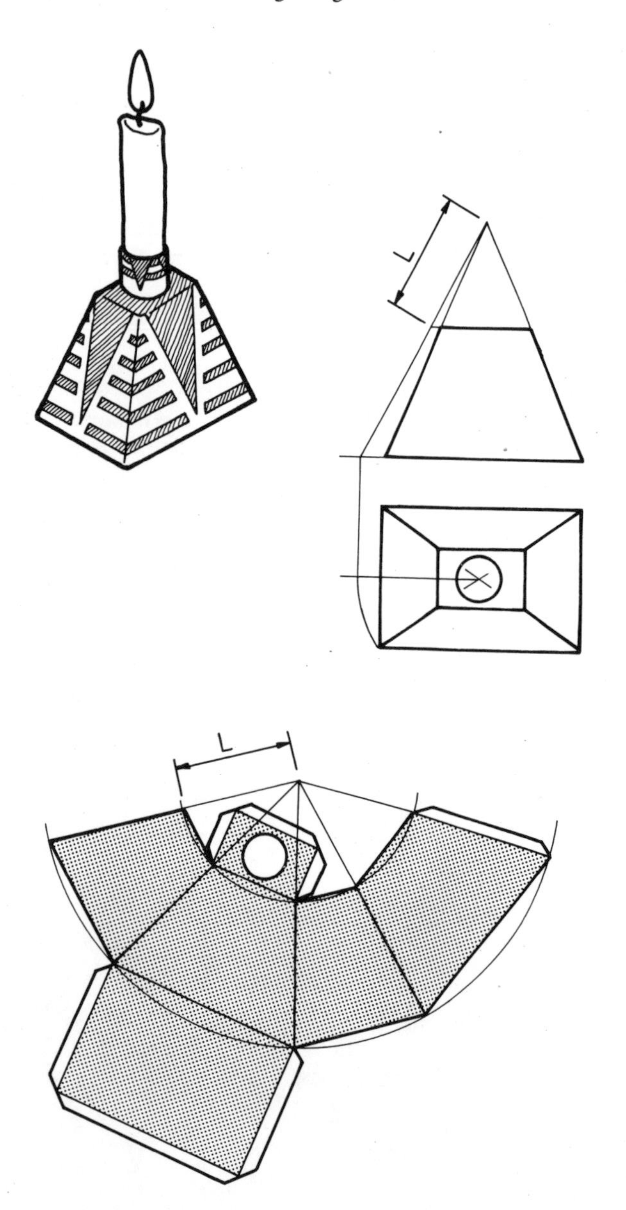

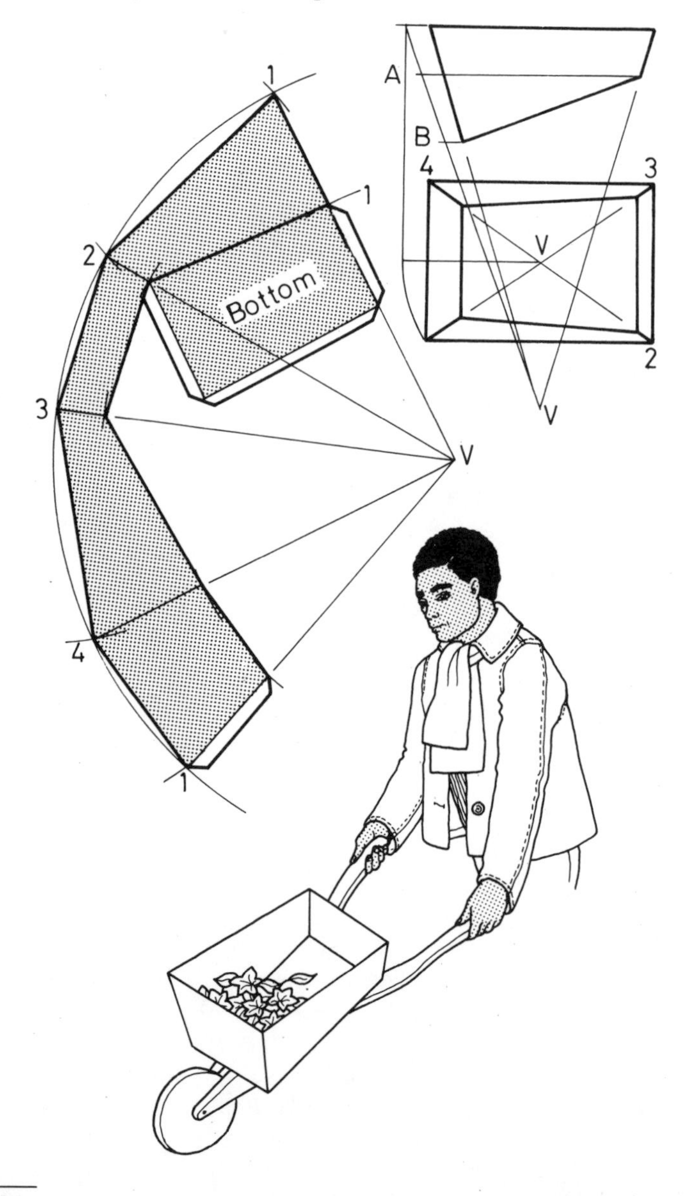

EXERCISES

1. (a) Construct the surface development for a rectangular prism 70 mm long by 40 mm wide by 30 mm high.

(b) Use your development as the basis for a design for a packet to hold 500 paper staples.

(c) Trace the development onto thin card, cut it out with scissors and make up the packet. Add a suitable label.

2. Construct a surface development for the body of a model lorry measuring 120 mm long by 55 mm square. Cut out your development and construct the model. 'Wheels' can be made from thick card glued to the body of the model. Design a 'cab' and add it to the body of your model lorry.

3. (a) Construct a surface development of the truncated hexagonal prism shown in the drawing. Work to a scale of 1:5.

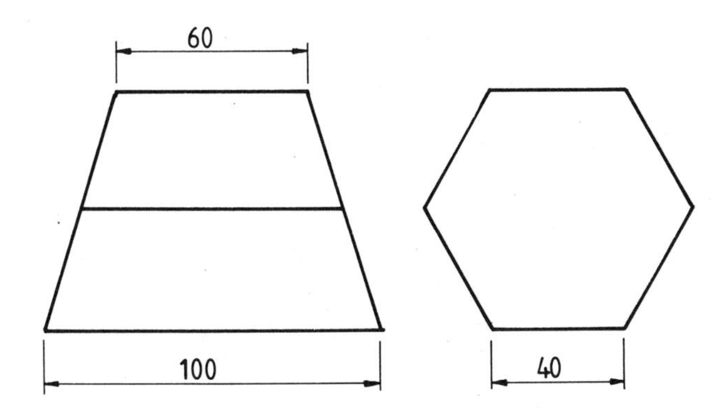

(b) Design a carrier bag based upon the shape of a regular *pentagonal* prism. Proceed as follows:

(i) make freehand drawings of suitable designs;
(ii) choose the best of your drawings and determine dimensions for your chosen design;
(iii) make notes on materials from which your design could be made; select one of these;
(iv) make notes on constructions by which your design could be made from your chosen material;
(v) construct a surface development for your design;
(vi) design a logogram for your design, which includes your own initial;
(vii) make a full size model of your design, with its logogram, from a suitable paper material.

4. Design a table lamp shade which is in the form of a truncated cylinder. Construct its development. Cut out your development with the aid of scissors and from it make a full size paper model of your design.

5. Draw, develop and make a conical party hat for yourself. Dimensions for the hat and the slope of its bottom edge should be judged from the size and shape of your head. Add suitable decorative details to the hat.

6. Construct a development for a *hexagonal* pyramid of base edge length 50 mm and height 80 mm.

7. You wish to make a hopper in the form of an inverted rectangular pyramid to contain 50 table tennis balls so that the balls are always easily available in a youth club hall.

Design a hopper of suitable size and shape, construct its development and make a model of your design from thin card to test whether it is suitable for its purpose.

Colour, shading, tinting

Colouring

In this section of the book, a number of examples of colouring items of graphic design work are included. A variety of colouring media has been used in these examples and the following six methods of applying colour have been adopted.

Colour pencils
These are in a good range of colours and can be purchased in many shops. They are easy to use. Mistakes can be erased, although some traces of colour may remain on your work after erasing.

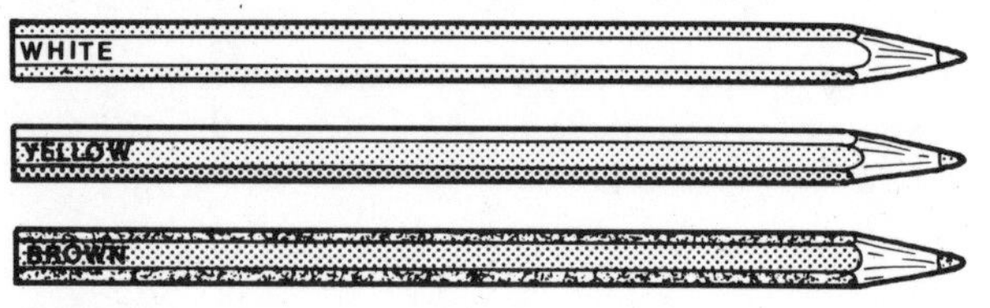

Colour pens
Colour pens are available in a large range of different types and colours: Biro pens; a variety of 'penstiks' with nylon or with felt nibs; coloured inks used in technical pens. The 'penstik' type of pen is made with a variety of nib widths from fine to broad. Many of the fluids in which the colour is held are spirit types and thus quick drying. Once applied it is difficult to erase mistakes except by covering with self-adhesive white paper or by painting out the error with white water-colour paint.

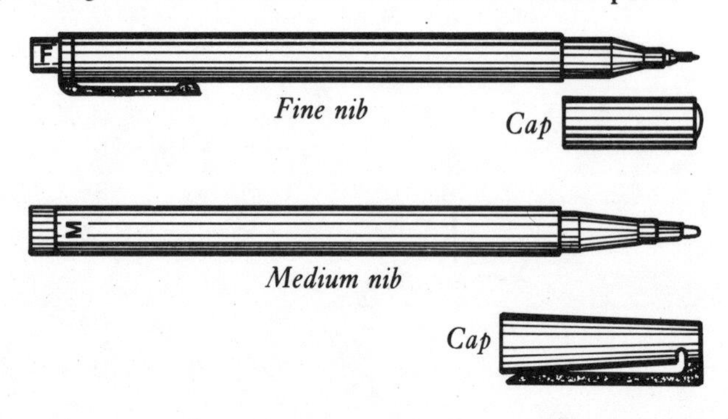

Water-colour
Water-colour paints are applied with an artist's paintbrush. They are available in block form, as powders or as pastes in tubes. It is advisable to use good quality tube paints if you wish to produce good colouring with water colours. White gouache water paint is of value for painting in highlights or for covering up mistakes in work involving other media.

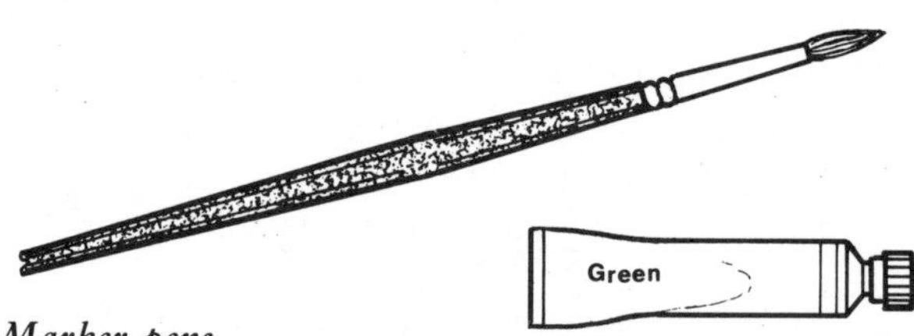

Marker pens
Marker pens are obtainable in a large range of colours. They consist of a tube container in which the colour liquid is held and to which a marker tip nib is fitted. Although the shapes of the nibs vary between makes, in general two types are available – broad and fine. It is advisable to use marker pens with spirit based colours. These dry rapidly after application. Care must be taken to avoid mistakes, because erasure is practically impossible except by covering the errors with self-adhesive papers.

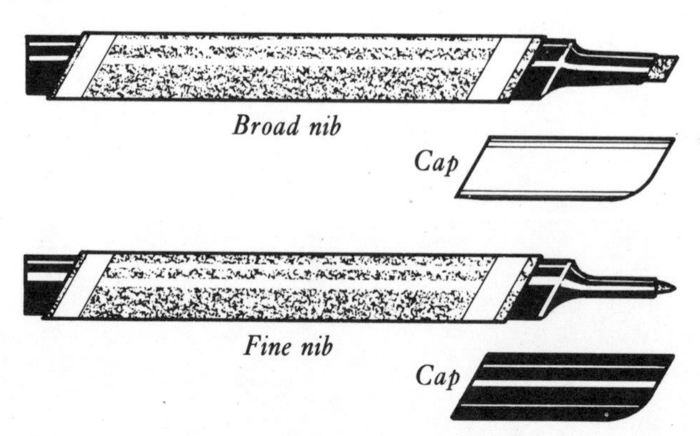

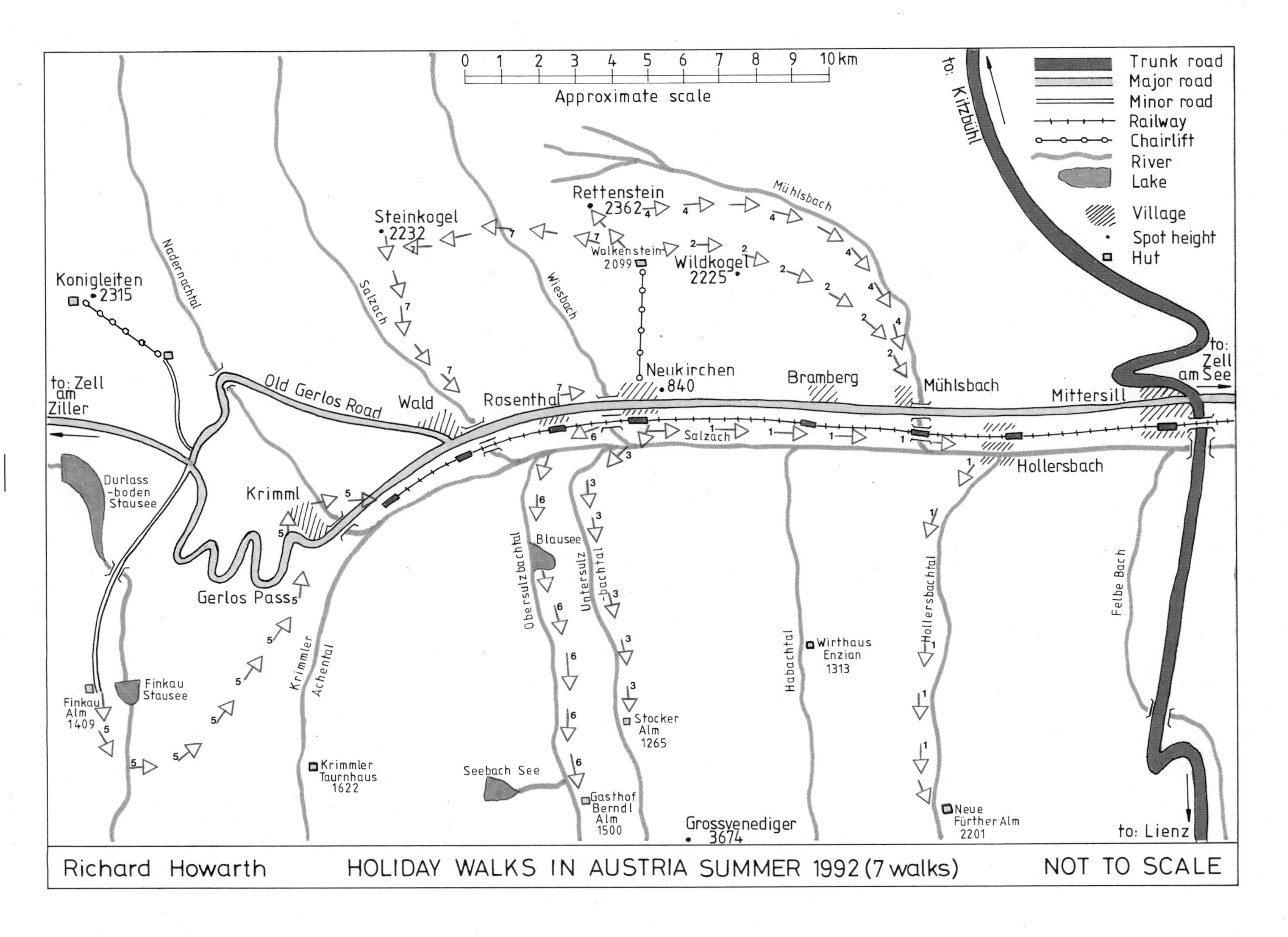
0 1 2 3 4 5 6 7 8 9 10 km
Approximate scale
to: Kitzbühl
Trunk road
Major road
Minor road
Railway
Chairlift
River
Lake
Village
Spot height
Hut
Rettenstein
2362
Mühlsbach
Steinkogel
2232
Wolkenstein
2099
Wildkogel
2225
Konigleiten
2315
Nadernachtal
Salzach
Wiesbach
Neukirchen
840
Bramberg
Mühlsbach
to: Zell am See
Mittersill
to: Zell am Ziller
Old Gerlos Road
Wald
Rosenthal
Salzach
Hollersbach
Durlass -boden Stausee
Krimml
Blausee
Gerlos Pass
Obersulzbachtal
Untersulz -bachtal
Hollersbachtal
Felbe Bach
Krimmler Achental
Habachtal
Wirthaus Enzian 1313
Finkau Stausee
Finkau Alm 1409
Stocker Alm 1265
Krimmler Taurnhaus 1622
Seebach See
Gasthof Berndl Alm 1500
Grossvenediger
3674
Neue Further Alm 2201
to: Lienz
Richard Howarth
HOLIDAY WALKS IN AUSTRIA SUMMER 1992 (7 walks)
NOT TO SCALE

Dry transfer colour
Sheets of self-adhesive film can be applied to drawings to produce a clean, flat colour. The best known is the 'Pantone' series made by Letraset.

Air-brushing
In air-brushing the coloured liquids are forced through an air-brush by compressed air onto the drawing surface. Some firms make special forms of air-brush which can be attached to their own make of marker pen. These use the coloured liquid in the pen as the colouring medium.

Project – Holiday walks in Austria

The sketch map on page 89 was drawn with the aid of the following:
Technical pens – To draw the margins, black lines and, with letter stencils, all lettering.
Red biro pen – The route arrows.
Red, blue and yellow fine-tip penstiks – Colouring of roads, rivers and lakes.
Dry transfer letters – All route numbering.

Pencil drawing

Some rules for pencil drawings

The rules given below apply to the use of all types of pencil, whether black or coloured.
1. To draw firm clean lines hold the pencil as shown in drawing 1. The depth of shade of the line depends upon the grade of pencil used and the pressure with which the lines are drawn. Line thickness depends upon the sharpness of the pencil.

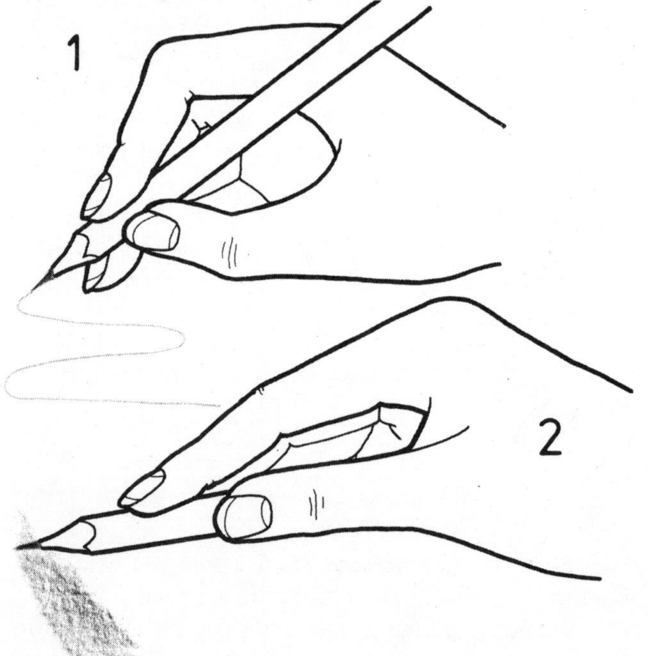

2. When shading areas with a pencil, hold the pencil as shown in drawing 2.
3. Shading with pencils usually produces a type of 'grain' effect – the shading strokes of the pencil. The effect is not unpleasant and may enhance the appearance of a drawing.
4. The 'grain' effect can be considerably reduced by going over the same shaded area with pencil strokes in different directions to the original shading strokes.
5. Deeper shades can be obtained by increasing the hand pressure with which the shading is applied.
6. Masking areas with low-tack masking film, or with masking tape, allows areas to be shaded with clean, sharp, clear-cut edges. Hand held pieces of paper can be used to produce the same results.
7. The intensity of pencil shading can sometimes be enhanced by going over a pencil-shaded area with a white pencil.

Examples of pencil shading

Shading with a grade B or BB pencil is an effective method of emphasising the form of an item being drawn. Five examples are shown.

Drawing 1 – Each of the flat surfaces is shaded to give a light, a medium or a dark depth of grey, as if the light shining on the object is from above and to the right.

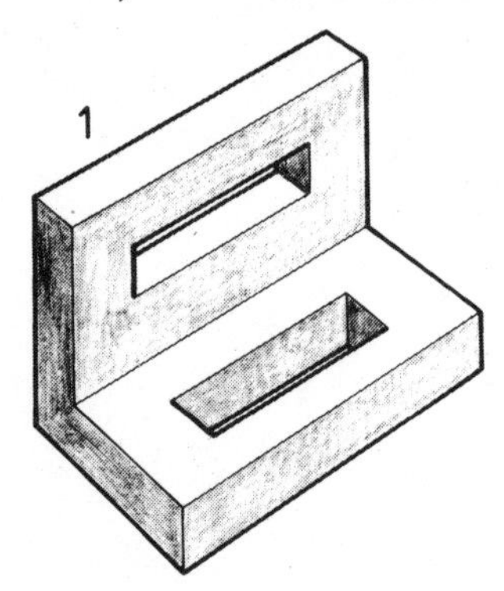

Drawing 2 – The depth of shading becomes heavier towards the sides of the drawing of both the outside and the inside of this short length of pipe.

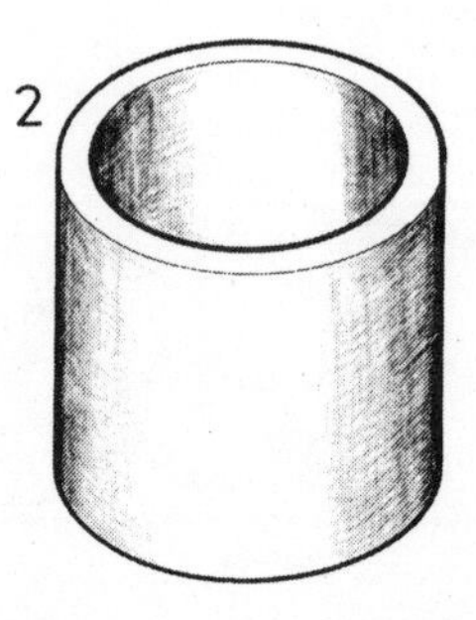

Drawing 3 – Shading alone, without outlining lines, can produce an effective piece of graphics.

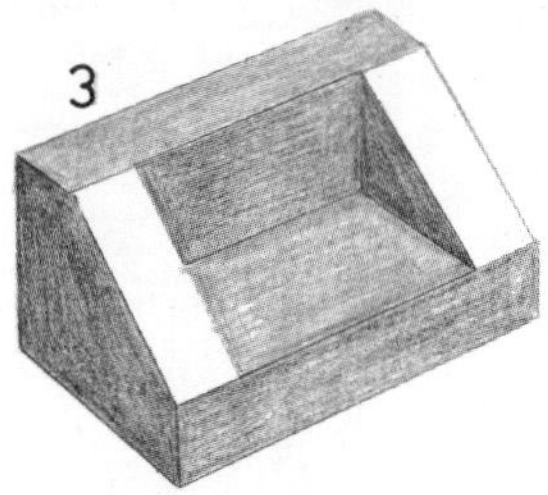

Drawing 4 – A sphere can be shaded as if a spot of light is in one position. The shading becomes deeper as it approaches the outline of the sphere.

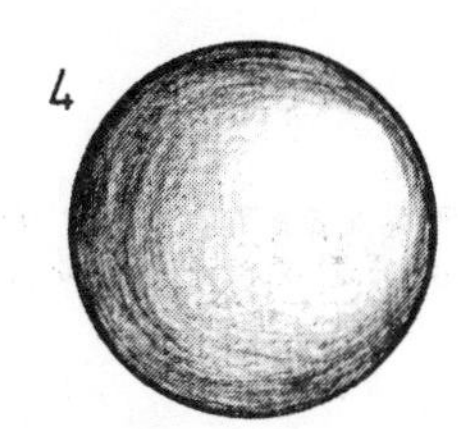

Drawing 5 – A freehand drawing of a telephone shaded by using a grade HB pencil for the lighter shades and a grade BB pencil for the darker shades. It is assumed that a light source is above and behind to the left of the telephone.

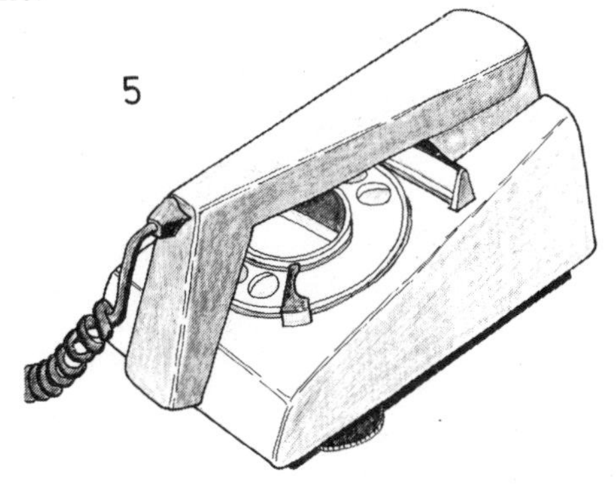

Examples of colour pencil shading

A number of drawings on pages 92 and 93 have been shaded with colour pencils.

Page 84, Drawing 1 – A small area lightly shaded with colour pencil. Note the 'grain' effect.

Drawing 2 – A similar small area shaded with heavier pressure applied to the pencil.

Drawing 3 – Shading from light to dark by gradually increasing the pressure on the pencil.

Drawing 4 – To produce this leaf shape, a piece of masking film was smoothed onto the sheet. A leaf shape was then cut in the film with a sharp knife. The leaf shape of film was then lifted from the sheet and the area it had covered was shaded. Finally the remaining masking film was lifted from the sheet. The leaf shape could also have been cut from a piece of drawing paper and the paper then held in position by hand as the area inside the cut-out part was shaded.

Drawing 5 – Part of this drawing has been drawn with thin lines, the other part shaded with the aid of masking tape.

Drawing 6 – An area shaded within an outline drawn with a grade B pencil.

Drawing 7 – An area shaded within an outline drawn with the same colour of pencil as the shading.

Drawing 8 – An area shaded within an outline drawn in black ink with a technical pen.

Drawing 9 – A drawing of a polished stone mounted on a piece of silver plate. Outline of the silver drawn in thin black ink lines, the shape of the stone emphasised with deeper shading towards its outline. A rectangular background formed by masking with masking tape.

Drawing 10 – A variety of coloured rectangles forming a neck pendant design within a pencilled rectangle. The shape of the pendant is emphasised with a background formed from straight coloured lines.

Drawing 11 – A letter A on a background of two colours.

Drawing 12 – A polished stone set on a wooden circle.

Drawing 13 – A design for a necklet.

Drawing 14 – Interlocking coloured arcs form this design for an item of jewellery.

Drawing 15 – A design for a tile surface.

Drawing 16 – A woodgrain effect produced with coloured pencil. A light coloured background emphasises the outline of the wooden shape.

Drawing 17 – A design for a house nameplate in position against a brick wall.

The coloured drawings on page 93 show some applications of colour work to the geometry of graphics.

Drawing 20 – Three different colours shaded with colour pencils emphasise the message of this histogram.

Drawing 21 – The bars of this graph have been coloured a deep blue with a colour pencil.

Drawing 22 – Pencil colours colour shade this pie graph.

Drawing 24 – Three different shapes of the answer to an area problem pencil shaded in different colours.

Drawings using colour pens

The remaining coloured drawings on pages 92 and 93 demonstrate some uses of coloured lines drawn with nylon tipped pens.

Drawings 18 and 19 – To enlarge the hen outline in drawing 18, a grid of squares was constructed around the outline. A grid of larger squares (drawing 19) was then drawn within which the larger outline of the bird could be constructed. The grid lines of both drawings greatly assist in obtaining an accurate copy.

Drawing 23 – A locus problem with the construction lines and the locus answer drawn with a red pen.

Drawing 24 – Fine construction lines in red.

Drawings 25 and 26 – The solutions to two vector problems, the answers (R in each case) drawn in red.

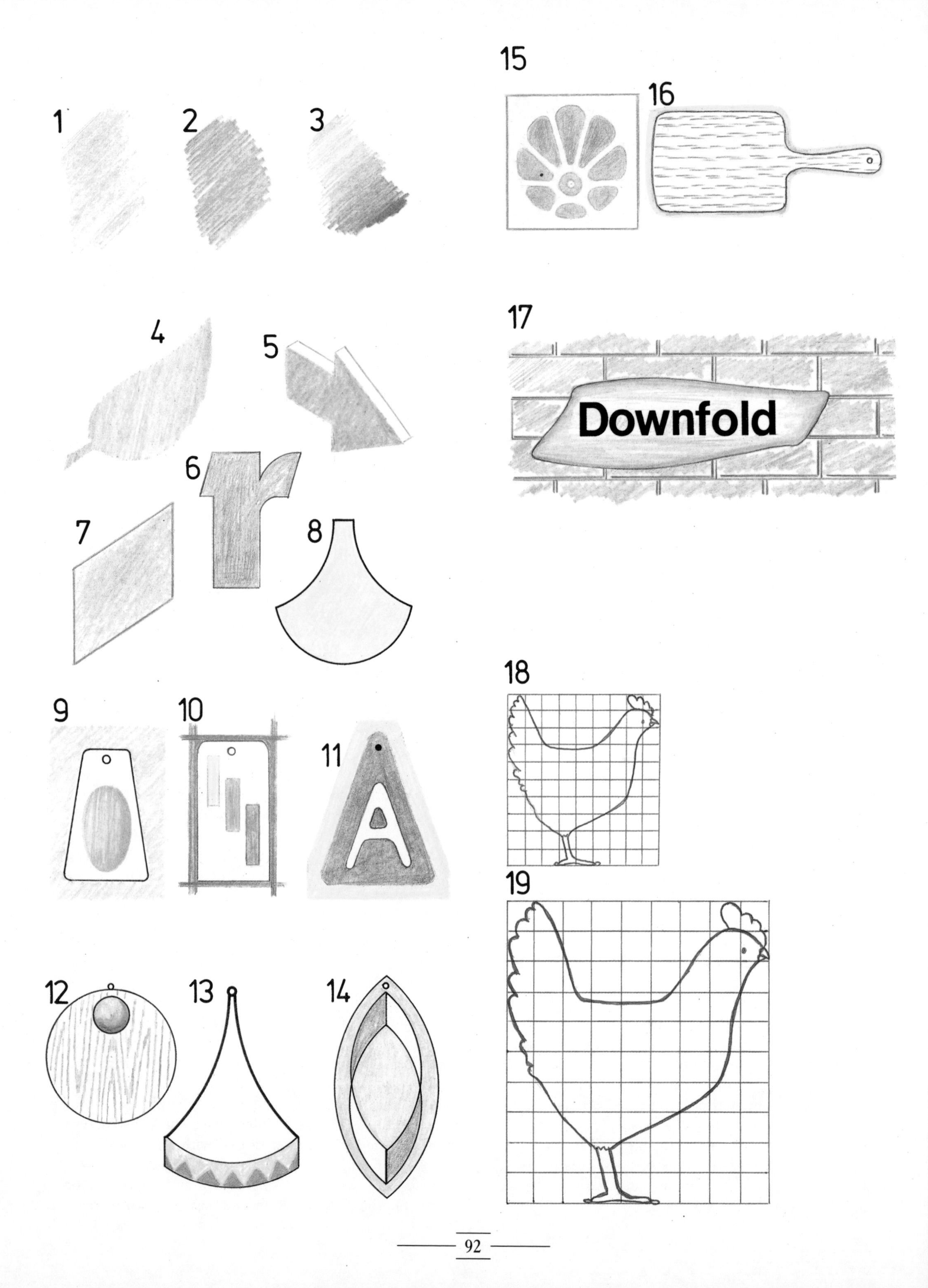
1
2
3
4
5
6
7
8
9
10
11
A
12
13
14
15
16
17
Downfold
18
19

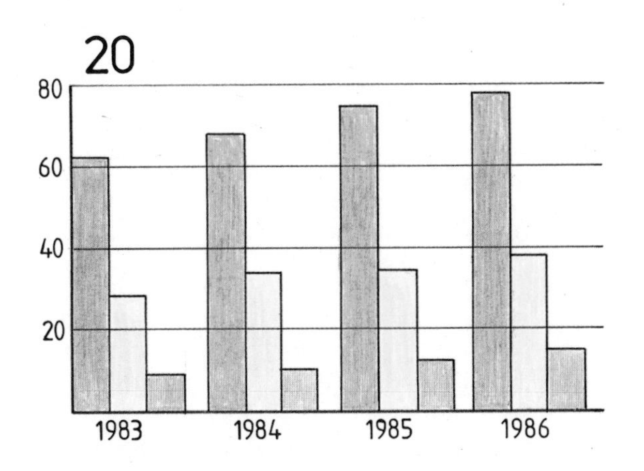
20
80
60
40
20
1983
1984
1985
1986

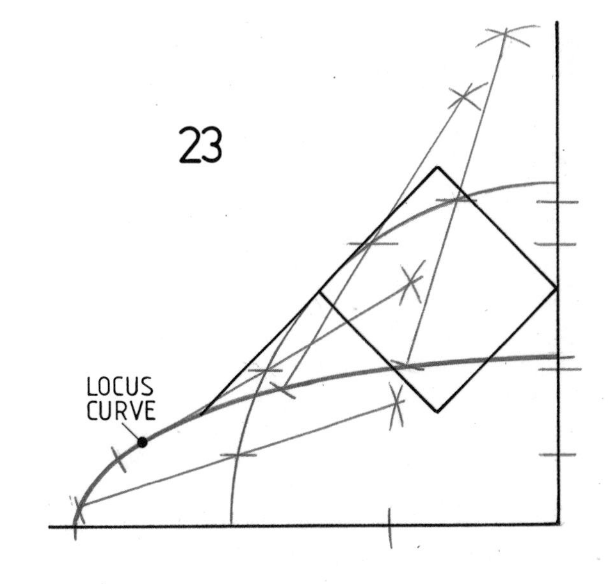
23
LOCUS
CURVE

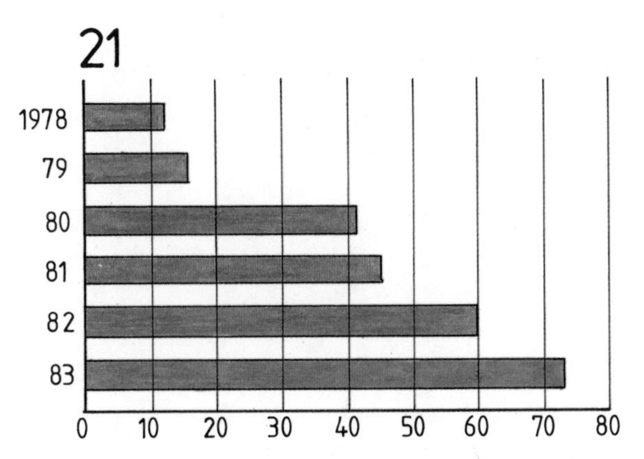
21
1978
79
80
81
82
83
0
10
20
30
40
50
60
70
80

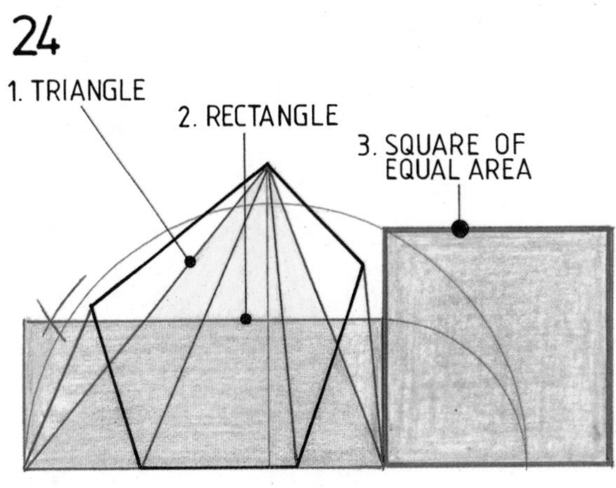
24
1. TRIANGLE
2. RECTANGLE
3. SQUARE OF
EQUAL AREA

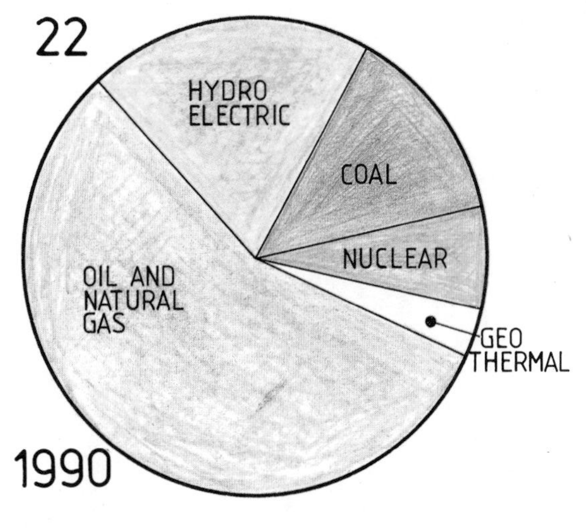
22
HYDRO
ELECTRIC
COAL
NUCLEAR
OIL AND
NATURAL
GAS
GEO
THERMAL
1990

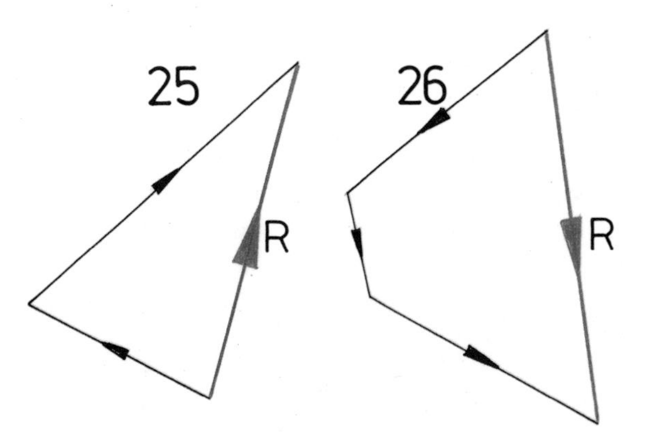
25
R
26
R

EXERCISES

1. Select a group of your school friends – about ten to twelve will be sufficient.

(a) Ask the members of the group for their home addresses.

(b) Ask them how they travel to and from school each day:

(i) on foot;
(ii) by bicycle;
(iii) in their parents' car;
(iv) by bus;
(v) other means of travel.

(c) Plot the positions of the homes of the members of the groups on a large-scale local road map.

(c) Design, draw and colour a sketch map which describes the routes taken by the group members in their travels to and from school. Your sketch map could include:

(i) different symbols for each mode of travel;
(ii) symbols for the routes covered;
(iii) differences between paths, minor roads, major roads;
(iv) symbols showing possible danger spots.

2. The outlines of a simple geometrical pattern are given in drawing 1. Copy this pattern with the aid of a straightedge and a set square. With the aid of coloured pencils shade in areas of the pattern to achieve a well designed blend of colours.

3. On a piece of isometric grid paper design a pattern based on the triangles of the grid. Two suggestions are included in drawing 2, but try to avoid copying these. Colour in areas of your design with colour pencils or colour pens to achieve a good balance in your pattern design.

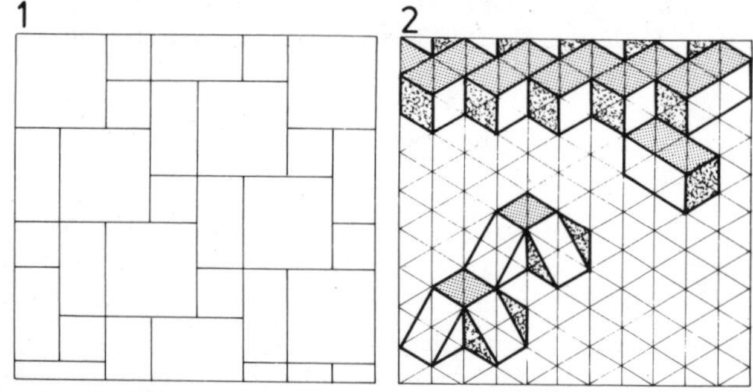

Straight line shading

Two examples of outlining a drawing with thick lines and shading with thin parallel lines are shown. In drawing 1, the bookcase has been shaded with parallel lines drawn with a fine pointed pencil. In drawing 2, brick outlines have been indicated as shadows with thick lines, as if light was falling from above and from the left, and freehand, thin, parallel lines indicate the brick surfaces.

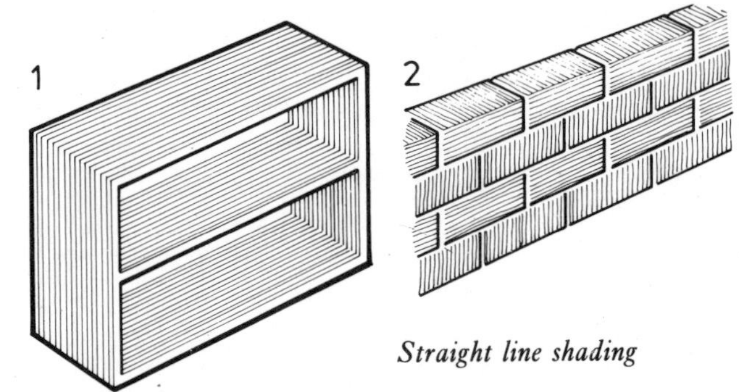

Straight line shading

Variations in line shading

Four examples of variation in depth of shading are shown.

Drawing 1 – Door chimes. Outlines are a mixture of thick and thin lines. Different depths of tone have been achieved by combining thin parallel lines at different spacings, thin crossing lines at different spacings and blacking in parts of the drawing.

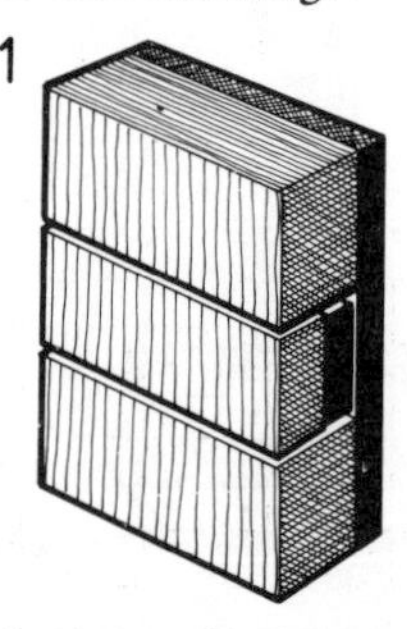

Drawing 2 – Paper handkerchief box. Outlines are thick lines. Different depths of shading have been drawn in with parallel lines at different spacings and thin crossing lines.

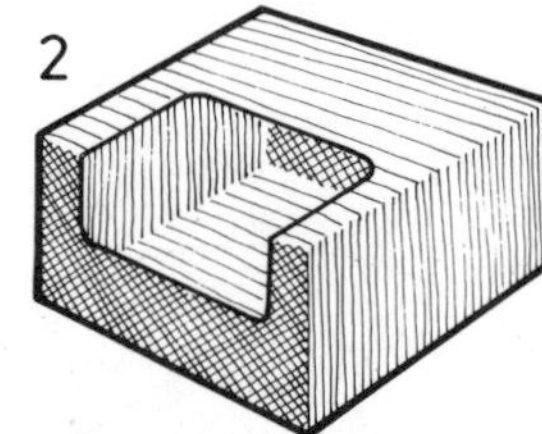

Drawing 3 – Stationery tray. Only thick outlines plus shading with dry transfer tones (see below).

Drawing 4 – Bedside cabinet. Only the lines of the outline have been drawn, all other surfaces being described by drawing wood graining, with varying spacings to suggest different depths of shade.

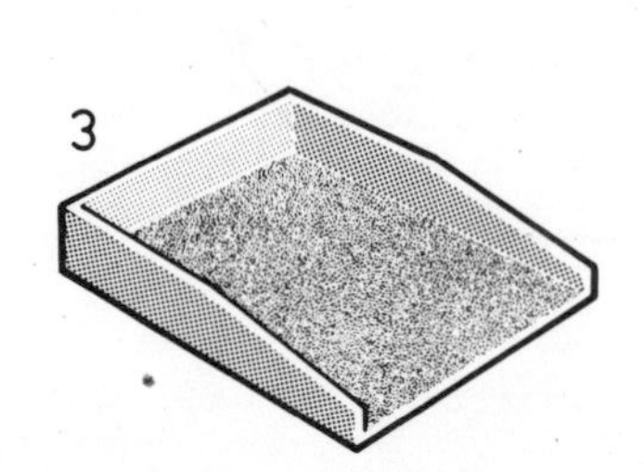

Shading with dry transfer tones

Dry transfer tones such as Letratone (produced by Letraset) are supplied as pre-printed, self-adhesive film mounted on a backing sheet. A very wide choice of dry transfer tones is available. The way in which the film is applied to a drawing is as follows.

1. Complete the drawing outlines in pen or pencil.
2. Select the sheet of tone suitable for the required shading effect.
3. Place the sheet over the drawing and cut out the rough shape of the part of the drawing to be shaded with a sharp knife or artist's scalpel. Peel the cut-out from the backing sheet.
4. Place the piece of self-adhesive film in position on the drawing and smooth it lightly in position – light finger pressure is all that is needed at this stage.
5. Trim the film to its exact shape with the sharp knife and peel any excess film from the drawing.
6. Smooth the film on to the drawing under pressure with a purpose-made burnishing tool.
7. When the film is firmly in place on the drawing, parts of the printed pattern on its surface can be removed by scraping with a sharp knife if required.

Four examples of drawings shaded with dry transfer tones are shown.

Drawing 1 – An adaptor plug. Two dot tones – 20% and 30% shading – were applied to the side and rear of the plug and a third speckled shading was applied over the fuse cover.

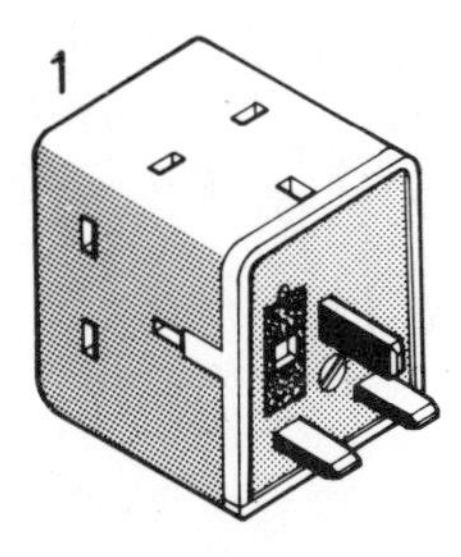

Drawing 2 – A shaft bearing. Most of the drawing was covered with a speckled tone shading to represent the surface of cast iron. The interior surface of the bearing hole was left free of tone. A second layer of the same tone was applied on the left hand edges. The shaping of the front rounded corner and along the filleted corner between base and upright was achieved by scraping off the surface of the transfer film with a knife.

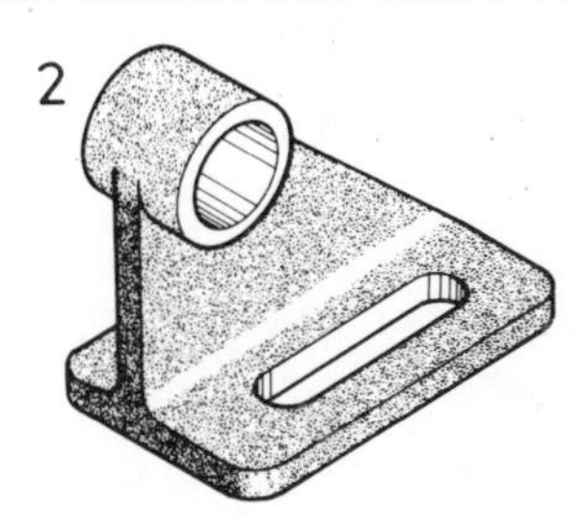

Drawing 3 – A length of sliding door track. Shaded with 20% dot tone.

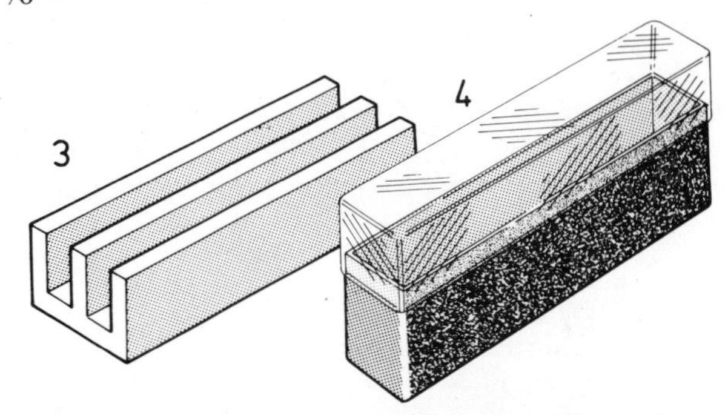

Drawing 4 – A 35 mm transparency case. Three shadings – a 10% dot, a 30% dot and a speckled tone – were used to shade this drawing.

Two freehand drawings compare dry transfer tone shading and shading using an ink pen. In the first, a hand micro-recorder, three tones were used. In the second, a camera flash gun, all shading has been applied freehand using a technical pen.

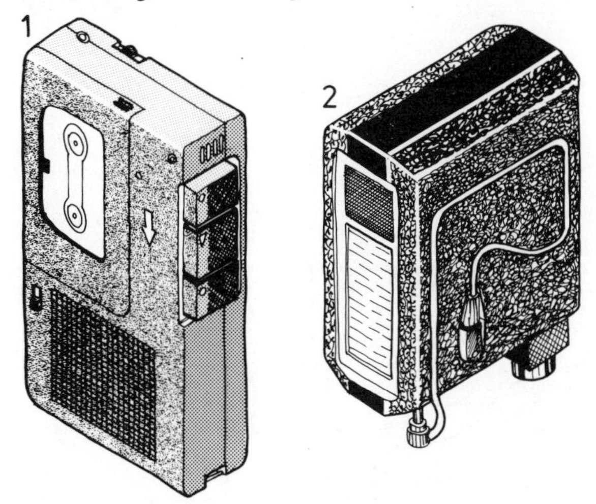

Freehand drawings comparing dry transfer tone shading (drawing 1) with pen shading (drawing 2)

Colour shading of 3D objects

Seventeen examples of the shading of 3D forms are given on pages 96 and 97. All have been shaded with colour pencils in such a way as to emphasise the depth of the object which has been drawn.

Drawing 1 – An isometric drawing of a cube. Colour pencil shading against the straight edges of a piece of paper. Top shaded as if it has a shiny surface.

Drawing 3 – Single-point estimated perspective drawing of a cube with its shadow.

Drawing 4 – A cylinder colour shaded against masking tape. Circularity emphasised by deeper colour towards outer surfaces.

Drawing 5 – A cylinder outlined in colour pencil lines. Circularity emphasised by lines parallel to the ellipse of the cylinder top.

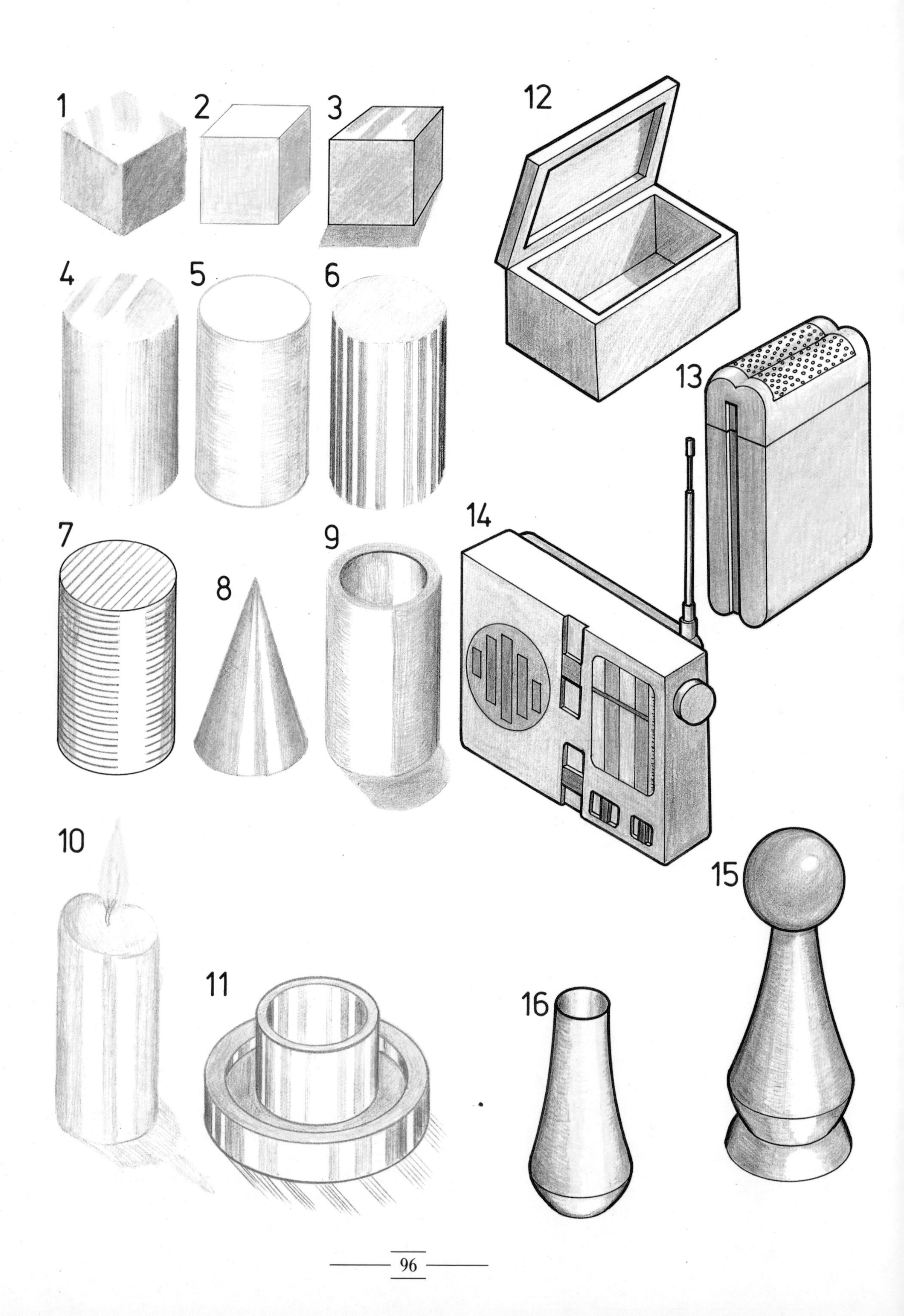
1
2
3
12
4
5
6
13
14
7
9
8
10
15
11
16

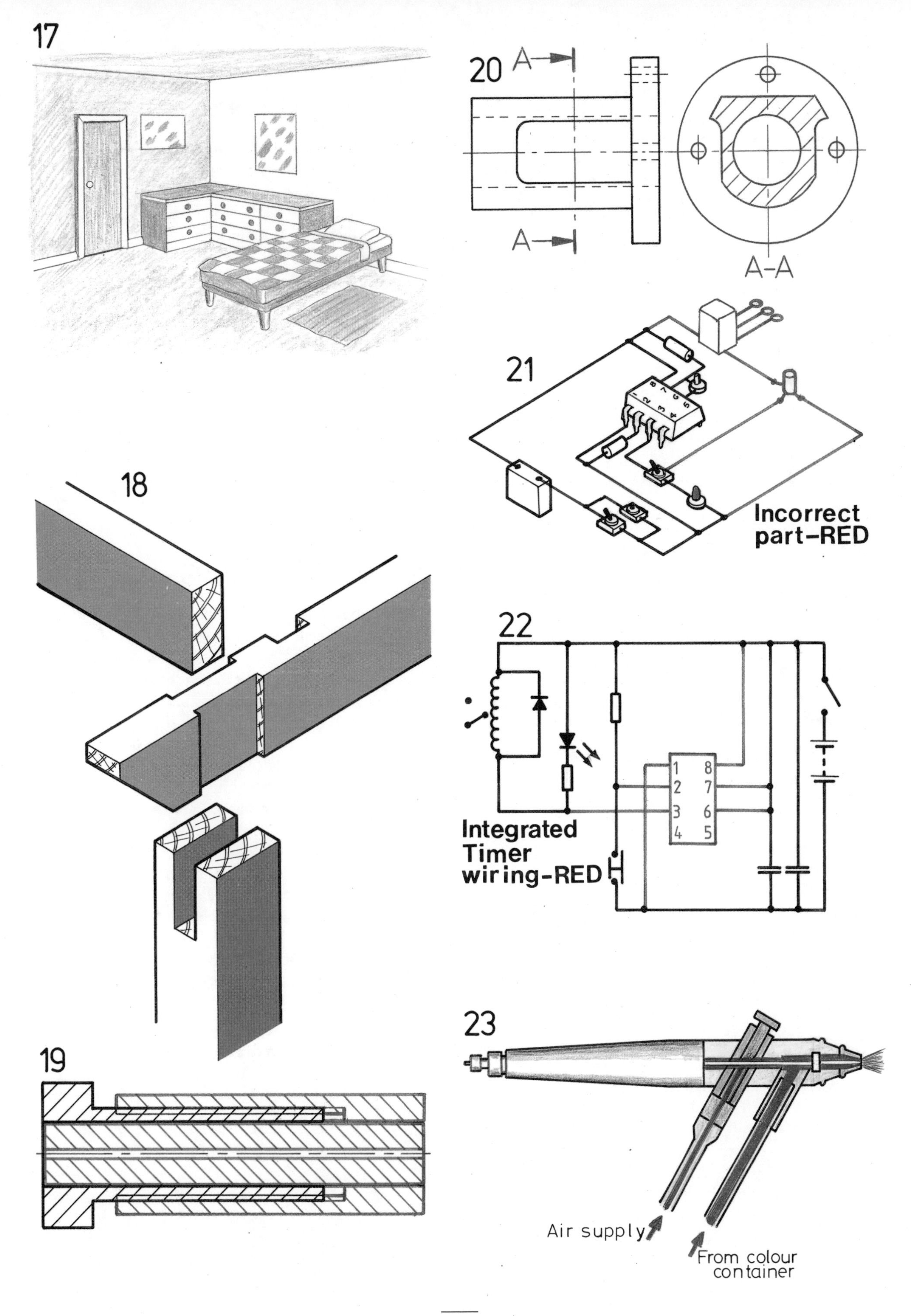
17
20
A
A
A-A
21
Incorrect
part-RED
18
22
1
2
3
4
8
7
6
5
Integrated
Timer
wiring-RED
19
23
Air supply
From colour
container

Drawing 6 – Another cylinder, drawn against a cut-out in masking film. Straight lines of colour deepening towards outside emphasise the shape of the cylinder.
Drawing 7 – Cylinder outline drawn with black ink lines. Parallel colour lines to emphasise cylinder shape.
Drawing 8 – A cone, colour shaded against masking film.
Drawing 9 – A pipe which has been colour shaded with black pencil but with top edges coloured pink.
Drawing 10 – The warmth of a candle emphasised by its red shading.
Drawing 11 – The hardness of a steel part emphasised by blue shading.
Drawing 12 – This box has been shaded by first drawing its outer edges with thick lines, then colouring with three tones, one of which is white, then adding a shadow to the box interior.
Drawing 13 – A shaver showing similar shading features – thick lines on outer edges and three depths of shading, but no shadow.
Drawing 14 – A transistor radio drawn with thick line outer edges and with several depths of colour.
Drawings 15 and 16 – Circular objects which have been colour shaded. Note the light spot on the sphere of 15, the bars of light catching the fronts of the shapes, and the way in which the colour pencil strokes follow the curves of the objects.
Drawing 17 – A two-point perspective of a bedroom interior with various details shaded with a variety of colour pencils.

Other examples of colour work

Further examples of the use of colour are given on page 97.
Drawing 18 – An isometric exploded drawing of a joint from a wooden framework with parts coloured with Pantone dry transfer sheet.
Drawing 19 – A section through a three part assembly with each of the three parts drawn in different colours of ink.
Drawing 20 – A First Angle orthographic projection in which the end view is a sectional view taken on A–A. Black, red and blue linework.
Drawing 21 – A pictorial electronics circuit drawing showing the parts which were wrongly assembled in the circuit in red.
Drawing 22 – An electronics circuit diagram which incorporates an integrated circuit timer. Timer connections drawn in red lines. Note the use of red dry transfer letters.
Drawing 23 – A part sectional drawing through an air-brush showing the routes of colour fluid (red) and compressed air (blue).
Note: This is a purely diagrammatic drawing. Correct, fuller details of a sectioned air-brush are given on page 107.

EXERCISES

On pages 112 and 113 and pages 116 and 117 you will find two mini graphics projects have been reproduced. The first deals with the designing of a pattern formed from tesserae. The second deals with the designing of an item of personal jewellery. Study the contents of the four pages closely and then write an evaluation of the two projects. Your evaluation could include comments as follows.
1. Do the final designs meet the requirements of the design brief?
2. Do the projects show original designing?
3. Do you consider the final designs are successful?
4. Are the materials, shape and form, constructions, colours, and fittings suitable to the designs?
5. Can you trace the design process sequence of design brief; investigation; ideas for solutions; chosen solution; development from the investigation; realisation of final design?

Letters and lettering

Many items of design graphics require the addition of letters and lettering to name, give details of, add notes about or explain drawings. Details about types of lettering for this purpose have already been given on pages 35 to 36. Examples of such lettering have also been included in the many design drawings throughout this book. You may wish, however, to design your own lettering for some particular purpose. A few examples are given on this page while further examples which have been partly colour washed are shown on page 100.
Drawing 1 Shaded letters – The letters in the words THE FLEET are particularly suitable for thick and thin line shading, because each letter is made up of straight lines. The effect can also be obtained on letters with curved lines.

1

THE FLEET

Drawing 2 Shaded letters – The letters of the word SHELTER have been drawn in perspective using only thin lines, the depth of the letters being obtained by imagining the faces being lit from above and from the front. The resulting shadows are drawn with thin vertical lines.

2

Drawing 3 Ghosted letters – The letters of A. YARWOOD give an appearance of depth even in those parts of the letters where nothing but space exists.

Drawing 4 An advertisement suggestion – The drawing of the telephone held in a hand is the result of a sketch made from observing a person holding a telephone. The resulting outline was then shaded with a dry transfer tone, parts of which have been scratched away with a knife. The words 'Phone today' were designed on the idea of a 'balloon' enclosing the words as if they were coming from the phone.

Logograms

As defined in the Oxford Concise Dictionary, a *logogram* is a 'sign or character representing a word in shorthand'; an *ideogram* is a 'character symbolising the idea of a thing without expressing its name'. Another word – *pictogram* – is now in common use in graphics to define a picture, symbol or sign representing an idea. Despite these clear definitions, some confusion in the use of the term logogram has arisen, with the result that nowadays the word has commonly come to be taken to define any sign, character, or piece of graphics which represents the name of a firm, organisation or association. A logogram is, in other words, now taken to include those trade marks and trade signs which many organisations employ to represent their name, even when they do not necessarily include letters or words.

Two sheets of drawings showing a project in designing a logogram have been included in this book – on pages 18 and 19. In this example of design work, the term logogram has its dictionary meaning. In the broader meaning of the term, logograms are interesting forms of modern graphical designs which are based on geometrical constructions with or without the significance of letters.

Water-colour washes

Water-colour washes applied with an artist's paintbrush are one of the most common methods of colouring used in graphic design drawing.

Some rules for applying colour washes

1. Use clean water. Throw away water that has become too deeply coloured with paint.
2. Use good quality artist's brushes. A Windsor and Newton number 3 brush is a good size, although a number 0 brush is of value for fine detail.
3. Either block or tube paints are suitable.
4. Aim at obtaining thin, pale and even tints.
5. Lightly painting an area to be coloured with clean water before applying colour assists in achieving an even wash.
6. Mix sufficient colour wash for the part to be coloured before commencing to paint. If you run out of colour wash while painting an area, it is difficult to cover up the drying line which will result.
7. Squeeze the brush dry to remove surplus paint at the end of colouring an area.
8. To achieve several depths of colour over one area, allow each coat to dry before applying another.
9. Work from side to side, or from top to bottom of the area to be painted.
10. Do not allow the working edge of your wash to dry until you have completely covered an area.

Examples of colour wash work

Fifteen examples of drawings which have been colour washed with water-colour paints are shown on pages 100 and 101.

Drawing 1 – A single-point perspective drawing of the initials for Craft, Design and Technology.
Drawing 2 – An oblique drawing of a year number.
Drawing 3 – An isometric drawing of the word METAL.
Drawing 4 – Two colours of wash – green to emphasise the letter e and light orange as a background.
Drawing 5 – A number emphasised by a coloured, shaped background.
Drawing 6 – A cabinet drawing of a nest of hexagonal boxes – two tones of green.
Drawings 7 and 8 – Two isometric drawings of metal parts shaded in two tones of blue.
Drawing 9 – A clock case and its battery. Reds and orange.
Drawing 10 – A sectional view with the shape of the section colour washed in blue.
Drawing 11 – A design for a bookrack with wooden parts coloured orange and plastic parts coloured blue.

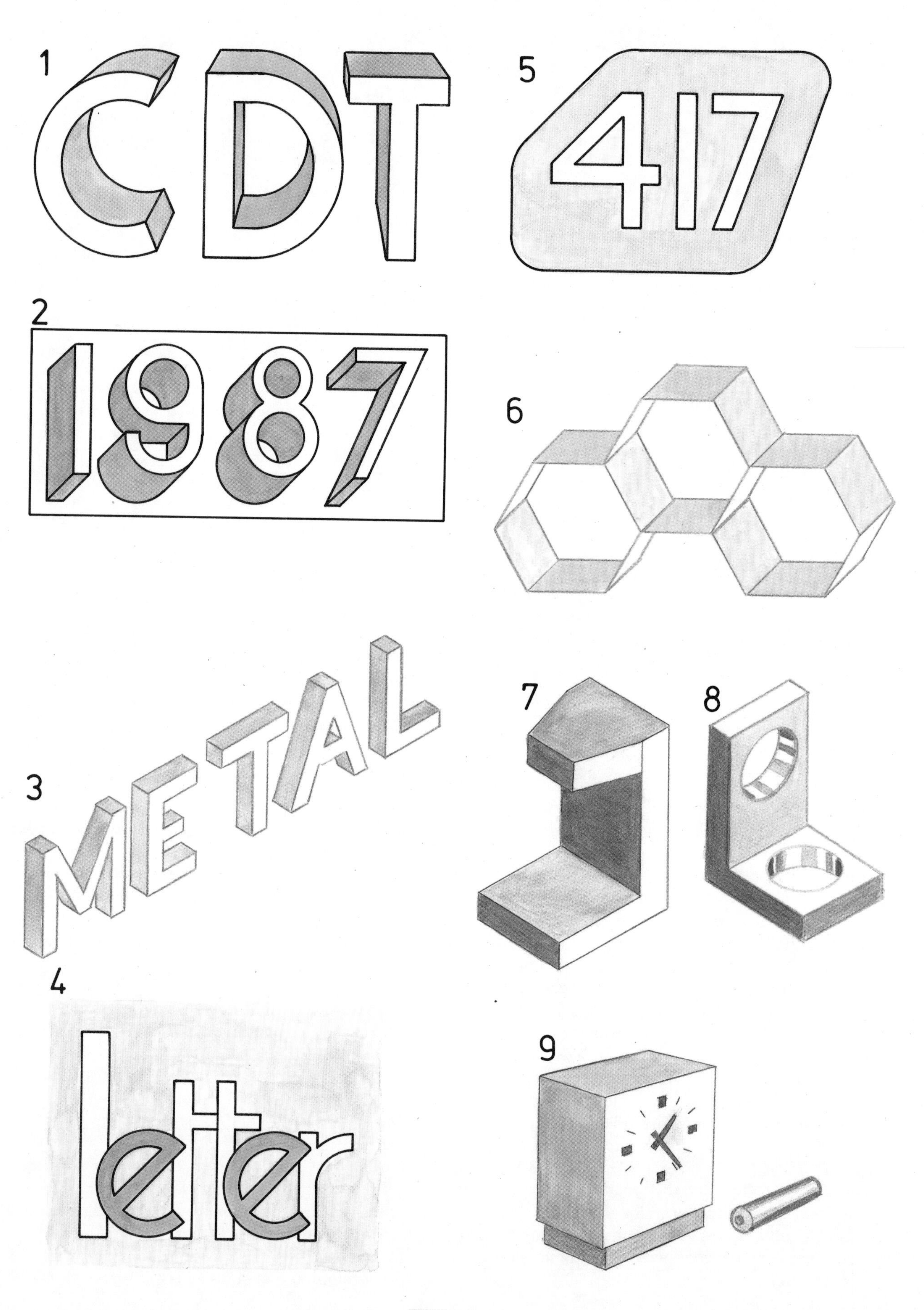
1
CDT
5
417
2
1987
6
3
METAL
7
8
4
letter
9

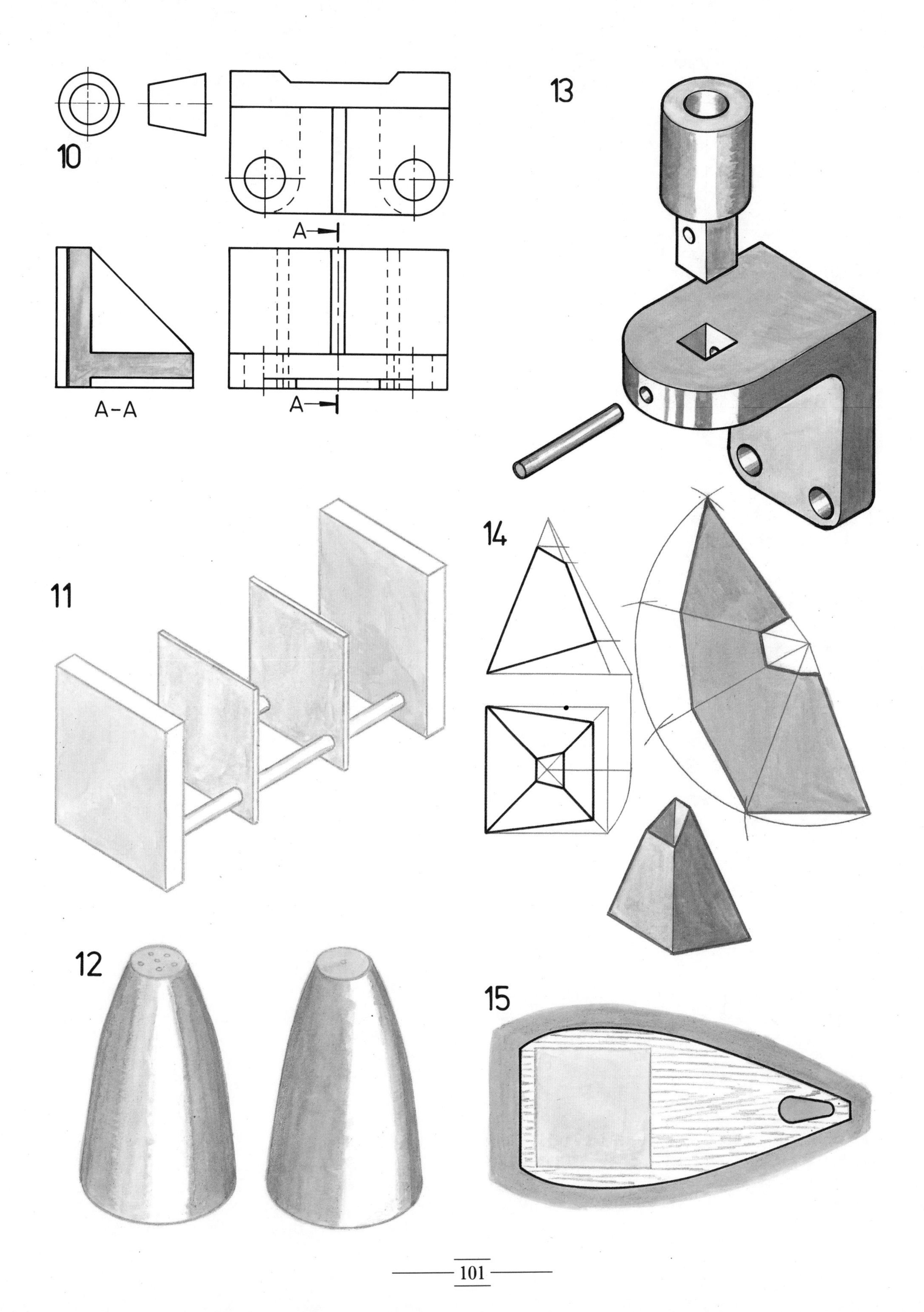
10
A
A–A
A
13
11
14
12
15

Drawing 12 – Salt and pepper pots coloured green – varying depths of shade.
Drawing 13 – An exploded isometric drawing. Each part colour washed with different colours.
Drawing 14 – The geometry for a development and the resulting development. Construction lines in red. Development colour washed red. A pictorial drawing of the assembled development in two shades of blue.
Drawing 15 – A cheese board outlined against a green background of colour wash. The ceramic tile of the board colour washed orange. Grain of the wood applied with brown colour pencil markings.

EXERCISES

Shade or colour all your designs in answer to the following exercises.

1. From the initials of your own name, design and draw:
either (a) an isometric drawing which uses your initials;
or (b) an oblique drawing including your initials;
or (c) a single point or a two-point perspective drawing of your initials.

2. Drawings 1 and 2 are examples of lettering designed to emphasise the meanings of the words they spell.

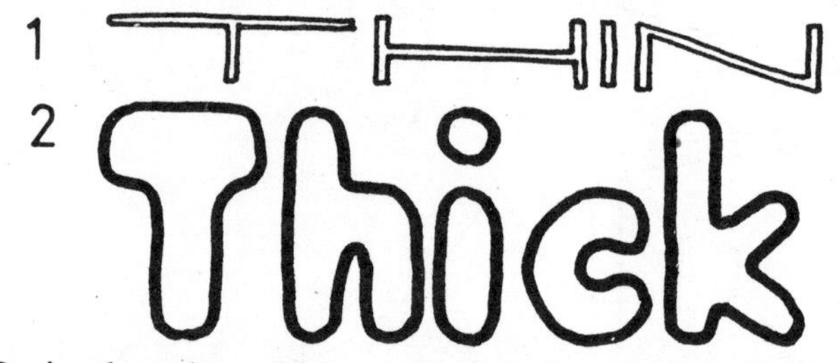

Design lettering which emphasises the words: tall; fat; heavy; slim; rough; smooth; tidy; untidy; round; square.

3. The letters E, H, L, M, N, T, V, W, X, Y, Z are made up of straight lines. Design three different shapes for each of these eleven letters. Then take the lower case letters e, h, l, m, n, t, v, w, x, y, z and draw these in the same three styles.

4. You wish to make a binder to hold forty sheets of design drawings of A4 size. Design and make the binder from cardboard. Design suitable wording for a cover to the binder.

5. Design a folder cover for your graphics design drawings to include the wording GRAPHICS DESIGN.

6. Drawing 3, on an isometric grid, shows a packet for containing tea. Copy the given drawing to the sizes indicated by the grid lines. On the three appearing faces of the packet drawing add your own designs for the wording 'FRESH TEA'.

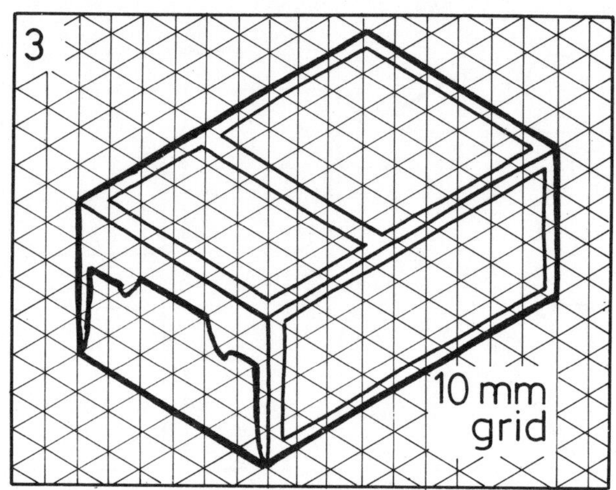

7. Drawing 4, on an isometric grid, is of a ceramic jar with a screw-top lid to hold hand cream. Design a label which would wrap around the bottom part of the jar and which includes the words 'HAND CREAM'. Then copy the given drawing to the sizes indicated by the grid lines and re-draw the label in position on the jar.

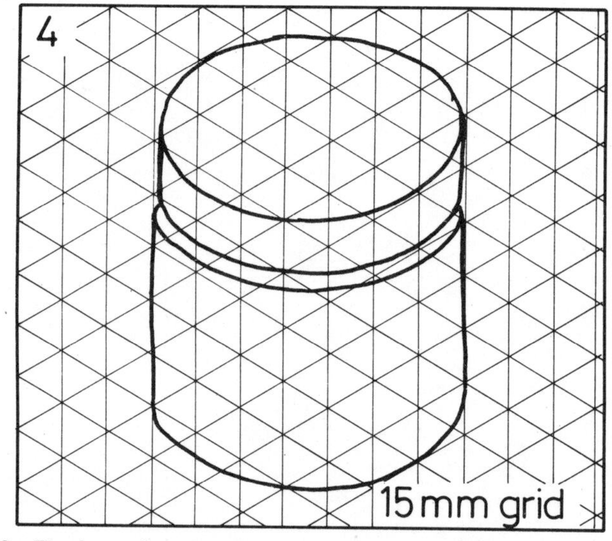

8. Design a house name or number plate for your own house. The letters/figures should be in a coloured plastic which can be pinned or screwed in position on a shaped wooden board.

9. Copy the perspective drawings given on page 62 with the aid of tracing paper. Enlarge each of the drawings to a satisfactory size. Colour the enlarged drawings with colour pencils, with water-colour washes or with markers.

10. Copy the given drawings (5) of a cube, a cylinder, a cone, a piece of piping and a bracket to the sizes indicated by the grid lines. Colour your drawings with either colour pencils, water-colour wash or markers.

11. Your school uses pieces of masking tape to fix papers to drawing boards. Design a holder for a 25 metre roll of 25 mm wide masking tape, which will allow short pieces of the tape to be cut from the roll on

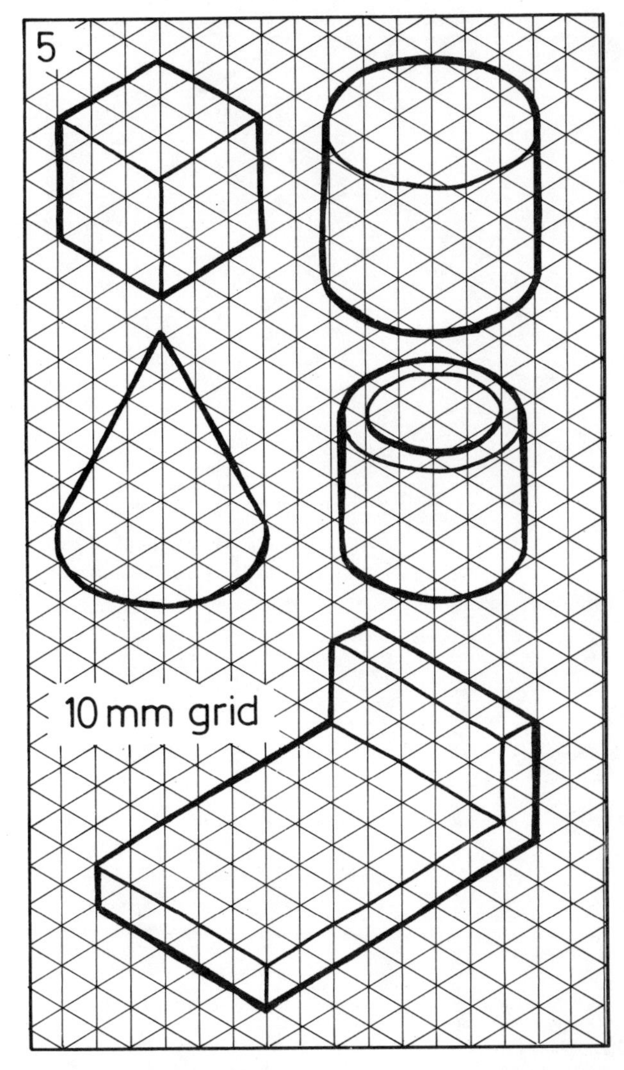

a sharp or serrated edge which is part of the design. Make a model of the design from cardboard and pieces of wood.

12. Copy the two given drawings (6 and 7) to the sizes given, using fine pencil lines. Design shapes for a pair of salad servers based on the two drawings. Name a suitable material from which the servers could be made.

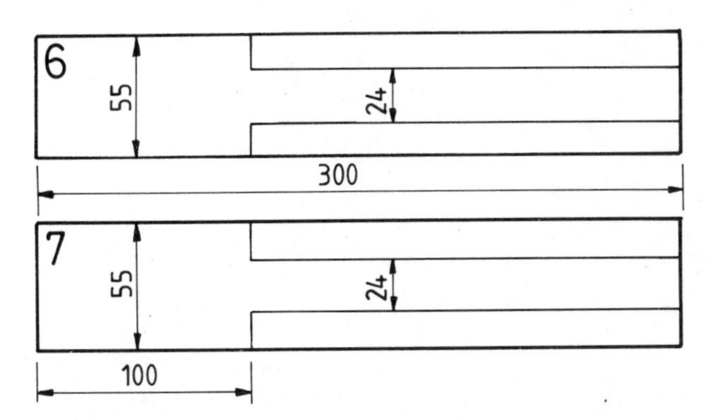

13. A kite project is given on pages 114 and 115. Using the information given in the project drawings, design, draw and colour suitable decorations for a 'diamond' shaped kite.

14. On page 120 you will see a colour photograph of some granite rocks. Study the photograph carefully. Could you design a pattern based upon the shapes, textures and colours suggested by the photograph? Design and then draw such a pattern. Add colour and suitable texture patterns to your design.

15. The drawings 9 to 14 on page 92 suggest designs for neck pendants or for lapel brooches. Design and draw a suitable hanging or clasp for each of the pendants/brooches.

16. Drawings 8 and 9 show dimensions for a paperback book and the overall dimensions for a desk bookrack for holding a number of these books. Drawing 11 on page 101 shows an outline for a bookrack suitable for holding paperbacks. The rack shown on page 101 is to be made from a hardwood with sheet plastic dividers. Copy the drawings of page 101 to the sizes given in the drawing on this page. Add any decorative features you consider might enhance the design.

Book 180 × 110 × 20 (to 40)

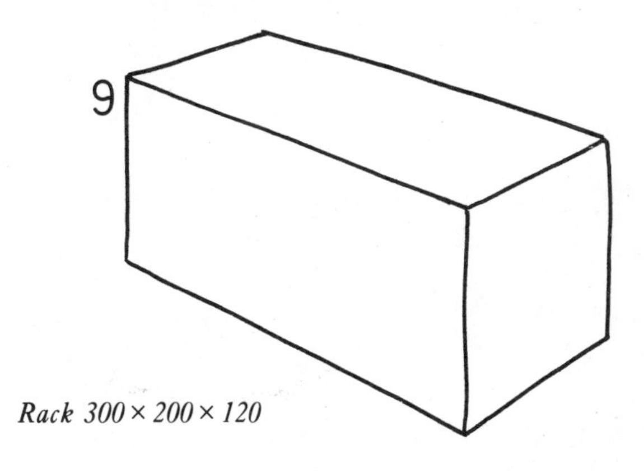

Rack 300 × 200 × 120

17. From your design in answer to Exercise 7, design an advertisement for the ceramic jar with your label design in place, a hand outline and the words 'Hand cream'.

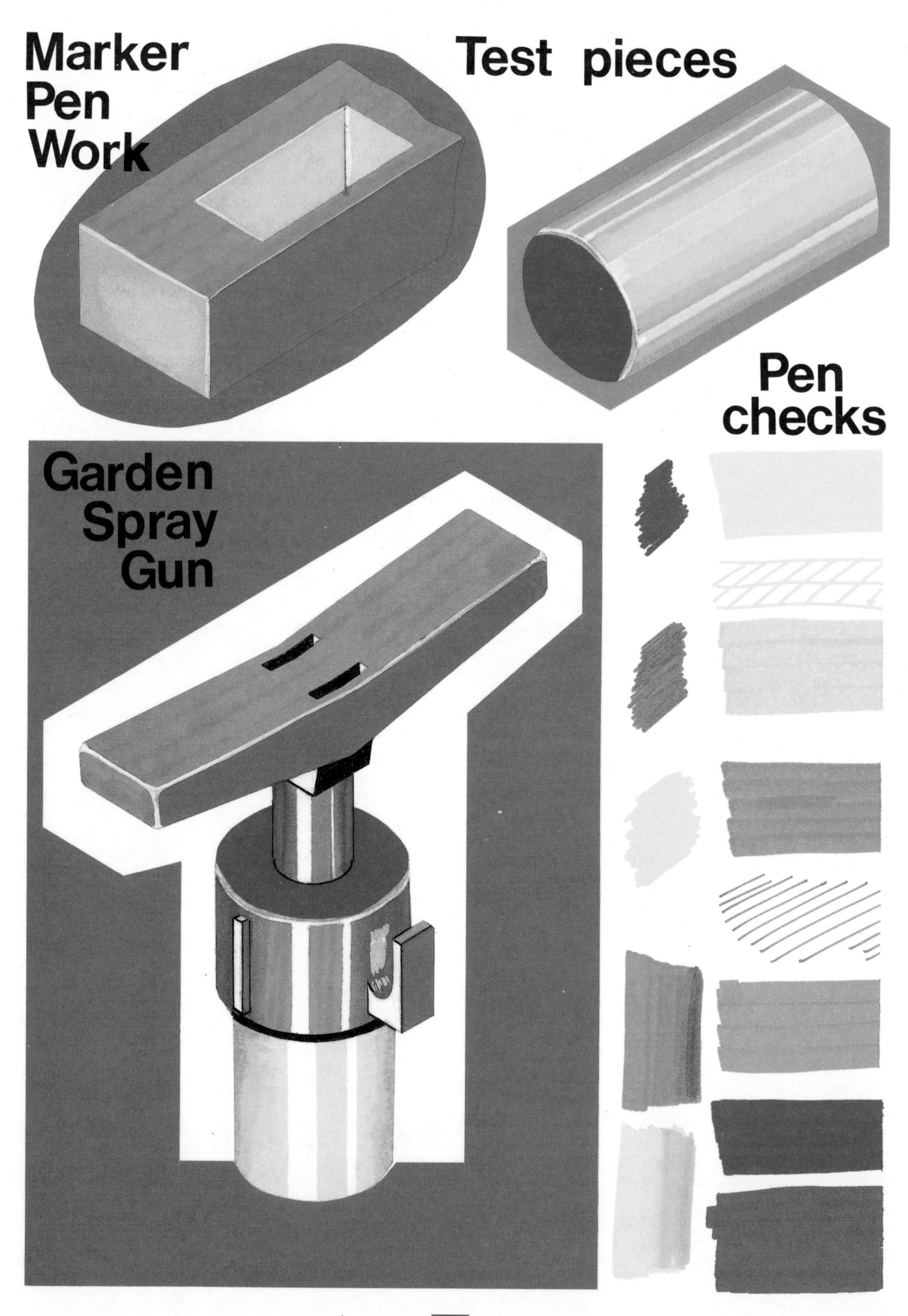
Marker
Pen
Work
Test pieces
Pen
checks
Garden
Spray
Gun

7 Model steam engine burner

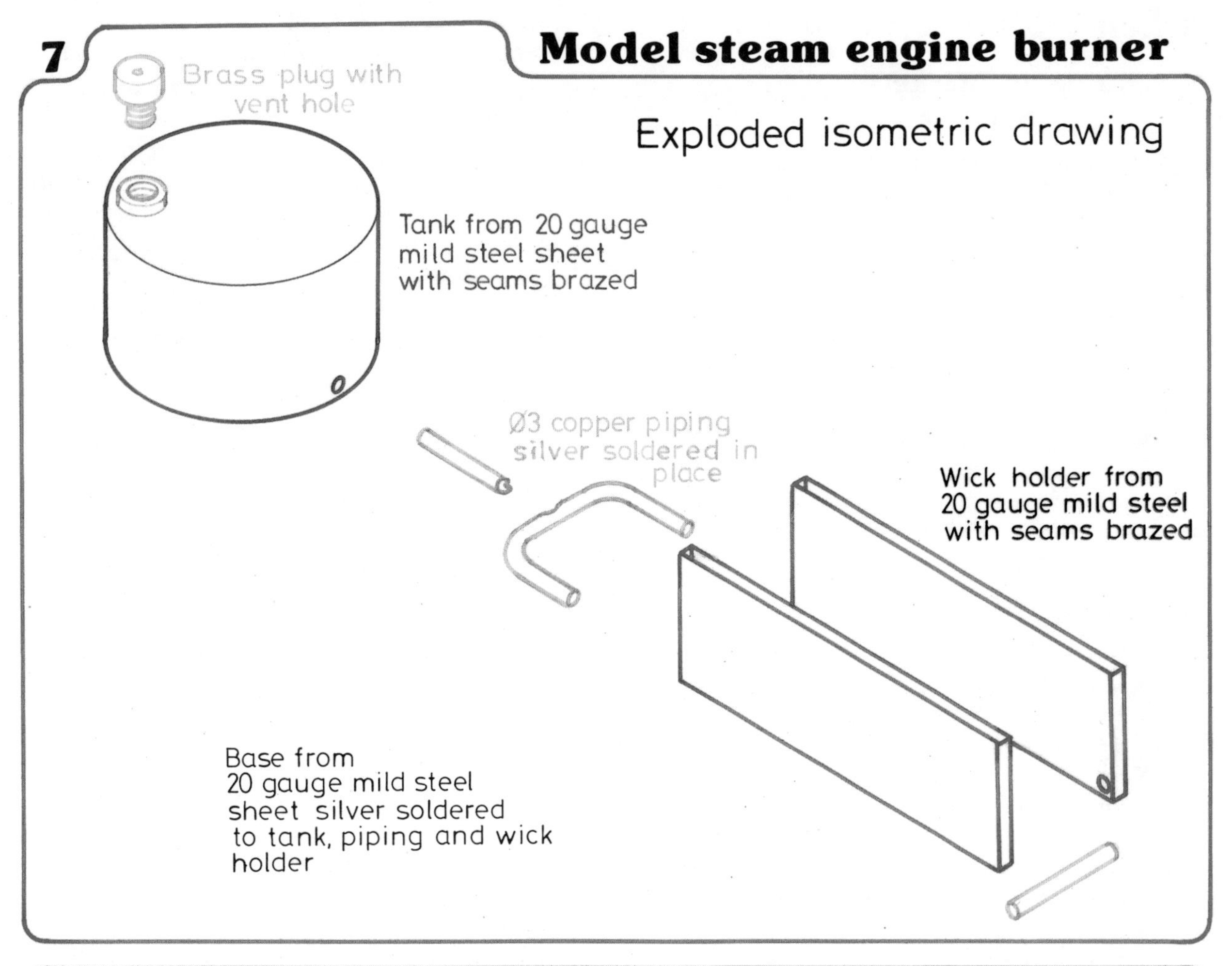

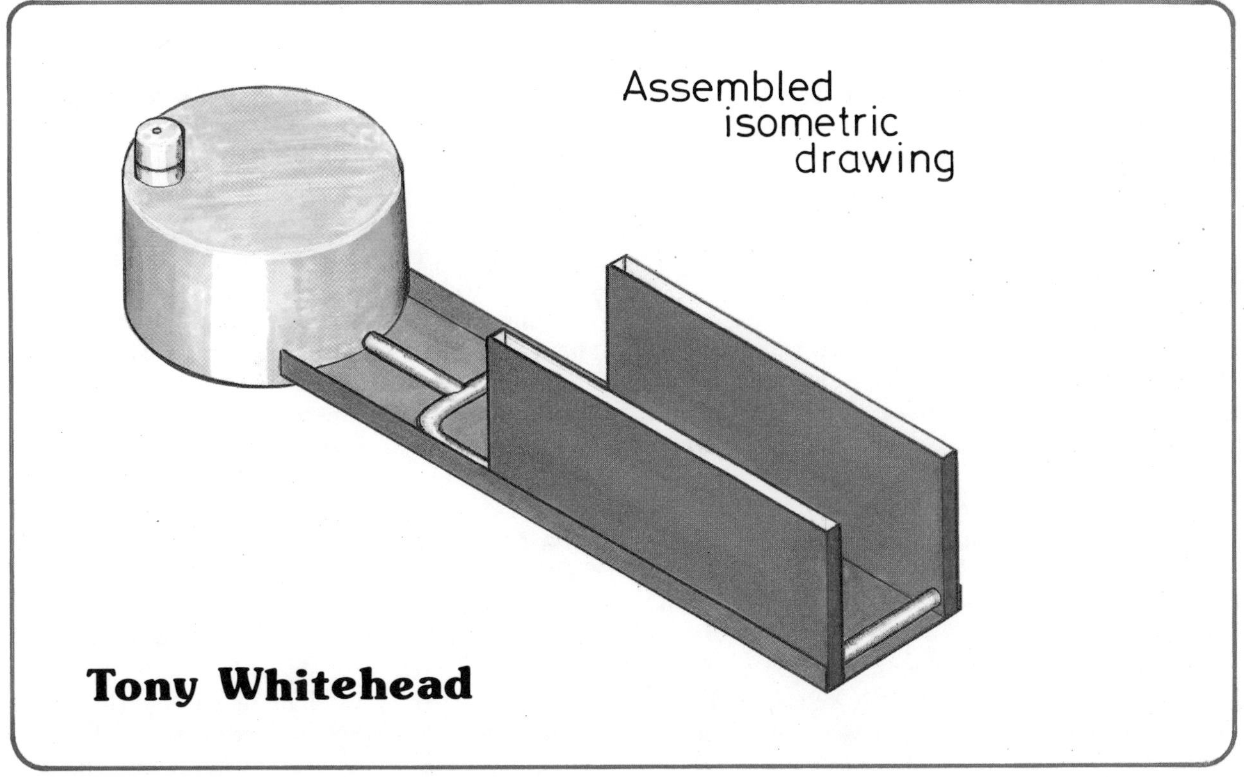

Tony Whitehead

Markers

Notes on using markers

1. Marker pens are made by a number of firms. Each firm makes them in a large range of colours.
2. Each firm makes each colour marker with two nib types – broad and fine. The broad nib types are for covering areas with colour, the fine are for filling in colour detail.
3. Deeper shades of the same colour can be obtained by going over the same area two or three times, but allow each coat to dry before applying another. Failure to observe this rule can result in the paper surface becoming damaged, with ill effects on your colouring.
4. Always replace the marker cap immediately after the marker has been used. This avoids evaporation of the colour liquid. Also, if a marker rolls across a drawing and its nib is covered, the pen will not mark the drawing.
5. Highlights can be painted in with white gouache water colour applied with a brush or they can be drawn in with a white colour pencil.
6. Practice on spare paper before using any particular colour of marker.
7. Use bleedproof or natural tracing papers. Even then have a backing sheet behind your drawing paper to take up any 'bleed' through the paper of marker colour.
8. Masking tape or low-tack masking film can assist in obtaining clean, sharp edges to your drawing, but there is a tendency for the colour to 'bleed' under the masking. Test before use.

Colouring a drawing with markers

The isometric drawing on page 104 of part of a garden spray gun and the assembled isometric drawing on page 105 were both coloured with marker pens. The garden spray gun handle has been shown as a freehand orthographic projection drawn on a square grid on page 59.

The procedure adopted for the drawing on page 104 was as follows:

1. It was made on bleedproof paper.
2. The drawing outline was constructed with the aid of a 30°, 60° set square and an isometric ellipse template and drawn with light pencil lines.
3. The colour of each marker was checked before an area was coloured.
4. All the blue areas were coloured first. Parts of the drawing were masked with masking film to achieve straight coloured edges. Those parts needing a deeper tone of blue were allowed to dry before second and third coats were applied.
5. The red area was then coloured.
6. The yellow areas were the last to be coloured.
7. Highlights were filled in with white water-colour paint applied with a fine (No. 0) artist's brush.
8. Shadow areas were filled in with a black pen.
9. A red Pantone film was applied to a sheet of white art paper and part of it cut away with a sharp knife to form a background to the marker-coloured drawing.
10. The coloured drawing of the spray gun was cut out from the bleedproof paper with a sharp knife working on a piece of hardboard, with the aid of a straightedge.
11. The drawing was fixed inside its Pantone background with stick adhesive.

The marker colour drawing of page 105 was coloured in a similar manner except that the cut-out drawing was fixed on the project drawing sheet number 7 within red borders. The previous six sheets for this project contained details of the investigation into the problems of designing the burner together with solution ideas drawings. This drawing – sheet 7 – shows the chosen solution drawing.

The exploded isometric drawing of sheet 7 was drawn with coloured pens, a different colour being used for each part of the burner. Notes in appropriate colours were added with the aid of lettering stencils. Note the use of dry transfer lettering for the sheet number, the project title and the pupil's name. The borders of the sheet were drawn with a red colour pen, the corners being drawn with the aid of a radius curve.

Air-brushes

Air-brushes are used for applying colour to surfaces. They operate on compressed air supplied at pressures of between 20 and 40 psi (pounds per square inch), an average working pressure being 30 psi. 20–40 psi is 1.4 to 2.8 kg/cm^2 (kilograms per square centimetre). The compressed air can be supplied from a compressor driven by an electric motor, from an air cylinder (or a carbon dioxide cylinder), from a can of compressed air or from a car spare tyre which has been inflated to about 40 psi. If a car tyre is used, it must be mounted on its rim and a special adaptor must be purchased to attach the air line to the brush.

The compressed air enters the brush from an air line attached between the source of compressed air and the brush. It is mixed with the colouring medium which is sucked from a colour container into the brush by the flow of air through the brush. The colour leaves the nozzle of the brush as a spray of droplets. The spray size can be adjusted in one of two ways, depending upon the type of air-brush being used.

There are two types of air-brush – single and dual-action. In a single-action brush, when the control trigger is pressed compressed air enters the brush,

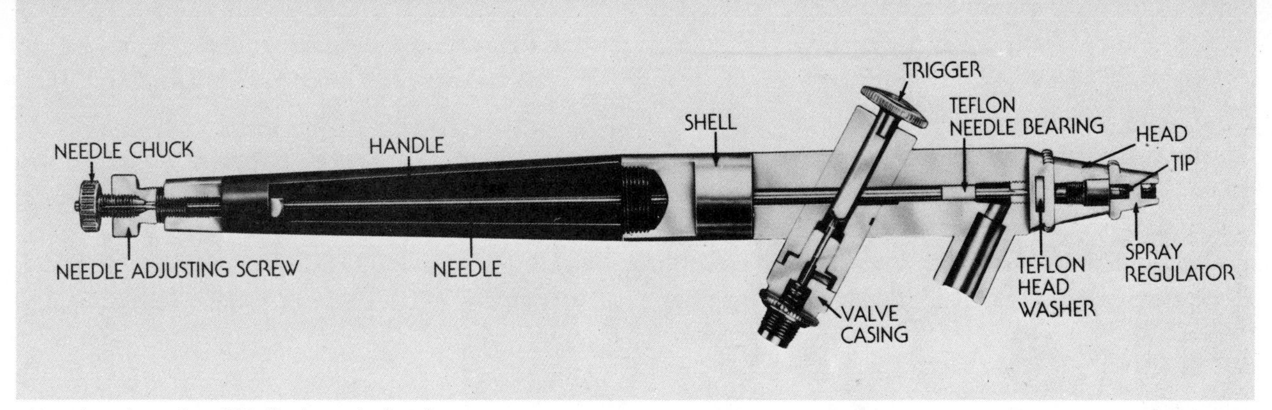

A section through a 200 Badger air-brush

forcing colour fluid from the brush nozzle. The quantity of fluid leaving the nozzle is controlled by an adjuster screw at the rear of the brush. In dual-action air-brushes, the control trigger is pressed down to allow compressed air through the brush and pressed backwards to allow colour fluid into the brush. The quantity of colour fluid, and hence the size of the spray, depends upon how far back the control trigger is pressed. With a dual-action air-brush, the user has complete trigger control over the flow of colour from the brush.

Because of their ease of trigger control, single-action air-brushes are much more common in schools than are dual-action brushes. Considerably more practice and skill are required to apply colour with a dual-action brush than with a single-action one.

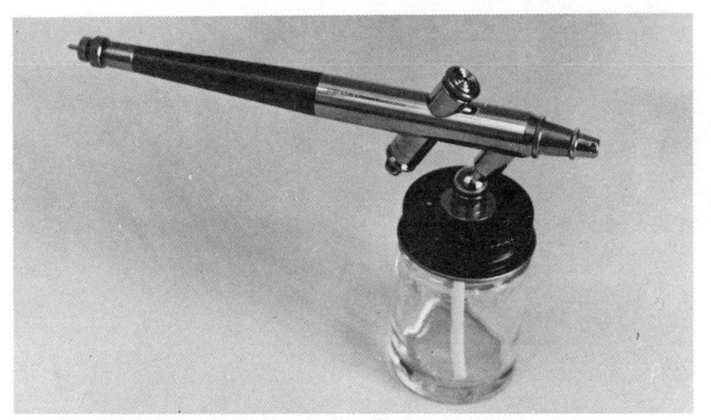

A 200 Badger air-brush – a single-action brush

The colour must always be a fluid, the consistency of which should be of a milk-like nature. If the fluid is too thick, the air-brush may become clogged or damaged. If too thin, the colour does not leave the brush in a suitable spray form. Various types of colour media can be used – water colours, inks, acrylic paints, various lacquers. In schools the most likely media to be applied by air-brush are inks and water colours. In single-action brushes, the colour fluid is supplied to the brush from a container fitted into the base of the brush.

Areas which are not to be coloured are often masked in some way to prevent the colour being applied over too wide an area. The masking can be done by low-tack masking film (see page 28), by masking tape, by newspaper held in place with masking tape or with hand held or taped pieces of shaped paper or card. Providing the colour medium will stick to the surface being coloured, any material can be coloured by air-brushing – papers, cards, woods, metals, ceramics.

1. Make a line drawing on tracing paper

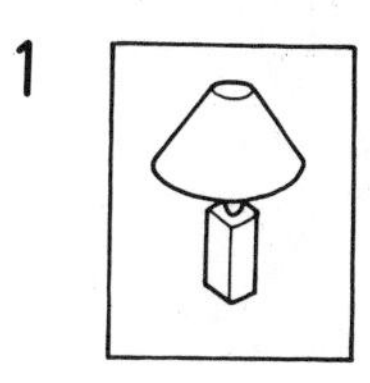

2. Smooth down low-tack masking film onto paper

3. Trace down onto masking film through carbon paper

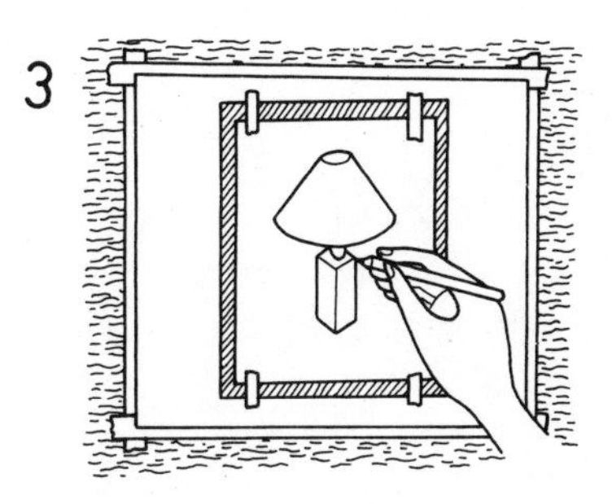

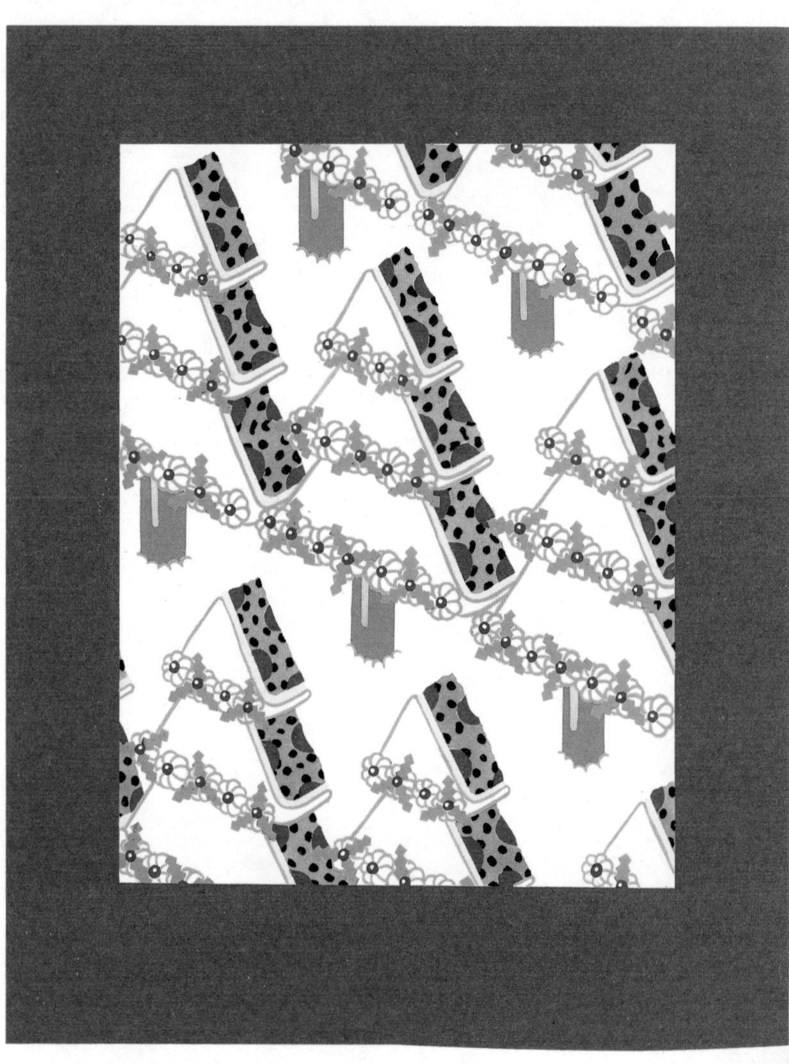

The finished Christmas card design from pages 4 to 7

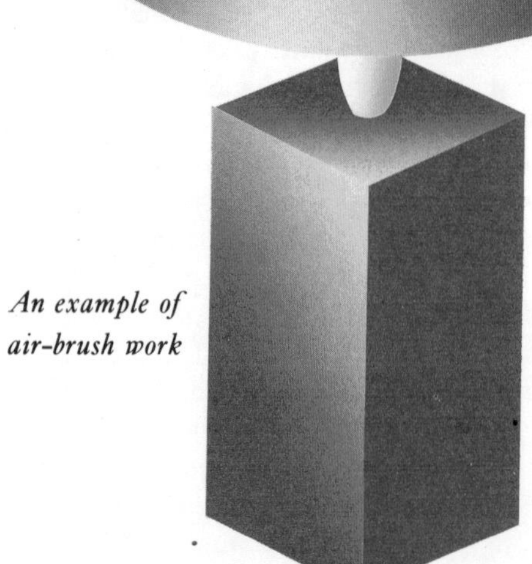

An example of air-brush work

Two examples of air brush work

4. Cut very lightly around all lines on masking film. Take care to cut the film, but not the art paper

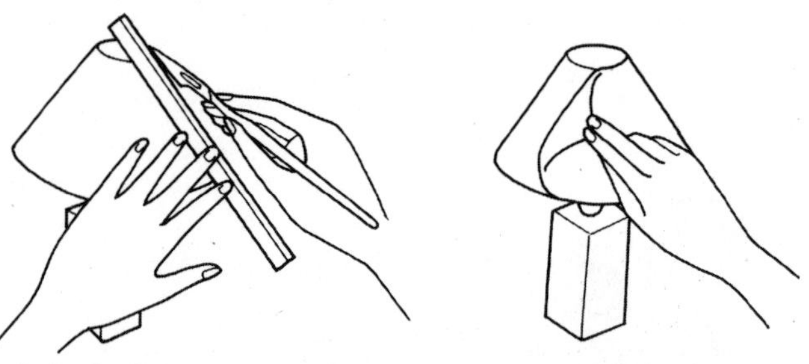

5. Peel off the first area to be air-brushed

6. Air-brush first part of drawing, turning the drawing as necessary

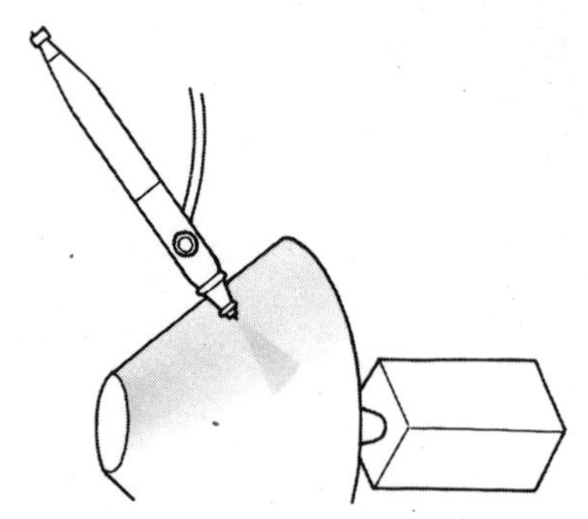

7. Replace first section of masking film. Remove film from second area to be air-brushed

8. Air-brush second area. Repeat stages 5 to 8 until air-brushing is completed

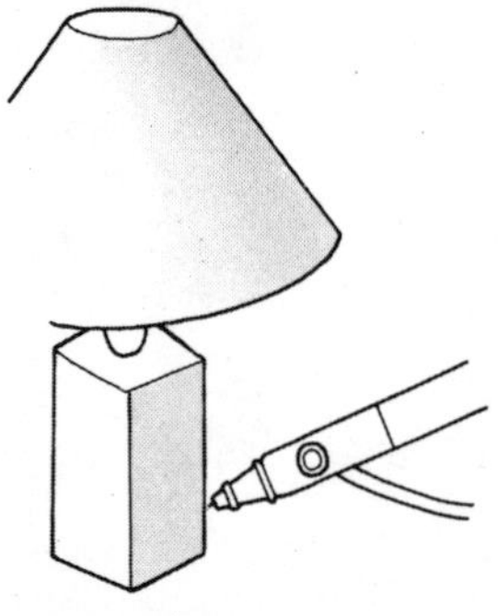

9. Remove all masking film to reveal completed artwork

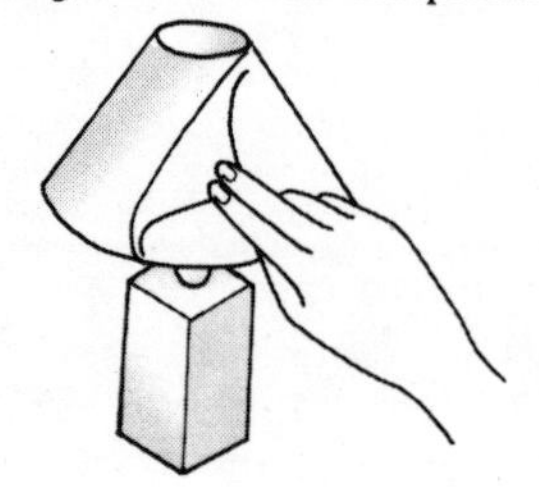

Note: The most important detail to be observed when using an air-brush is that the internal passages through the brush must be thoroughly cleaned of colouring medium *immediately* you have finished working with the brush. Immediately after use, the brush must be cleaned by passing through it, under air pressure, several containerfuls of the solvent for the colour which has been employed. This should be clean water for ink and water colours and the appropriate solvents for other types of colouring media, such as thinners for lacquers, white spirit for some paints. When spraying anything but water or water based colours through an air-brush, great care must be exercised. Ensure that the area you are working in is well ventilated and wear a filter mask.

Method of holding air-brush

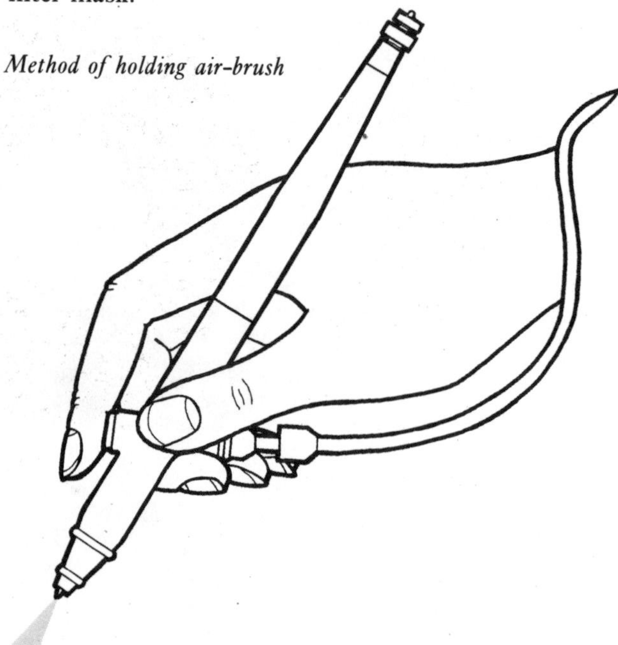

Examples of air-brush artwork

Four examples of air-brush work are given on pages 100 and 101, together with some air-brush practice to test the air-brush equipment before the finished artwork was started.

Christmas card

The finished artwork resulting from the project given on pages 5, 6 and 7 is reproduced on page 108. To the right of the Christmas card and below it are some practicę 'doodles' with an air-brush to test the working of the equipment. These 'doodles' show different types of colour effects which can be obtained by air-brushing.

Reading lamp

The reading lamp artwork on page 108 is the result of the procedures described on this page. Note in particular the complete absence of outline lines because low-tack masking film is used to determine the areas to be painted.

Record 'sleeve'

The record sleeve on page 109 results from air-brush work with details painted in with artist's brushes using gouache water-colour paints. The fine black lines have been added with a technical pen with the aid of a ruler.

Poster

The poster on page 109 is another example of air-brush work with small details added with artist's brushes.

Graphics projects

The next pages include four graphics projects. These are:

Pages 112 and 113 – A pattern project
Pages 114 and 115 – A kite project
Pages 116 and 117 – A jewellery project
Pages 118 and 122 – A stage scenery project.

These projects are not complete. The following items were excluded because of the limitations of space in this book.

1. models;
2. evaluations;
3. in the kite and stage scenery projects, some of the sheets of drawings and photographs;
4. photographs of the realised designs.

Project – A pattern

The colour work on pages 112 and 113 is a graphic design project in answer to the design brief:

'Design a geometric pattern based on tesserae.'

The project follows the design sequence emphasised throughout this book – design brief: investigation; ideas for solutions; chosen solution; realisation.

All the black linework and the notes were drawn with the aid of a technical pen with a 0.25 mm nib, or with a black biro pen. The blue margin lines were drawn with a penstik. Four different methods of colouring have been employed – colour pencils, water-colour washes, markers and air-brushing. The marker colour work was drawn on bleedproof paper, which was then cut to shape and mounted in place on the art paper containing the other drawings. The final artwork – the realised design – is an example of air-brush colour work.

Note how ideas based on geometric shapes are attempted and discarded, sometimes to be picked up again later as the project develops. Note also how, when a firm idea of a chosen solution appears to be selected, it is developed until a final design is accepted. Both sheets for this project were of A3 size.

Project – A kite

Four of the sheets of graphics from this project are given on pages 114 and 115. Two further sheets, not included here, develop the ideas for 'Other shapes for kites come to mind' as stated at the bottom of Sheet 4. Note these details in the four sheets of drawings.

1. *Sheet sizes* – All four sheets are of A3 size with drawings placed in a 'portrait' format.
2. *Borders* – 5 mm wide coloured borders at top and bottom of each sheet are joined vertically with thin lines. Water-colour wash of a blue tint was used to colour the wide borders.
3. *Printing and notes* – The title block containing the project title and the designer's name, together with all sheet numbers, have been added from dry transfer lettering sheets. All notes have been typed on typing paper, cut out with scissors and glued in position on the drawing sheets.
4. *Drawings* – The following drawing techniques have been employed.

(a) Most of the drawings are single-view orthographic projections (front views) drawn either freehand or with instruments.
(b) Some freehand perspective drawings showing construction details have been included, together with freehand perspective drawings on Sheet 4 for further ideas of kite shapes.
(c) Some of the drawings have been water-colour washed with tints of blues or reds.

Project – Jewellery

The two sheets of this project have been reproduced on pages 116 and 117. The project contains examples of four types of colouring – colour pencil work, marker colour work, water-colour wash and air-brush colour. The marker colour work was made up on bleedproof paper which was then mounted in position on an A3 size sheet of art paper. All lines and notes were drawn with either a technical pen or a biro. The project again follows the sequence – design brief; investigation; ideas for solutions; chosen solution; realisation. In this example, note that the following details were investigated while ideas for solutions were explored – shape and form; materials (wood, metals, plastics); construction; shaping and forming; colour; finish; fittings. The design brief for the project was simply:

'Design an item of personal jewellery'.

Project – Stage play 'The Rivals'

Six of the sheets of graphics from this project are given on pages 118 to 122. Two of the colour photographs associated with the project are included on page 120. The complete folio also contained the following which have not been shown because of lack of space:

1. numerous freehand sketches of ideas for the scenery together with notes
2. sketches and notes about stage lighting
3. notes and sketches suggesting scenery colours.
4. notes explaining the project. See page 118.

GRAPHIC PROJECT: Design a Geometric Pattern based on Tessarae.

Basic Shapes:

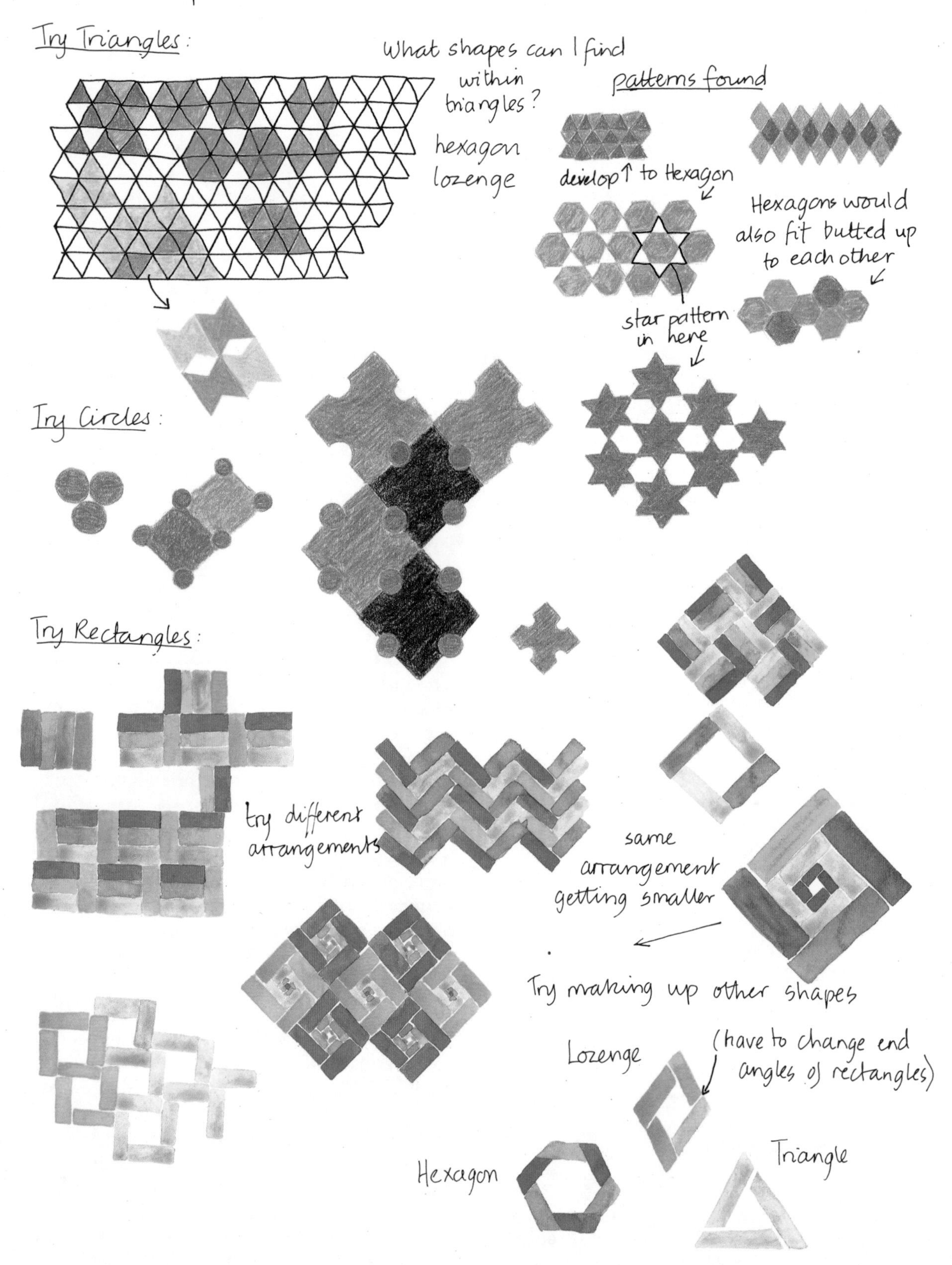

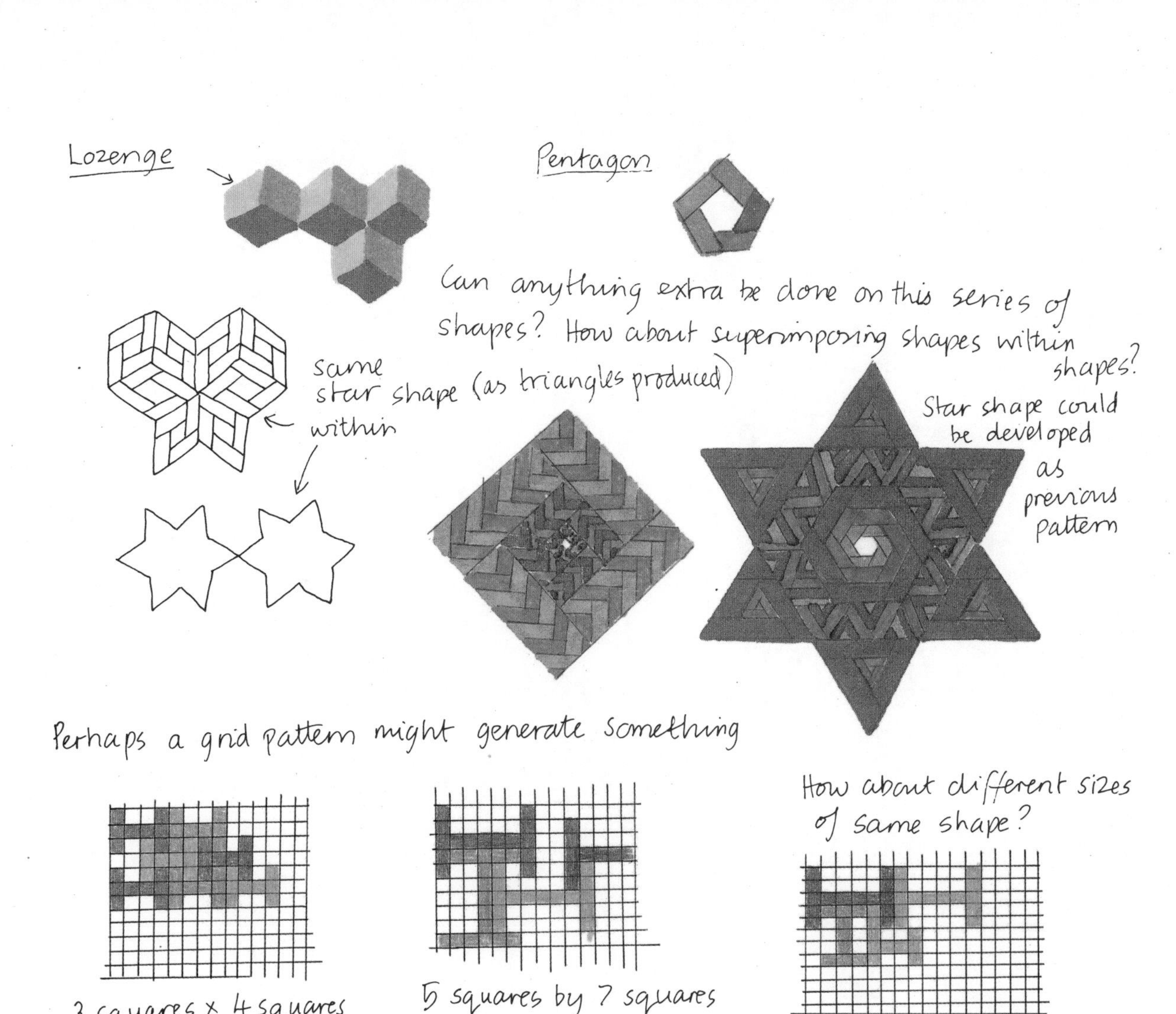

How about trying 7×9 squares

leaves this gap – can I get rid of this by adding to the shape? Yes. Add one square here

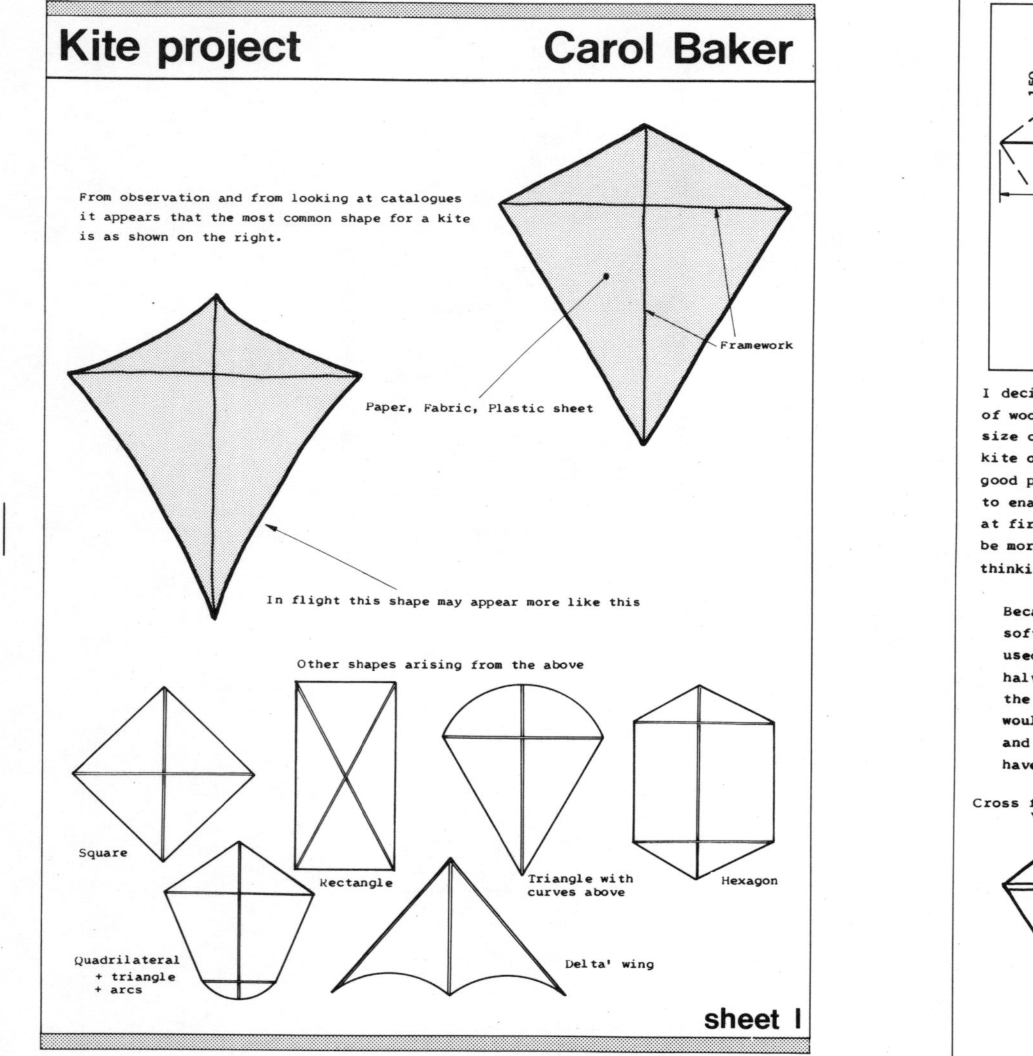

Kite project

Carol Baker

From observation and from looking at catalogues it appears that the most common shape for a kite is as shown on the right.

sheet 1

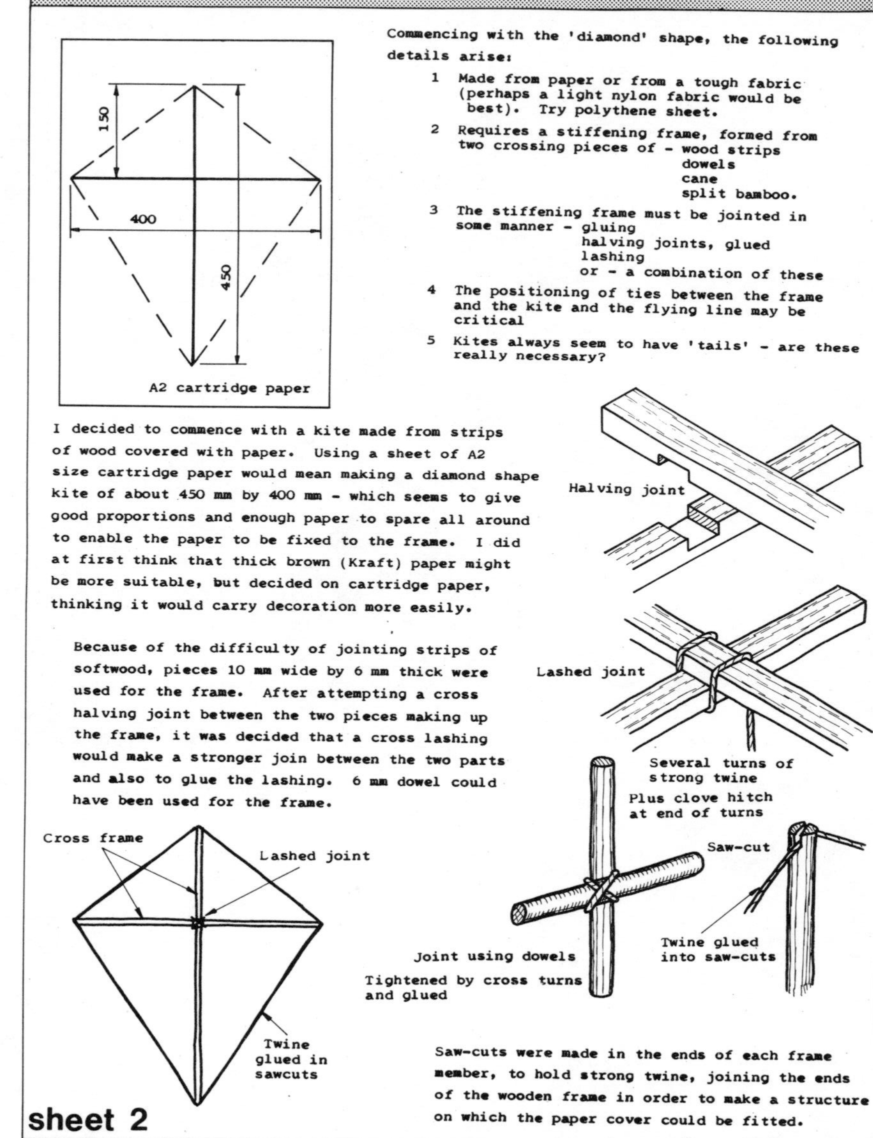

Commencing with the 'diamond' shape, the following details arise:

1. Made from paper or from a tough fabric (perhaps a light nylon fabric would be best). Try polythene sheet.
2. Requires a stiffening frame, formed from two crossing pieces of - wood strips, dowels, cane, split bamboo.
3. The stiffening frame must be jointed in some manner - gluing, halving joints, glued lashing, or - a combination of these
4. The positioning of ties between the frame and the kite and the flying line may be critical
5. Kites always seem to have 'tails' - are these really necessary?

I decided to commence with a kite made from strips of wood covered with paper. Using a sheet of A2 size cartridge paper would mean making a diamond shape kite of about 450 mm by 400 mm - which seems to give good proportions and enough paper to spare all around to enable the paper to be fixed to the frame. I did at first think that thick brown (Kraft) paper might be more suitable, but decided on cartridge paper, thinking it would carry decoration more easily.

Because of the difficulty of jointing strips of softwood, pieces 10 mm wide by 6 mm thick were used for the frame. After attempting a cross halving joint between the two pieces making up the frame, it was decided that a cross lashing would make a stronger join between the two parts and also to glue the lashing. 6 mm dowel could have been used for the frame.

Saw-cuts were made in the ends of each frame member, to hold strong twine, joining the ends of the wooden frame in order to make a structure on which the paper cover could be fitted.

sheet 2

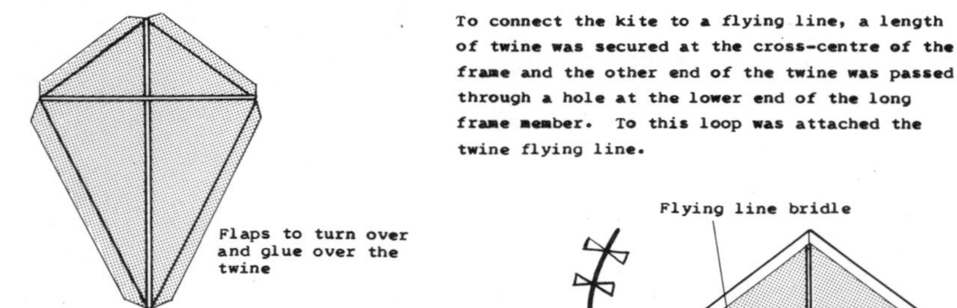

To connect the kite to a flying line, a length of twine was secured at the cross-centre of the frame and the other end of the twine was passed through a hole at the lower end of the long frame member. To this loop was attached the twine flying line.

From reading about kites, making this kite and talking to others who have made them, the following arose:

1 Yes – this kite needs a 'tail'
2 The positioning of the flying line on the 'bridle' can be critical to good flying of the kite

Thus a tail was fitted.
The first was made from a length of thin paper about 25 mm wide and 2 m long.
I tried flying the kite with its tail with the flying line in various positions along the bridle line. Eventually I got the kite to fly quite well. However, after some time, the tail broke away and the kite fell to the ground. This meant a new kite tail, which was made up from strips of paper tied at equal intervals along a nylon twine.

Flying line bridle

Tail

Paper

After gluing the flaps over the twine

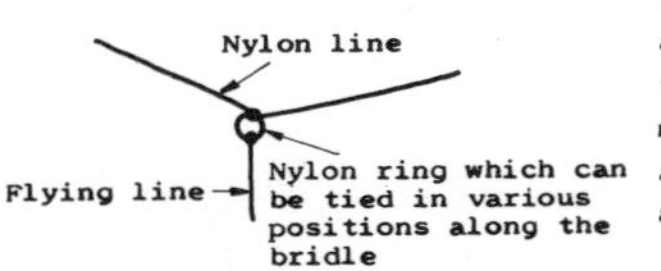

The position of the flying line on the bridle appears to depend partly on wind strength or on the speed with which the kite is flying. This means there may be a need for some method of adjustment between the position of the flying line and its bridle on the kite.

Next, I decided to attempt decorating the kite and the following decorations came to mind.

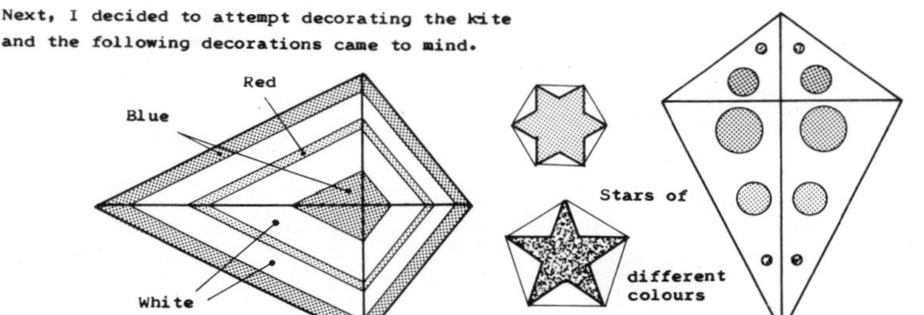

sheet 3

Here is a drawing of my first successful kite.

Wood 150 x 30 x 10

Dowels ∅13 x 150 long

I quickly found it to be necessary to have to make some form of equipment on which the flying line could be wound and unwound. This was made from 2 pieces of wood 150 mm by 30 mm by 10 mm with ∅13 mm dowel 'handles'.

Red

Lashed and glued

Blue

Red

6 mm dowel

Red

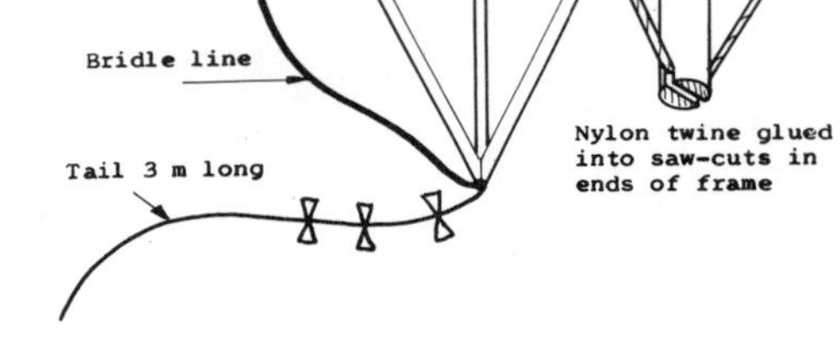

After flying this kite for some hours, the cartridge paper began to tear, so my next kite will be made from a light-weight man-made fibre cloth – probably nylon. This will however give rise to some difficulties when gluing the fabric to the frame, because nylon does not glue well with most adhesives.
I shall also make the next kite to a larger size.
Another type of kite I should like to experiment with is a BOX kite. I should also like to try making a DELTA WING kite. Neither of these kites appear to require tails and they could possibly be made to carry equipment – e.g. a camera, wind testing equipment etc. to some height.

Other shapes for kites come to mind – Geometric shapes
Animal shapes
Aeroplane shapes
Bird shapes

sheet 4

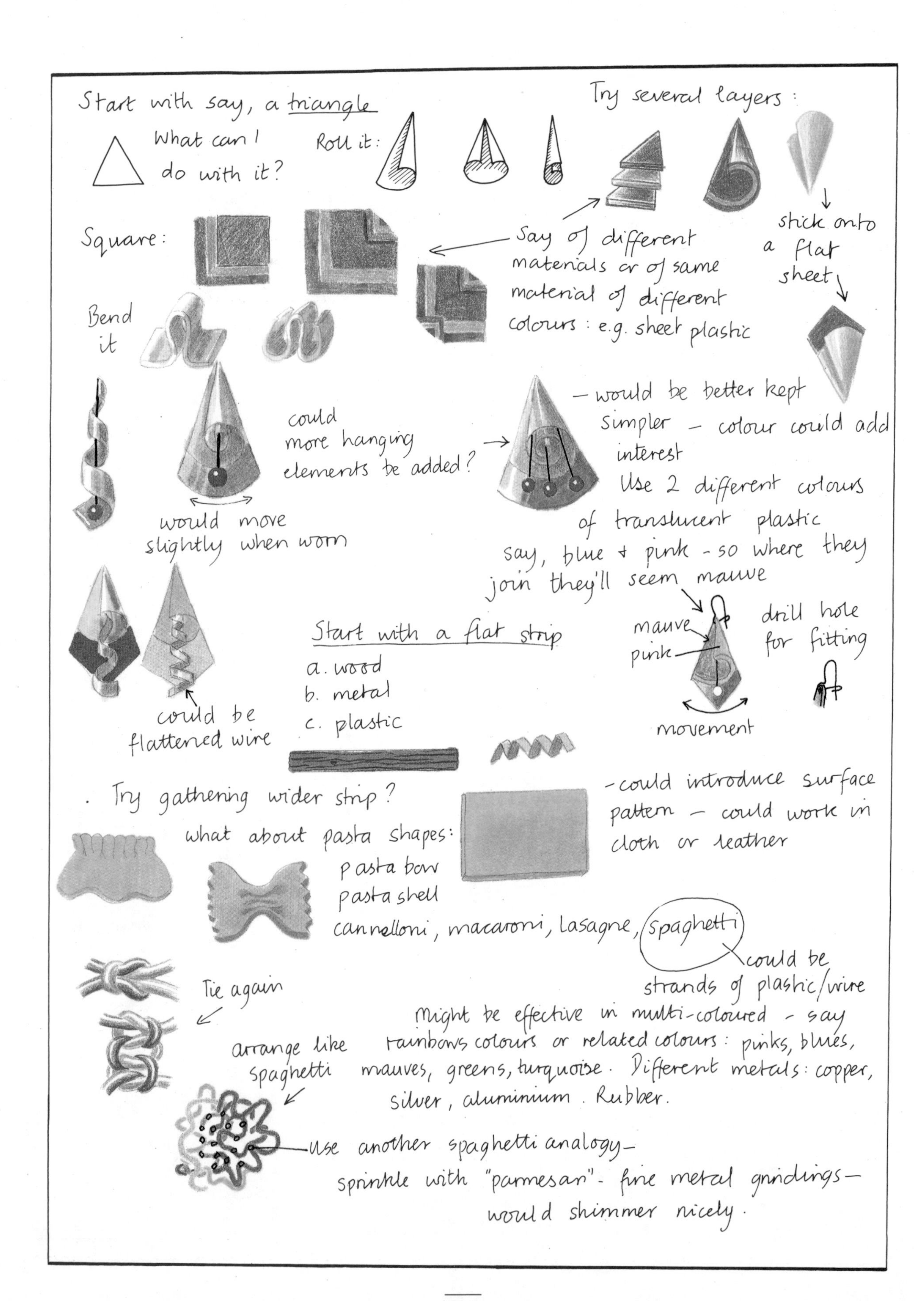

Start with say, a triangle
What can I do with it?
Roll it:
Try several layers:
Square:
Say of different materials or of same material of different colours: e.g. sheet plastic
stick onto a flat sheet
Bend it
could more hanging elements be added?
- would be better kept simpler - colour could add interest
Use 2 different colours of translucent plastic say, blue + pink - so where they join they'll seem mauve
would move slightly when worn
mauve
pink
drill hole for fitting
movement
could be flattened wire
Start with a flat strip
a. wood
b. metal
c. plastic
. Try gathering wider strip?
what about pasta shapes:
pasta bow
pasta shell
cannelloni, macaroni, lasagne, spaghetti
could be strands of plastic/wire
-could introduce surface pattern - could work in cloth or leather
Tie again
Might be effective in multi-coloured - say rainbows colours or related colours: pinks, blues, mauves, greens, turquoise. Different metals: copper, silver, aluminium. Rubber.
arrange like spaghetti
use another spaghetti analogy - sprinkle with "parmesan" - fine metal grindings - would shimmer nicely.

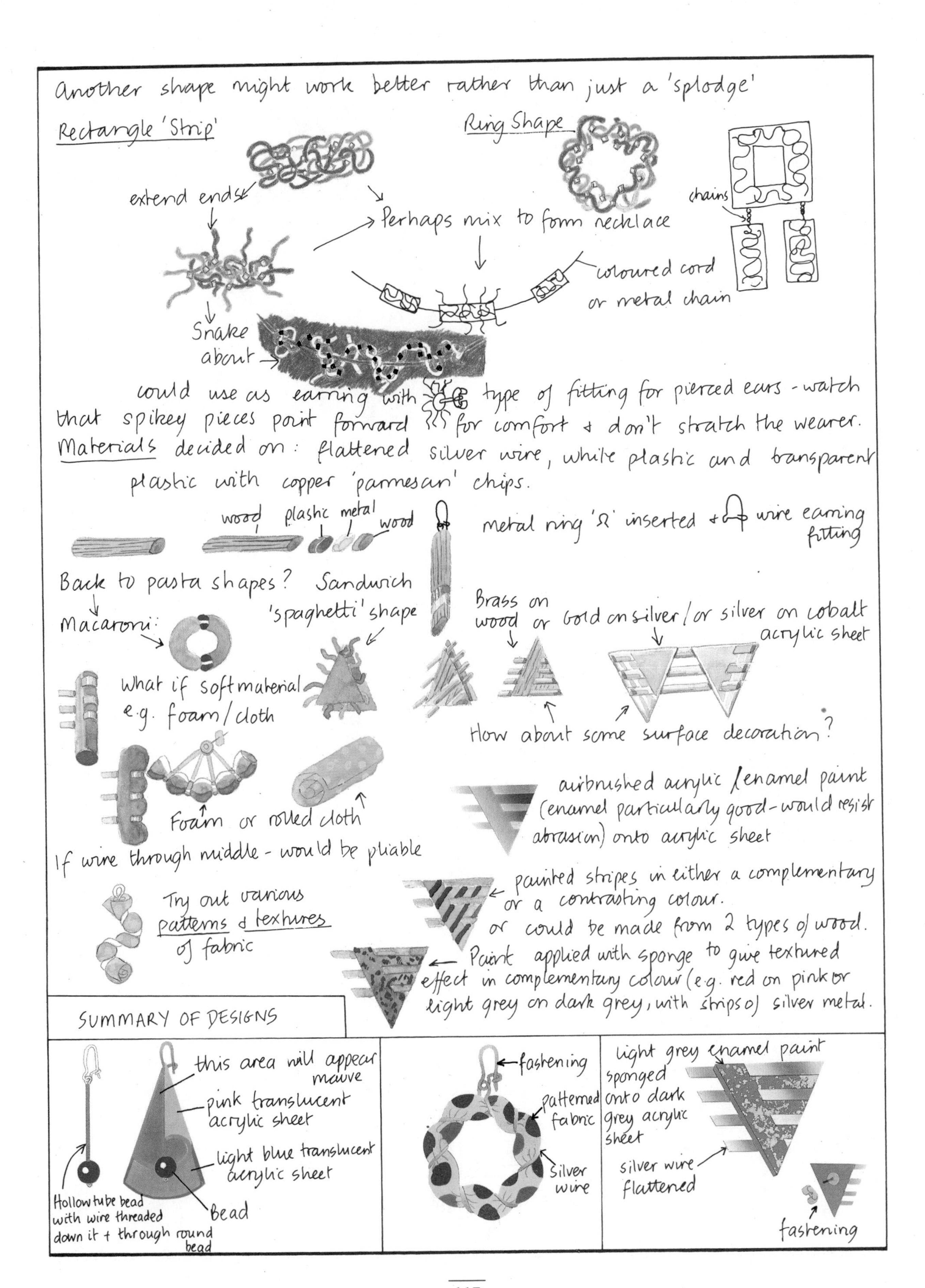
Another shape might work better rather than just a 'splodge'
Rectangle 'Strip'
Ring Shape
extend ends
Perhaps mix to form necklace
chains
coloured cord or metal chain
Snake about
could use as earring with type of fitting for pierced ears - watch that spikey pieces point forward for comfort & don't stratch the wearer.
Materials decided on: flattened silver wire, white plastic and transparent plastic with copper 'parmesan' chips.
wood
plastic
metal
wood
metal ring 'Ω' inserted & wire earring fitting
Back to pasta shapes? Sandwich 'spaghetti' shape
Macaroni:
Brass on wood or Gold on silver / or silver on cobalt acrylic sheet
What if soft material e.g. foam/cloth
How about some surface decoration?
Foam or rolled cloth
airbrushed acrylic / enamel paint (enamel particularly good - would resist abrasion) onto acrylic sheet
If wire through middle - would be pliable
Try out various patterns & textures of fabric
painted stripes in either a complementary or a contrasting colour.
or could be made from 2 types of wood.
Paint applied with sponge to give textured effect in complementary colour (e.g. red on pink or light grey on dark grey, with strips of silver metal.
SUMMARY OF DESIGNS
this area will appear mauve
pink translucent acrylic sheet
light blue translucent acrylic sheet
Hollow tube bead with wire threaded down it + through round bead
Bead
fastening
patterned fabric
silver wire
light grey enamel paint sponged onto dark grey acrylic sheet
silver wire flattened
fastening

'The Rivals'

An eighteenth century play set in Bath, written by Richard Brinsley Sheridan.

The school is attempting to stage this play in the school hall during the latter weeks of the Easter term. I have been asked to attempt designing scenery for the play as a graphics project and then to pass on details of the project to the school's art department.

The play is made up of the following acts and scenes:

Prologue	In front of curtain	
Act 1	*Scene 1*	A street in Bath
	Scene 2	Mrs Malaprop's dressing room
Act 2	*Scene 1*	Captain Absolute's dressing room
	Scene 2	The North parade ground
Act 3	*Scene 1*	The North parade ground
	Scene 2	Julia's dressing room
	Scene 3	Mrs Malaprop's lodgings
	Scene 4	Acre's lodgings
Act 4	*Scene 1*	Acre's lodgings
	Scene 2	Mrs Malaprop's lodgings
	Scene 3	The North parade ground
Act 5	*Scene 1*	Julia's dressing room
	Scene 2	The South parade ground
	Scene 3	Kingsmead Fields
Epilogue	In front of curtain	

Details of the dimensions of the school's stage are given in drawing 1.

The school possesses a number of stage 'flats'. The photograph shows some of these 'flats' as used in the last school play. The flats are either 1.25 m or 2 m wide and some include doors and/or windows. Because of the limitations of:

(a) the number of flats available,

(b) the shortage of time and the costs incurred in making new flats or painting much scenery,

it has been decided to restrict the number of scenes to four only, as follows.

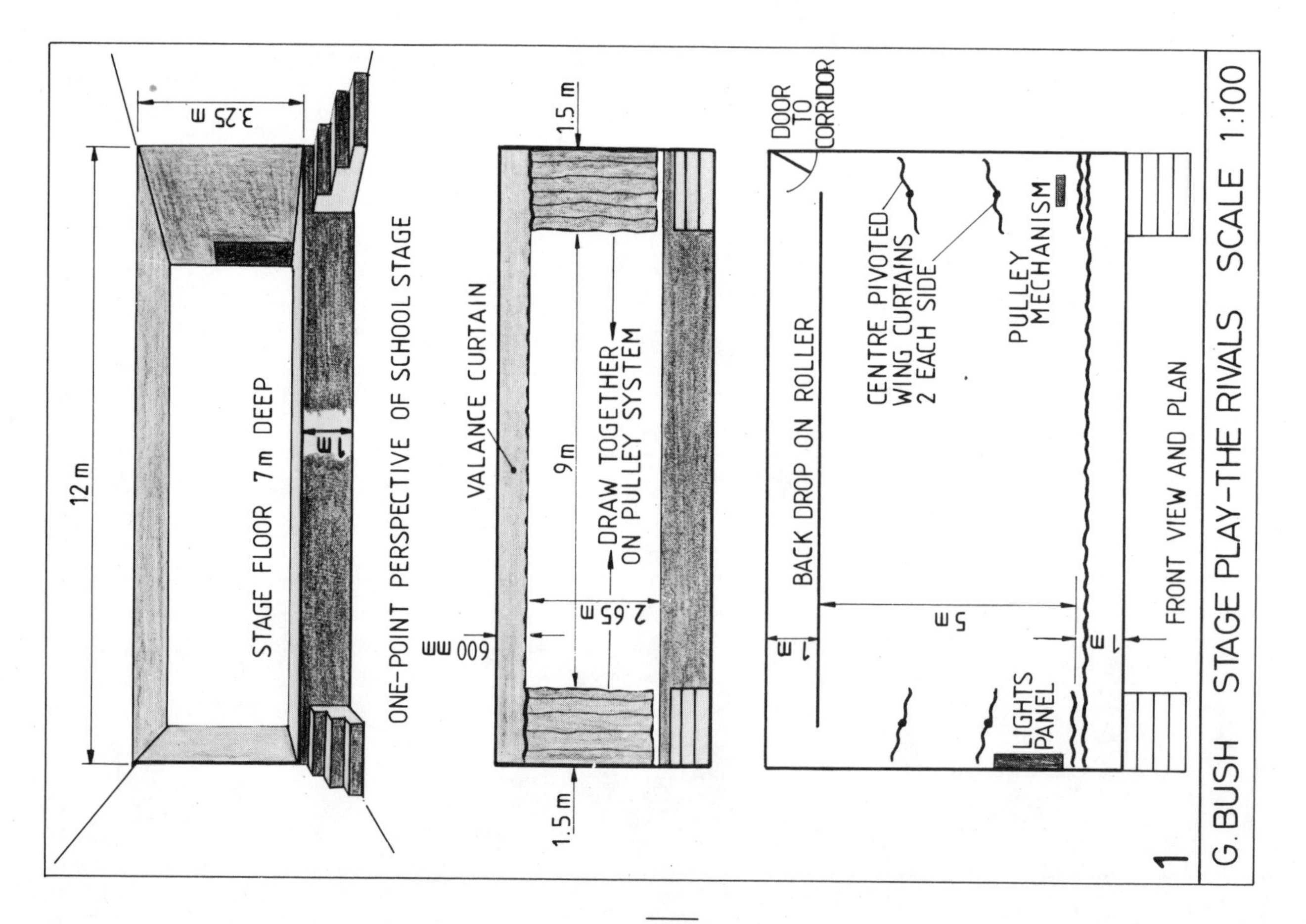

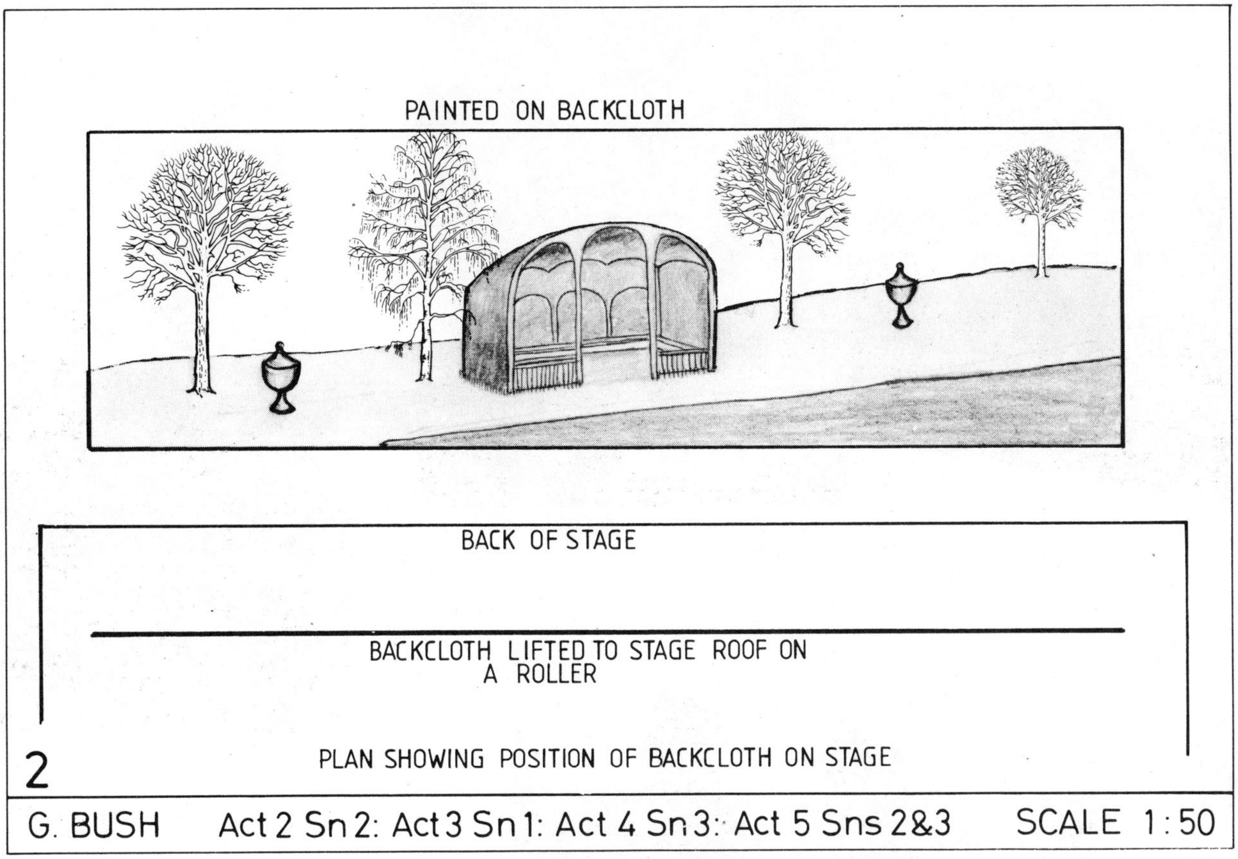

PAINTED ON BACKCLOTH

BACK OF STAGE

BACKCLOTH LIFTED TO STAGE ROOF ON
A ROLLER

2

PLAN SHOWING POSITION OF BACKCLOTH ON STAGE

G. BUSH Act 2 Sn 2: Act 3 Sn 1: Act 4 Sn 3: Act 5 Sns 2&3 SCALE 1:50

1. *Backcloth*
Drawing 2: North Parade Ground. To be also used as:
South Parade Ground
Kingsmead Fields
Thus the backcloth covers Act 2, Scene 2; Act 3, Scene 1; Act 4, Scene 3; Act 5, Scene 2; Act 5, Scene 3.

2. *A Street in Bath*
Drawing 3: Act 1, Scene 1.

3. *Mrs Malaprop's dressing room*
Drawing 4: Act 1, Scene 2. To be also used as:
Julia's dressing room, with appropriate changes of furniture – Act 3, Scene 2; Act 5, Scene 1.
Mrs. Malaprop's lodgings with appropriate changes of furniture – Act 3, Scene 3; Act 4, Scene 2.

4. *Captain Absolute's dressing room*
Drawing 5: Act 2, Scene 1. To be also used as:
Acre's lodgings with changes of furniture – Act 3, Scene 4; Act 4, Scene 1.

Photographs taken in Bath are to be used to assist in designing the scenery and painting it.

It may be necessary to make new flats. Details of how these can be constructed are given in drawing 6.

Six colour photographs are given on page 120. Two of these are associated with the project on stage scenery for '*The Rivals*', as noted earlier. The other four photographs are as follows.
1. A model associated with the lamp design project, one drawing from which is given on page 55.
2. A photograph of three models associated with a home furnishing project developed by a group of three girls.
3. A photograph of granite rocks. Ideas derived from this type of photograph showing shape, form and texture of natural features could form the basis for part of a design project.
4. A photograph showing a model made of 'Corriflute' sheet material. This photograph was kindly supplied by Commotion Technology Supplies of Enfield, Middlesex.

Two photographs taken in Bath on which 'The Rivals' scenery was based

An example of natural shapes

'Corriflute' sheet used for model making

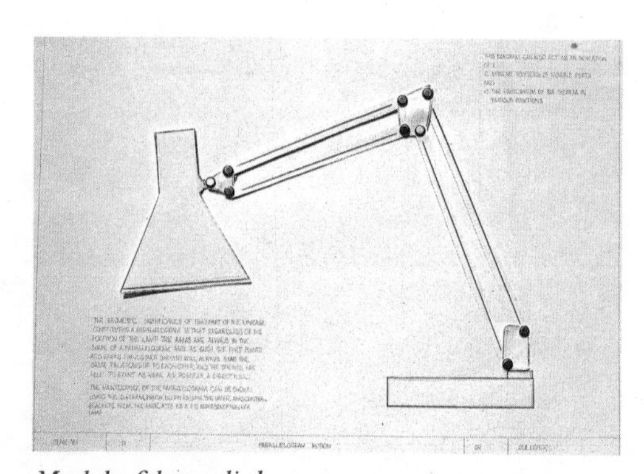

Model of lamp link system

Models of furniture

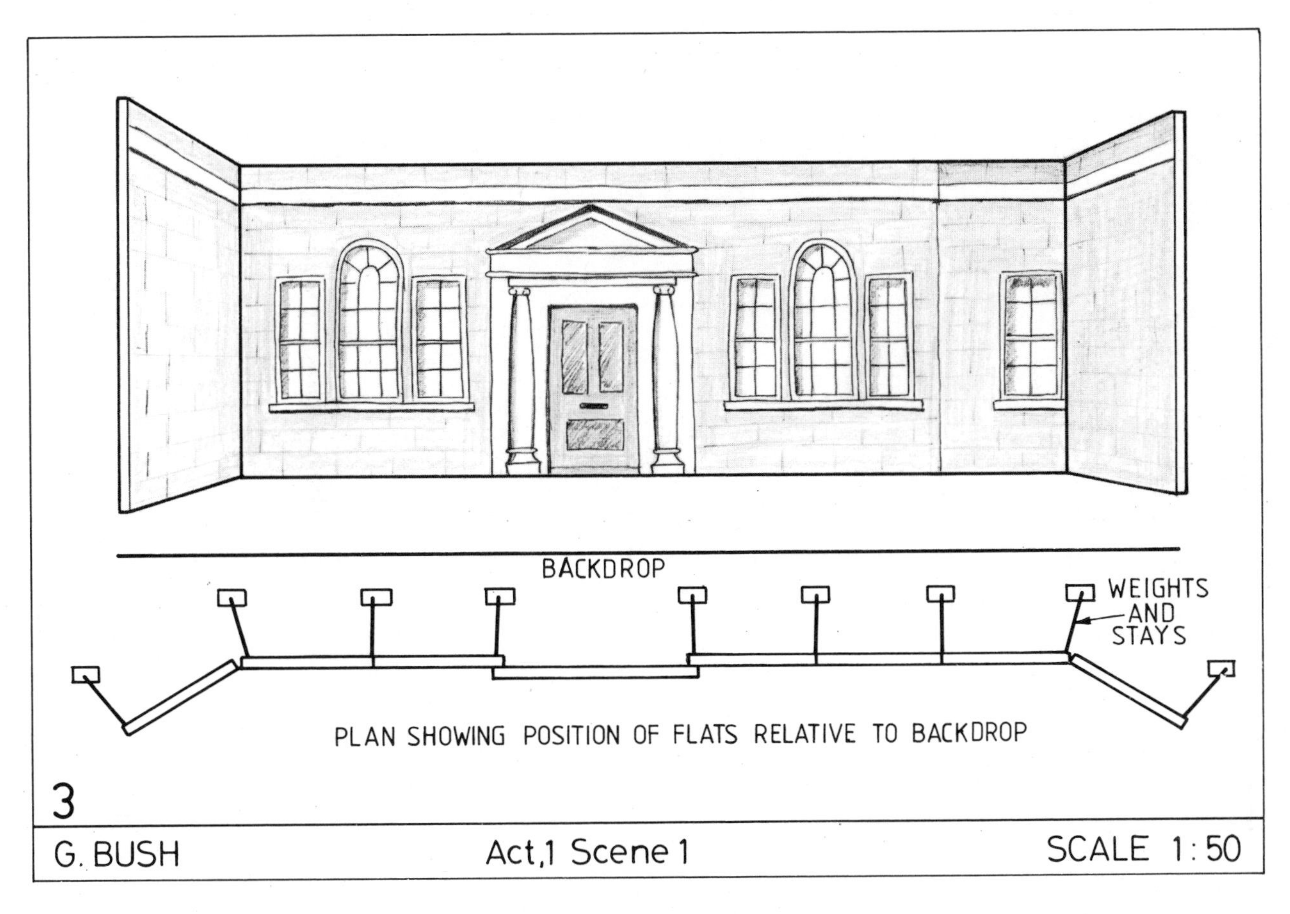
BACKDROP
WEIGHTS AND STAYS
PLAN SHOWING POSITION OF FLATS RELATIVE TO BACKDROP
3
G. BUSH
Act,1 Scene 1
SCALE 1: 50

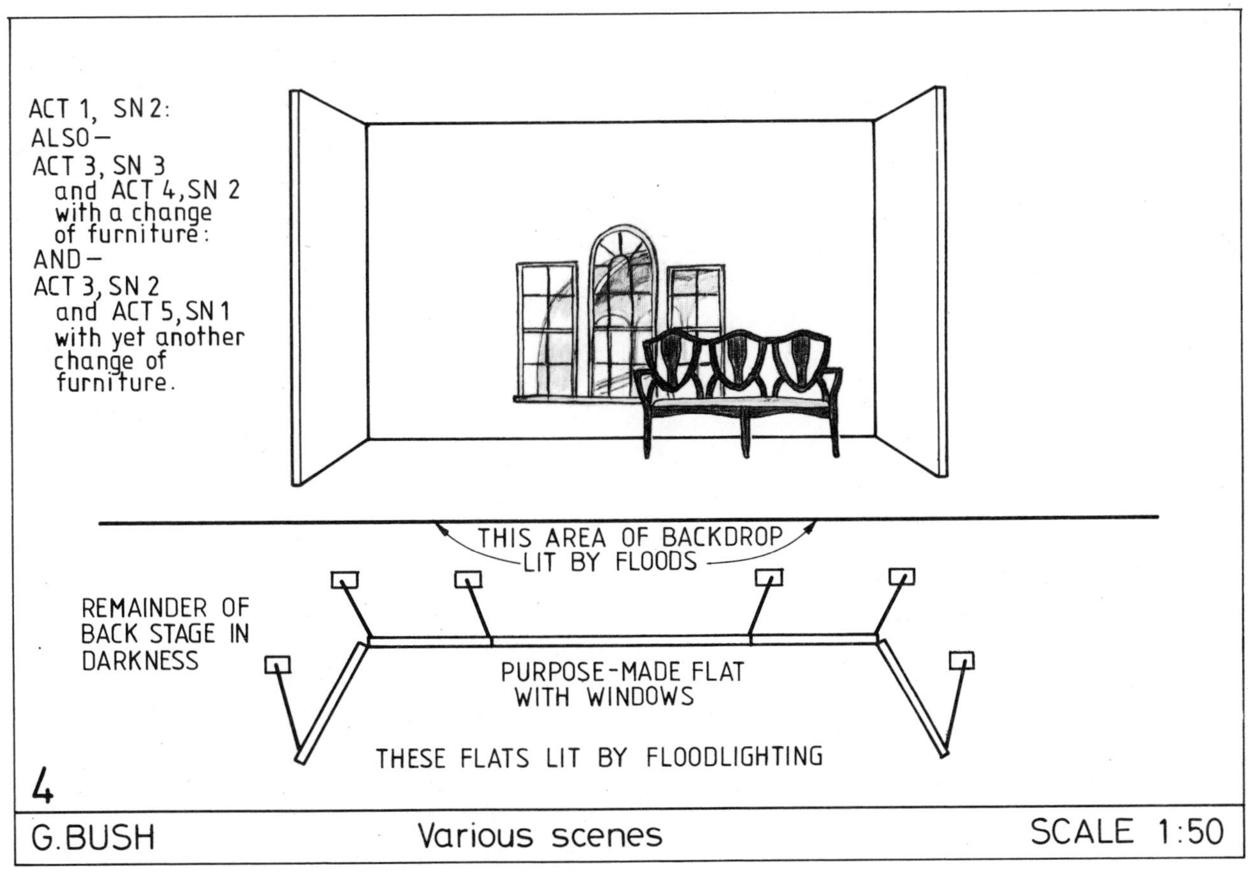
ACT 1, SN 2:
ALSO–
ACT 3, SN 3
and ACT 4, SN 2
with a change
of furniture:
AND–
ACT 3, SN 2
and ACT 5, SN 1
with yet another
change of
furniture.
THIS AREA OF BACKDROP
LIT BY FLOODS
REMAINDER OF
BACK STAGE IN
DARKNESS
PURPOSE-MADE FLAT
WITH WINDOWS
THESE FLATS LIT BY FLOODLIGHTING
4
G.BUSH
Various scenes
SCALE 1:50

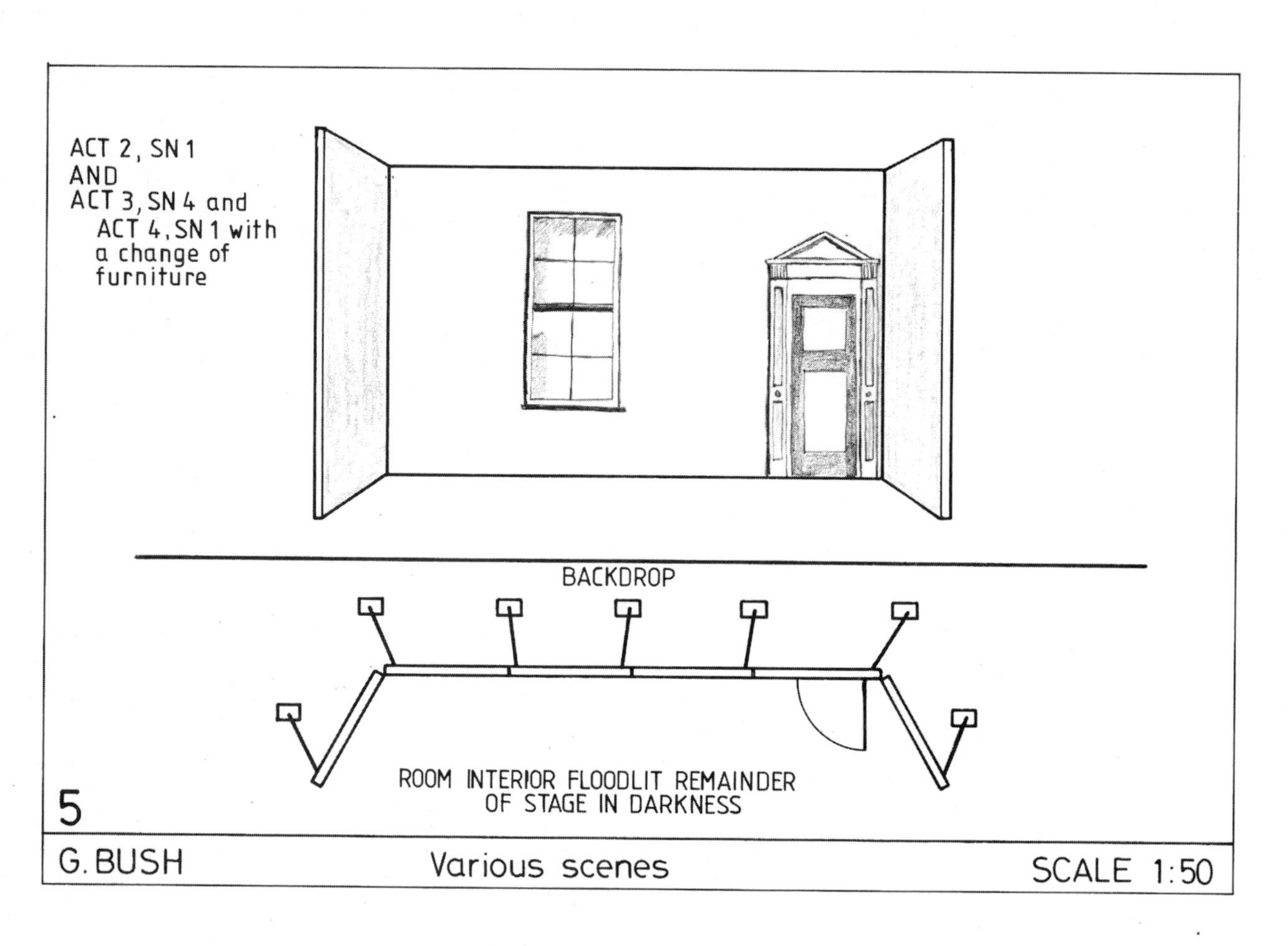
ACT 2, SN 1
AND
ACT 3, SN 4 and
ACT 4, SN 1 with
a change of
furniture
BACKDROP
ROOM INTERIOR FLOODLIT REMAINDER
OF STAGE IN DARKNESS
5
G. BUSH
Various scenes
SCALE 1:50

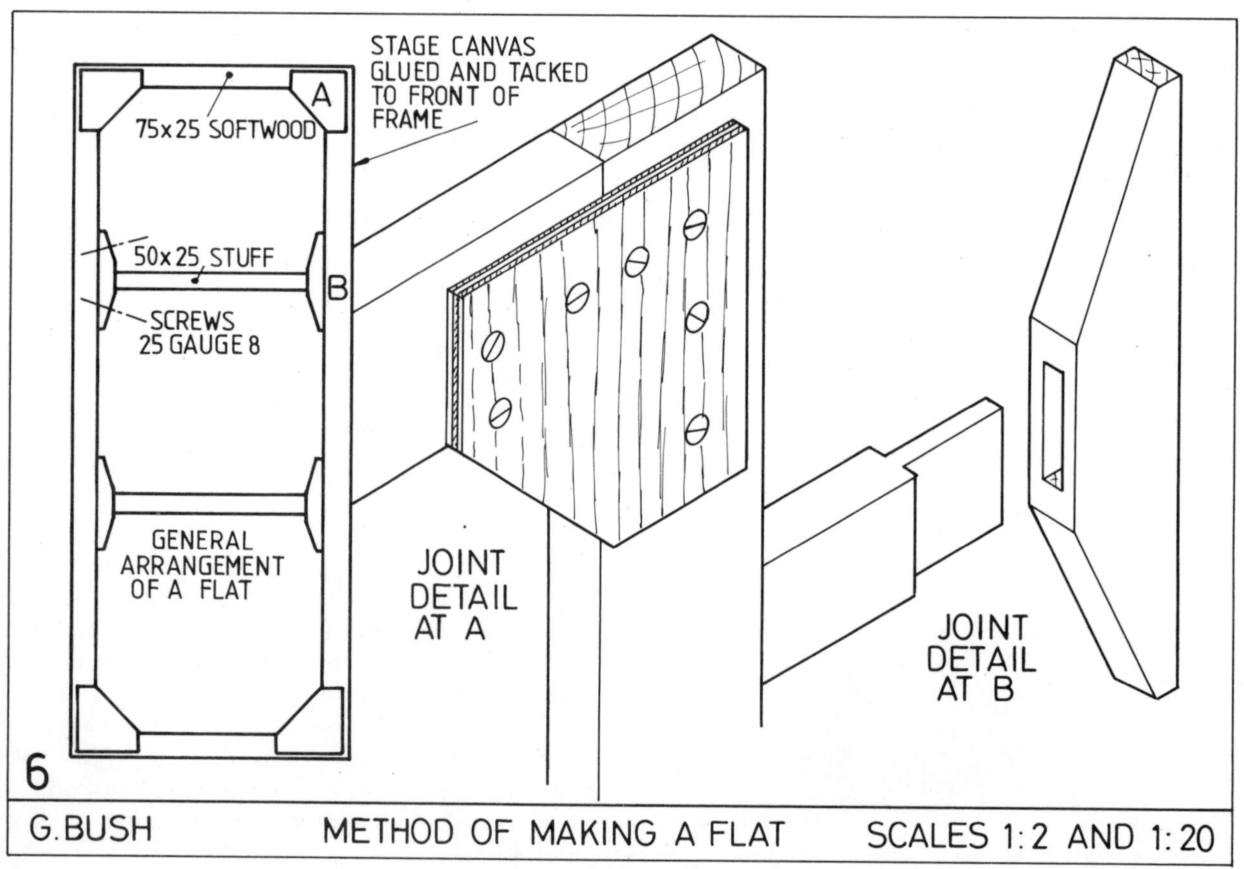
STAGE CANVAS
GLUED AND TACKED
TO FRONT OF
FRAME
A
75x25 SOFTWOOD
50x25 STUFF
B
SCREWS
25 GAUGE 8
GENERAL
ARRANGEMENT
OF A FLAT
JOINT
DETAIL
AT A
JOINT
DETAIL
AT B
6
G. BUSH
METHOD OF MAKING A FLAT
SCALES 1:2 AND 1:20

EXERCISES

1. The photograph above shows part of the scenery for a school play. The scenery was designed by pupils at the school which produced the play. The piece of the scenery shown was made from offcuts of softwood cut to length and glued inside a wooden frame.

   (a) What does this piece of scenery represent?

   (b) At what time of the year do you think the school staged a play?

   (c) Make a freehand sketch of the piece of scenery.

   (d) Design and draw a similar piece of scenery suitable for a school play in which a witch is shown riding on her broom stick with her black cat sitting on the other end of the stock.

2. The photograph shows a Sony Walkman. Make a freehand sketch of the outline of the cassette player. Shade your drawing using any of the methods shown on pages 94 and 95.

3. Copy the two drawings of a hand holding a pencil on page 90 on to tracing paper. Transfer the drawings on to a sheet of drawing paper. Then enlarge them to twice their size using the grid method shown on page 92. Shade your enlarged drawings to give the hands a three-dimensional appearance.

4. Using drawing 6 on page 92 as an example, design and draw the following letters to a similar design. Make each of your letters 60 mm high:

   b e h j l n t u z

   Shade these letters with colour pencils.

5. Assume you own a firm which makes and sells computers and that the firm is called:

   Your Name Limited.

   Design a logo for this firm based on the initials of your name.

6. (a) Draw a regular hexagon of side length 30 mm (see page 40);

   (b) draw a number of the same size hexagons touching each other as shown below;

   (c) design a shape, based on the hexagons, which will interlock with other identical shapes to form a tesselated pattern (see pages 112 and 113).

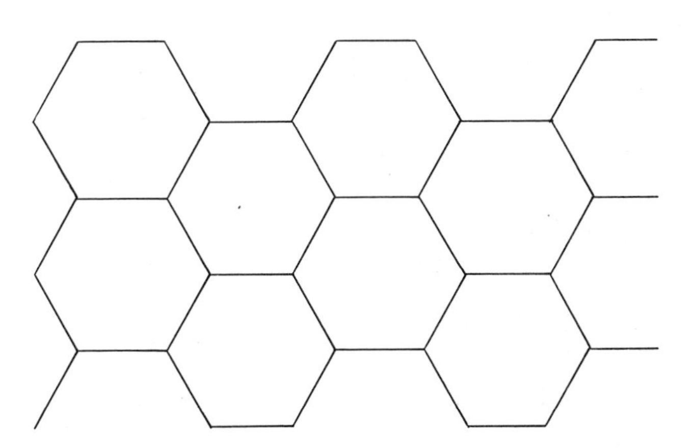

7. Study the three specimen projects given on pages 112 to 117.

   Evaluate the three projects in note and drawing form. State which details in the designs you think could have been improved and suggest your own improvements.

8. Study the *Rivals* project which starts on page 118.

   Draw up an evaluation of the designs for the various acts and scenes in note and sketch form. Show how you would include the improvements you might think are necessary.

SECTION 5 Craft and technology

The design process

A flow diagram defining the stages involved in a process for graphic design is given on page 8. A second diagram listing the stages involved in the process of designing for craft and technology projects is given on page 125 opposite. The reader is advised to study both these flow diagrams and to make comparisons between them.

In considering craft and technology design it is important to understand that graphics and notes play a very important role in the process. This is clearly demonstrated in the chart opposite. Every stage in the design process requires some form of graphics and notes. This dependence upon graphics is again demonstrated in the six sheets of drawings reproduced on pages 128 to 130 in which the design process is applied to the designing of a TV and video stand – a craft project.

The major similarity between the processes of designing graphics and designing a craft or technology project lies in the sequence which has been repeatedly stated throughout this book – design brief; investigation; ideas for solutions; chosen solution; realisation; evaluation.

The major differences lie in the fact that the methods of graphics involved in the designing of a craft or technology project are the media by which ideas are communicated. These ideas should eventually lead to the actual making or realisation of the design.

Because a craft or technology project is aiming at the actual realisation of a successful design, the investigation must include an analysis of details such as the following.

* What ergonomics problems are involved?
* Which dimensions are most suitable?
* What are the most desirable shapes and forms?
* Which are the most appropriate materials?
* How will the design be constructed? Can some of the ideas for solutions in fact be constructed?
* Will fittings have to be purchased or made?
* Is the design sufficiently strong to withstand the stresses and strains which will be imposed upon it when it is in use?
* Is the design economical to make as regards time spent in making; materials involved; economical use of line, shape and form?
* Will a surface finish have to be applied? Which is the most suitable finish with regard to aspects such as weather; wear and tear; corrosion?
* Will the completed design be safe to use?

When designing technology projects, different types of control mechanisms may have to be considered – electrical; electronic; pneumatic; hydraulic; mechanical.

Different types of structures may require investigation and testing.

In particular, in technology project design, consideration will need to be given to testing circuits, mechanisms and structures by means of models, made either from materials at hand or from specially designed kits.

Outline flow diagram

The abridged flow diagram in circular form given on page 126 shows that even when a design has been realised, the design process has not necessarily been completed. If an evaluation of the design suggests that faults are apparent or that modification might improve the resulting design, then the design process has not been completed. Hence the broken line in the diagram which tells the designer he or she may have to go back to the design brief and again follow the route – investigation; ideas for solutions; model; working drawing; realisation – to discover whether the realised design can be improved or not.

CRAFT AND TECHNOLOGY DESIGN

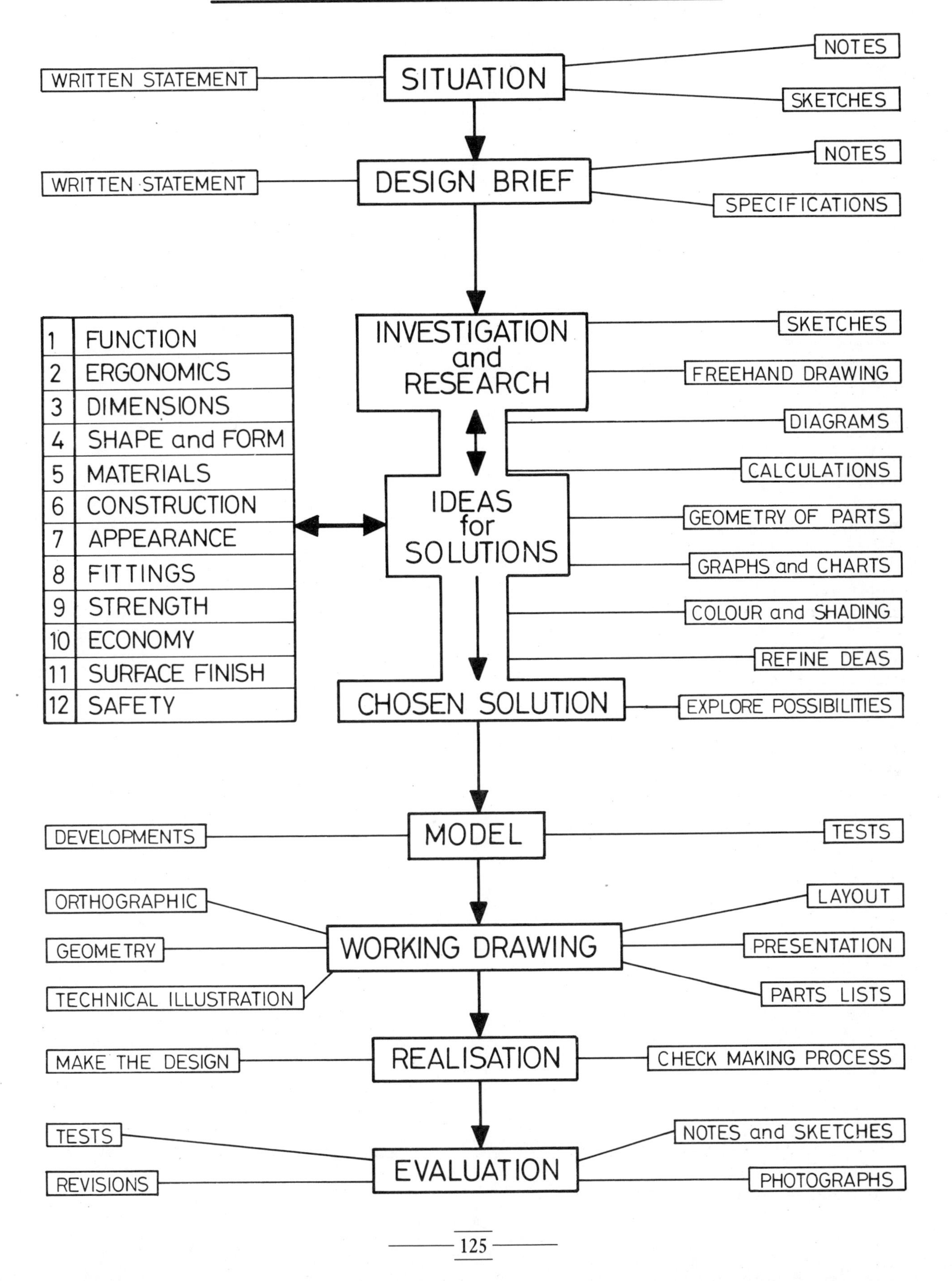

Ergonomics

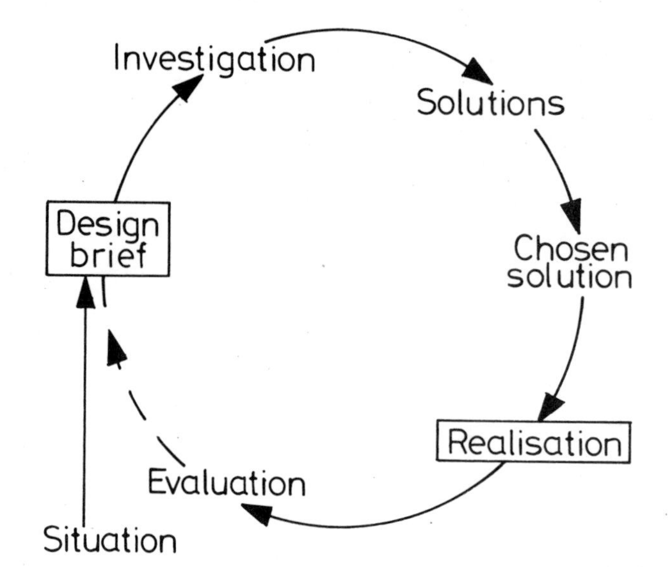

A flow diagram showing an outline process for craft and technology design

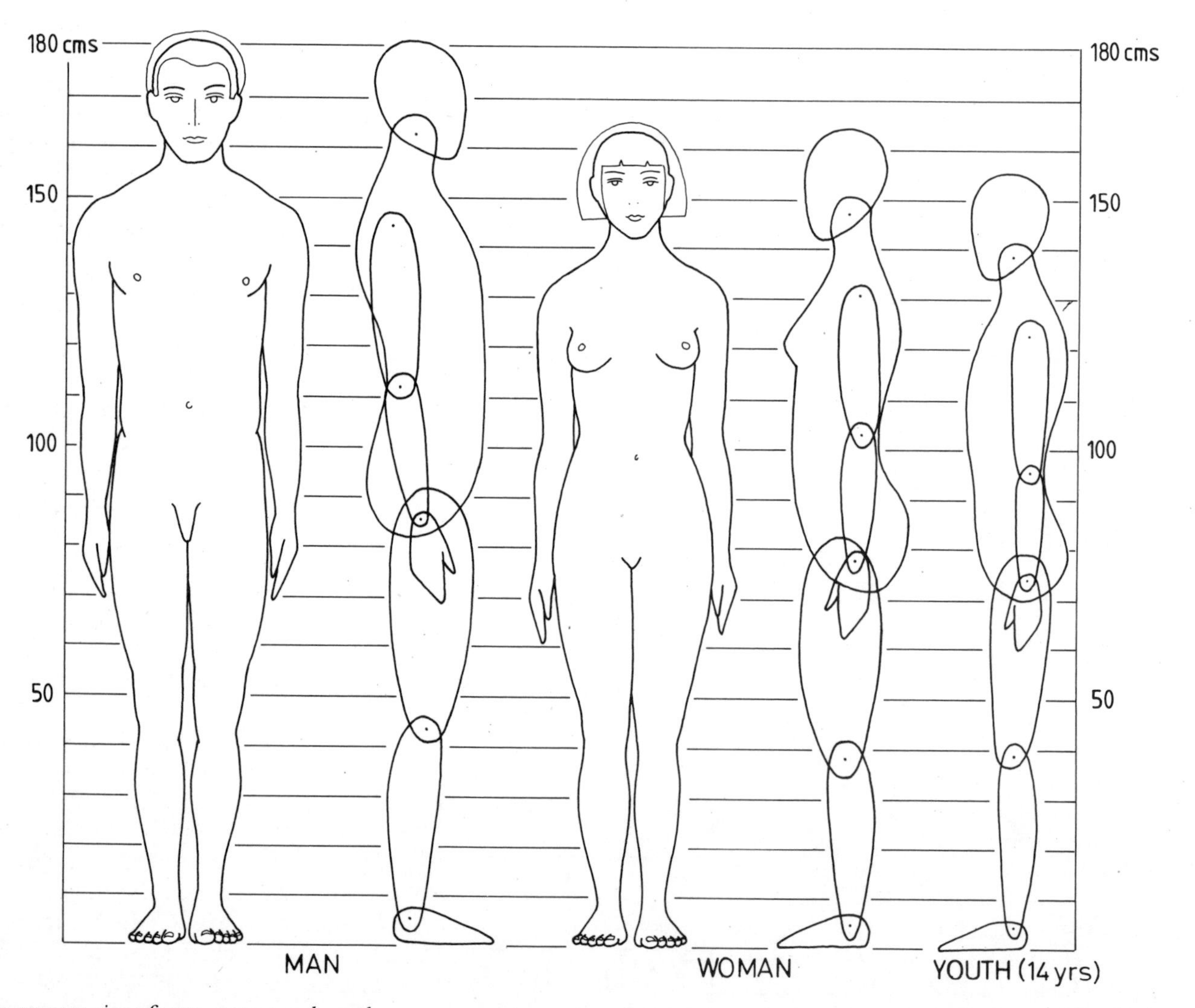

The average size of men, women and youths

A study of measurements associated with human bodies is known as *anthropometry*. The word *ergonomics* (from *ergos*, work, and *nomos*, natural laws) was first coined, by what came to be known as the Ergonomics Research Council, in 1949. The word ergonomics is now taken to refer to the relationship between designs and human beings. Ergonomics is not only concerned with sizes but also with shape, form, comfort, movement, vision, hearing, smell, taste and touch.

The drawings opposite of a man, a woman and a 14-year-old youth indicate the average sizes of men, women and youths. The side views of a puppet man, puppet woman and puppet youth have been included here for you to copy and, if needed, to make up into jointed flat figures which you can then use for the purpose of deciding sizes in your design work. The figures can be scaled down or up in size if necessary. They can be made from card, plywood, sheet plastic or sheet aluminium and jointed with pins, paper fasteners, thin rivets or thin nails. The photograph on this page shows part of a project in which a pupil made a study of the ergonomics of the more common designs used by humans.

EXERCISES

1. With the assistance of another person, measure and state your own dimensions of: height; inside leg with feet flat on floor; elbow to finger tips; shoulder to elbow; knee to foot with foot flat on floor; your own most comfortable chair and table heights.
2. Make the same measurements with a group of, say, six members of your class. Work out averages of size from the figures you have obtained.
3. Measure the following and state whether you think the measurements you make are ergonomically sound. The height of a doorway; the width of a room door; the height of your school desk; the height of a chair arm rest above the seat of the chair; the height of a light switch on your bedroom wall.
4. Measure the handles of the following items and write down their dimensions in millimetres – a carving knife; a pair of scissors; a pair of garden shears; a pair of pliers; a pair of carpenters pincers. Now measure and make a note of the size of your hands. Compare the two sets of measurements.
5. Measure and write down the dimensions of the decks and chairs in your classroom; heights, widths, lengths, depths. Compare these with those you noted in working through Exercise 1. Are the items of school furniture you have measured comfortable? If not, can you make suggestions for making them more comfortable.

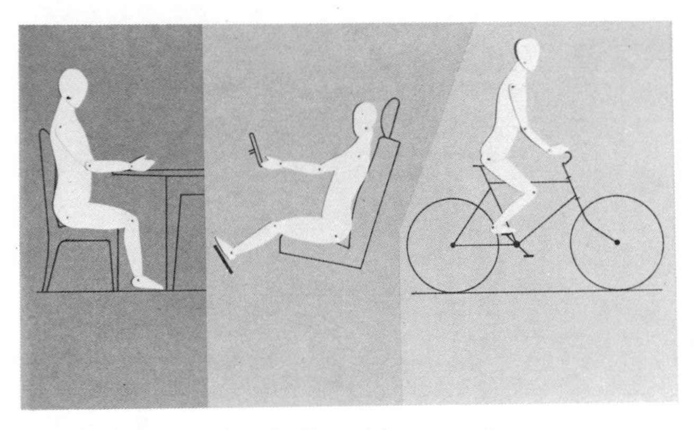

Models from a project dealing with ergonomics

Project – A TV and video stand

Pages 128 to 130 illustrate part of a typical project involving the craft and technology design process defined in the flow diagram of page 125. The six sheets of this project are pencil and technical pen work. Some of the drawings are freehand, some have been drawn with the aid of instruments and drawing aids. The arrows of sheet 5 were coloured in with a water-colour wash. The drawings involve methods of estimated perspectives, isometric drawing and orthographic projection or a mixture of orthographic and isometric. Line shading has been added to many of the drawings to emphasise shape and form.

Study the three pages closely. Note how details from the design process investigation have been included in note form among the drawings of ideas for solutions.

Because of the limitations of space in this book, it has not been possible to include the full project. The following have not been included here:

1. a model – the reader could attempt making a small scale model, say to a scale of 1:5, of the design from strips of wood and cardboard;
2. the realisation of the design;
3. an evaluation of the project – the reader could attempt his or her own evaluation of the project.

Note that, despite the inclusion of a cutting list (Sheet 6), the costs of each item have not been included. Why is this?

The graphics shown in this project are typical examples of drawings required when producing folios for design projects connected with school craft work. The student producing the graphics needs a good working knowledge of the craft he or she is learning, in order to be able to produce the required design drawings. The student may have to refer to books, or to sheets of work prepared by teachers to produce some of the required drawings, but avoid just copying form such sources.

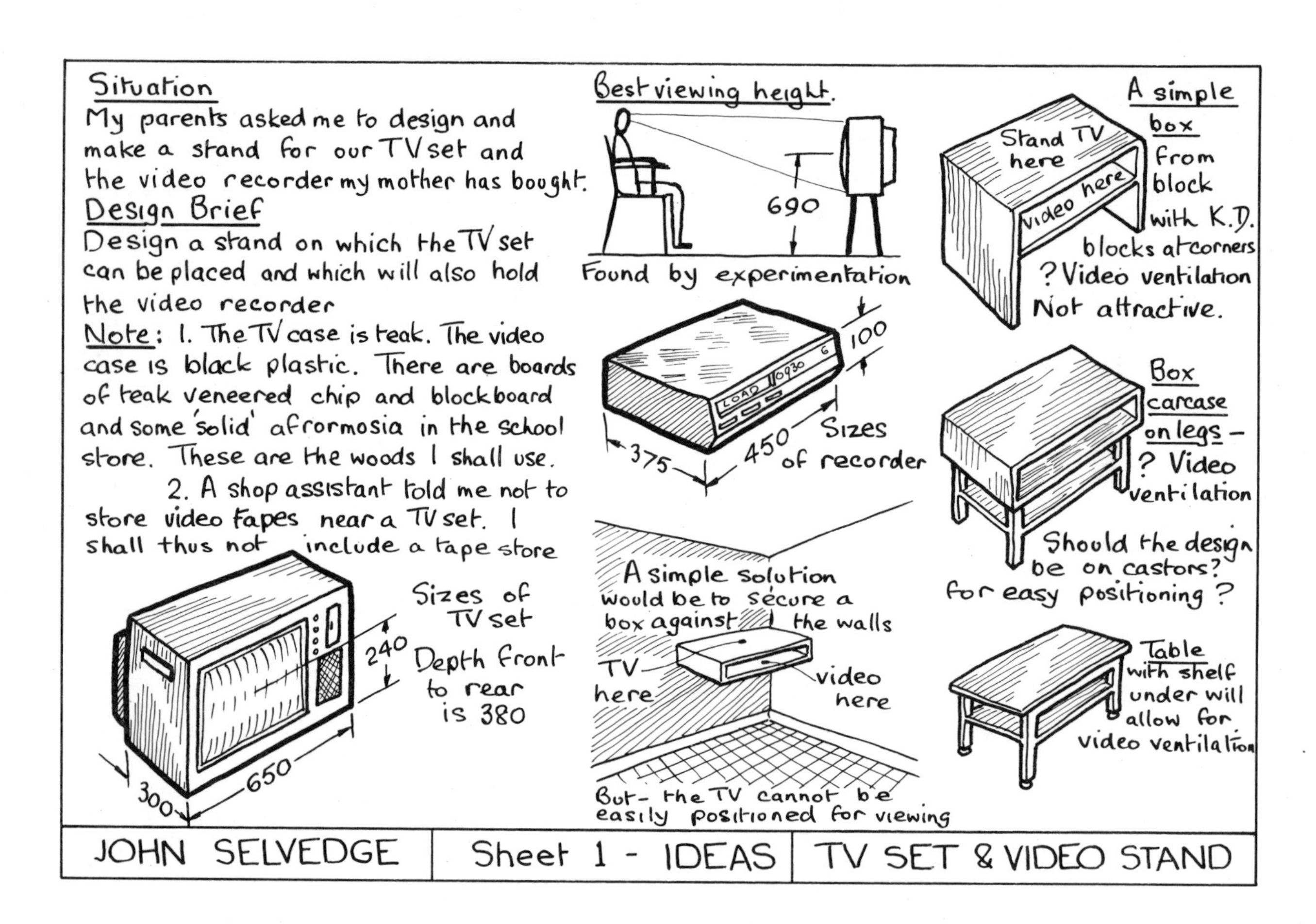
Situation
My parents asked me to design and make a stand for our TV set and the video recorder my mother has bought.
Design Brief
Design a stand on which the TV set can be placed and which will also hold the video recorder
Note: 1. The TV case is teak. The video case is black plastic. There are boards of teak veneered chip and blockboard and some 'solid' afrormosia in the school store. These are the woods I shall use.
2. A shop assistant told me not to store video tapes near a TV set. I shall thus not include a tape store
Sizes of TV set
240
Depth front to rear is 380
300
650
Best viewing height.
690
Found by experimentation
100
375
450
Sizes of recorder
LOAD
A simple solution would be to secure a box against the walls
TV here
video here
But - the TV cannot be easily positioned for viewing
A simple box from block with K.D. blocks at corners
Stand TV here
video here
? Video ventilation
Not attractive.
Box carcase on legs - ? Video ventilation
Should the design be on castors? for easy positioning?
Table with shelf under will allow for video ventilation
JOHN SELVEDGE
Sheet 1 - IDEAS
TV SET & VIDEO STAND

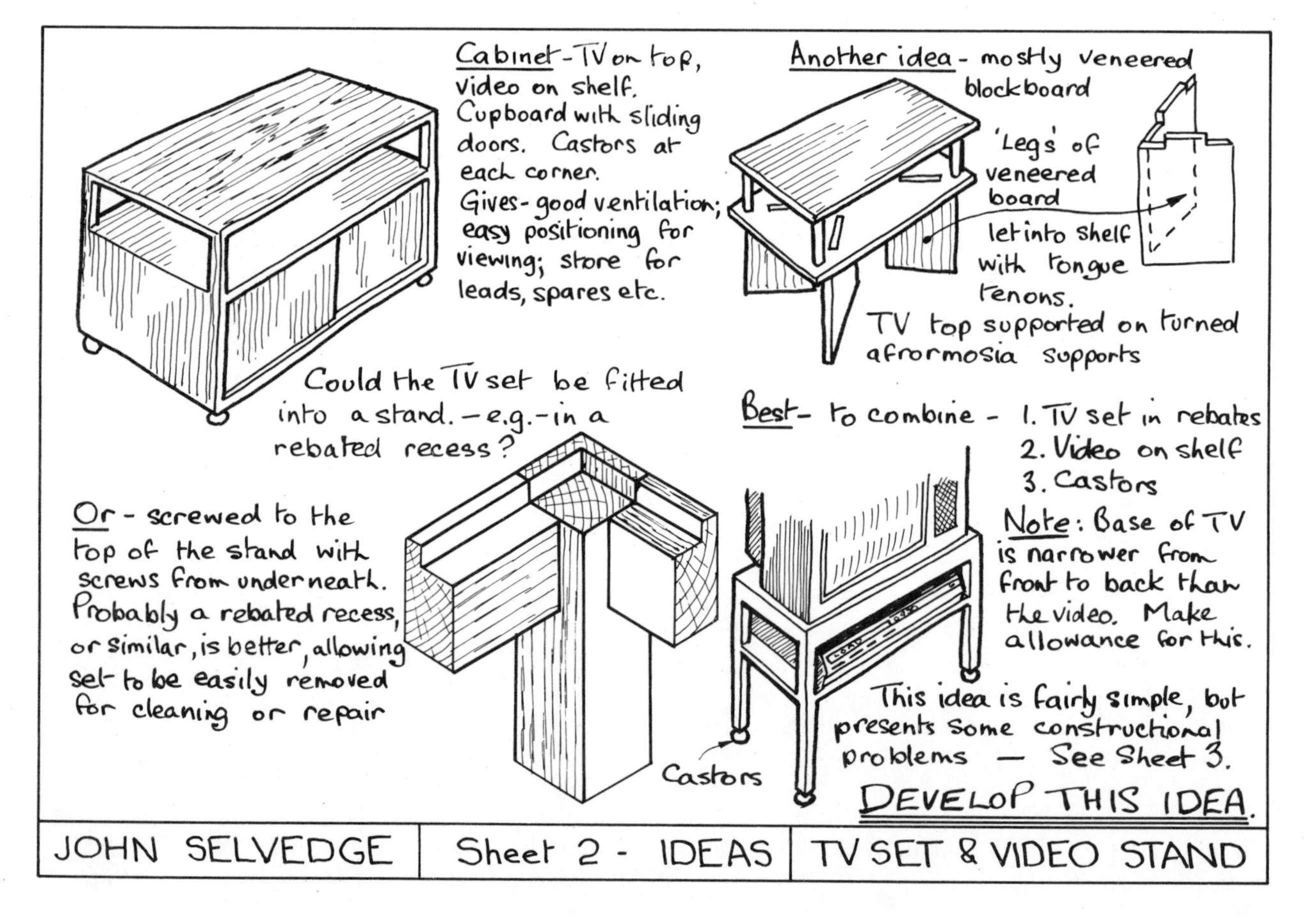
Cabinet - TV on top, video on shelf. Cupboard with sliding doors. Castors at each corner.
Gives - good ventilation; easy positioning for viewing; store for leads, spares etc.
Another idea - mostly veneered blockboard
'Legs' of veneered board let into shelf with tongue tenons.
TV top supported on turned afrormosia supports
Could the TV set be fitted into a stand. - e.g. - in a rebated recess?
Or - screwed to the top of the stand with screws from underneath. Probably a rebated recess, or similar, is better, allowing set to be easily removed for cleaning or repair
Best - to combine - 1. TV set in rebates
2. Video on shelf
3. Castors
Note: Base of TV is narrower from front to back than the video. Make allowance for this.
This idea is fairly simple, but presents some constructional problems — See Sheet 3.
Castors
DEVELOP THIS IDEA.
JOHN SELVEDGE
Sheet 2 - IDEAS
TV SET & VIDEO STAND

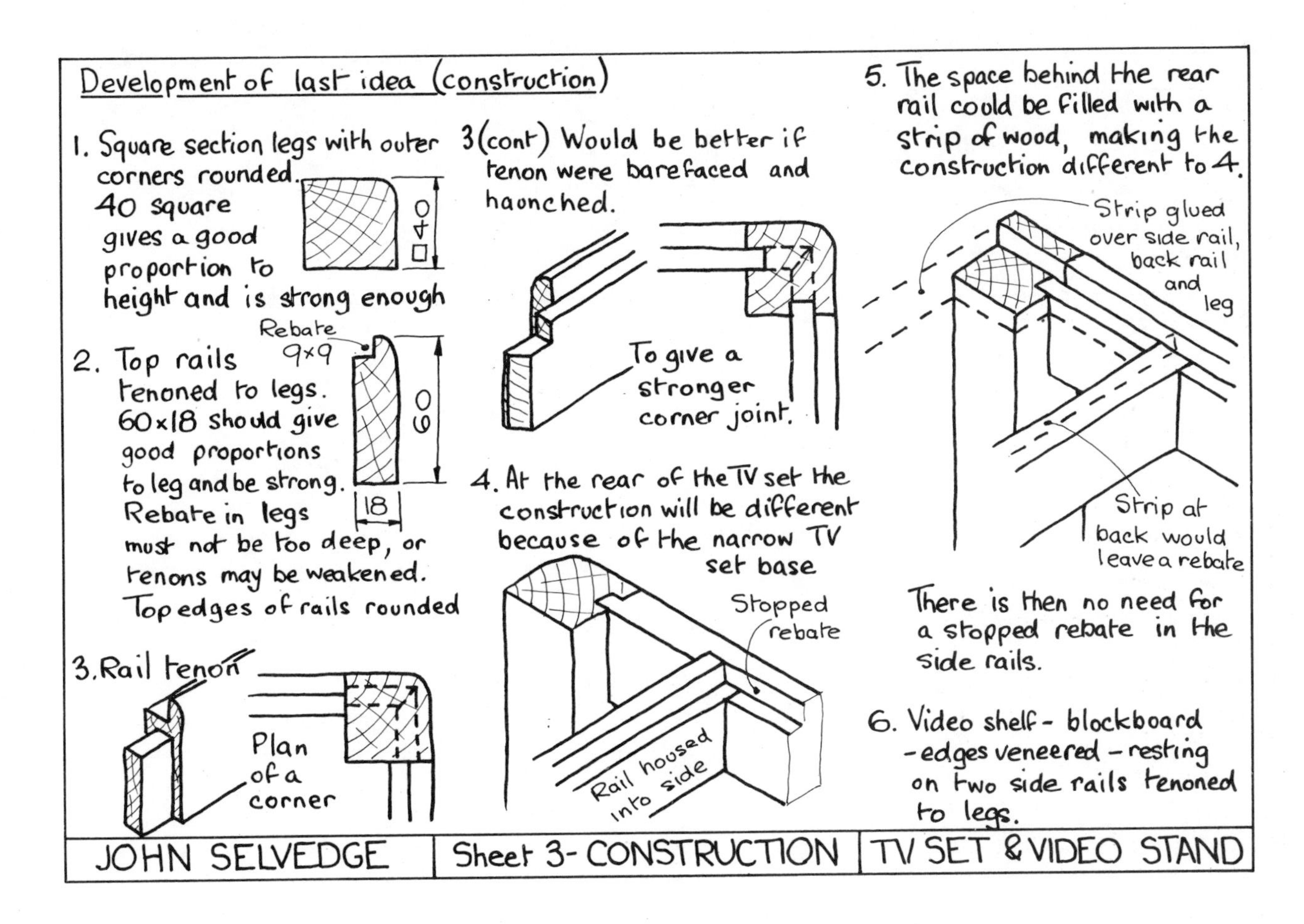

Development of last idea (construction)
1. Square section legs with outer corners rounded. 40 square gives a good proportion to height and is strong enough
□40
2. Top rails tenoned to legs. 60×18 should give good proportions to leg and be strong. Rebate in legs must not be too deep, or tenons may be weakened. Top edges of rails rounded
Rebate 9×9
60
18
3. Rail tenon
Plan of a corner
3 (cont) Would be better if tenon were barefaced and haunched.
To give a stronger corner joint.
4. At the rear of the TV set the construction will be different because of the narrow TV set base
Stopped rebate
Rail housed into side
5. The space behind the rear rail could be filled with a strip of wood, making the construction different to 4.
Strip glued over side rail, back rail and leg
Strip at back would leave a rebate
There is then no need for a stopped rebate in the side rails.
6. Video shelf - blockboard - edges veneered - resting on two side rails tenoned to legs.
JOHN SELVEDGE
Sheet 3- CONSTRUCTION
TV SET & VIDEO STAND

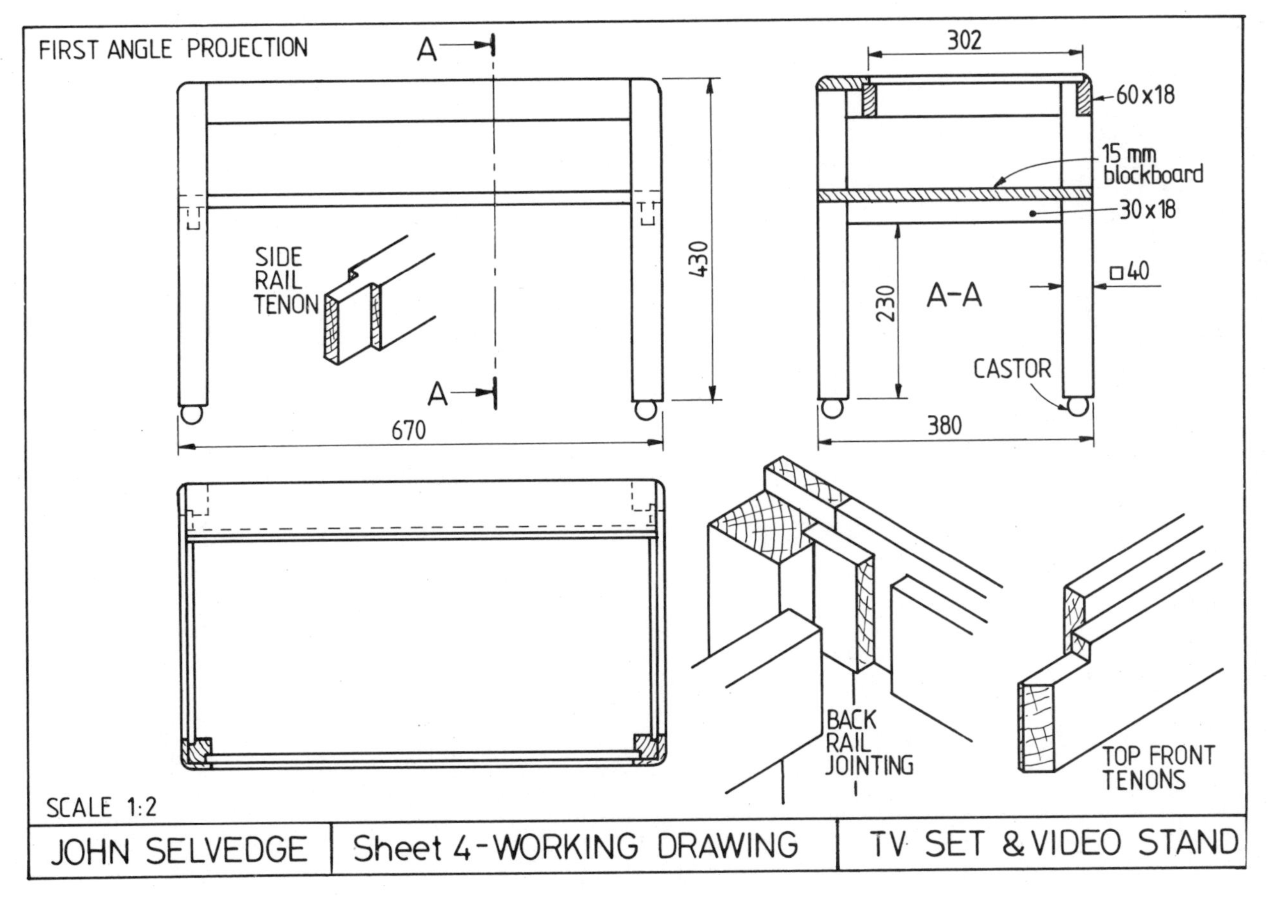

FIRST ANGLE PROJECTION
A
A
SIDE RAIL TENON
430
670
302
60x18
15 mm blockboard
30x18
□40
230
A-A
CASTOR
380
BACK RAIL JOINTING
TOP FRONT TENONS
SCALE 1:2
JOHN SELVEDGE
Sheet 4-WORKING DRAWING
TV SET & VIDEO STAND

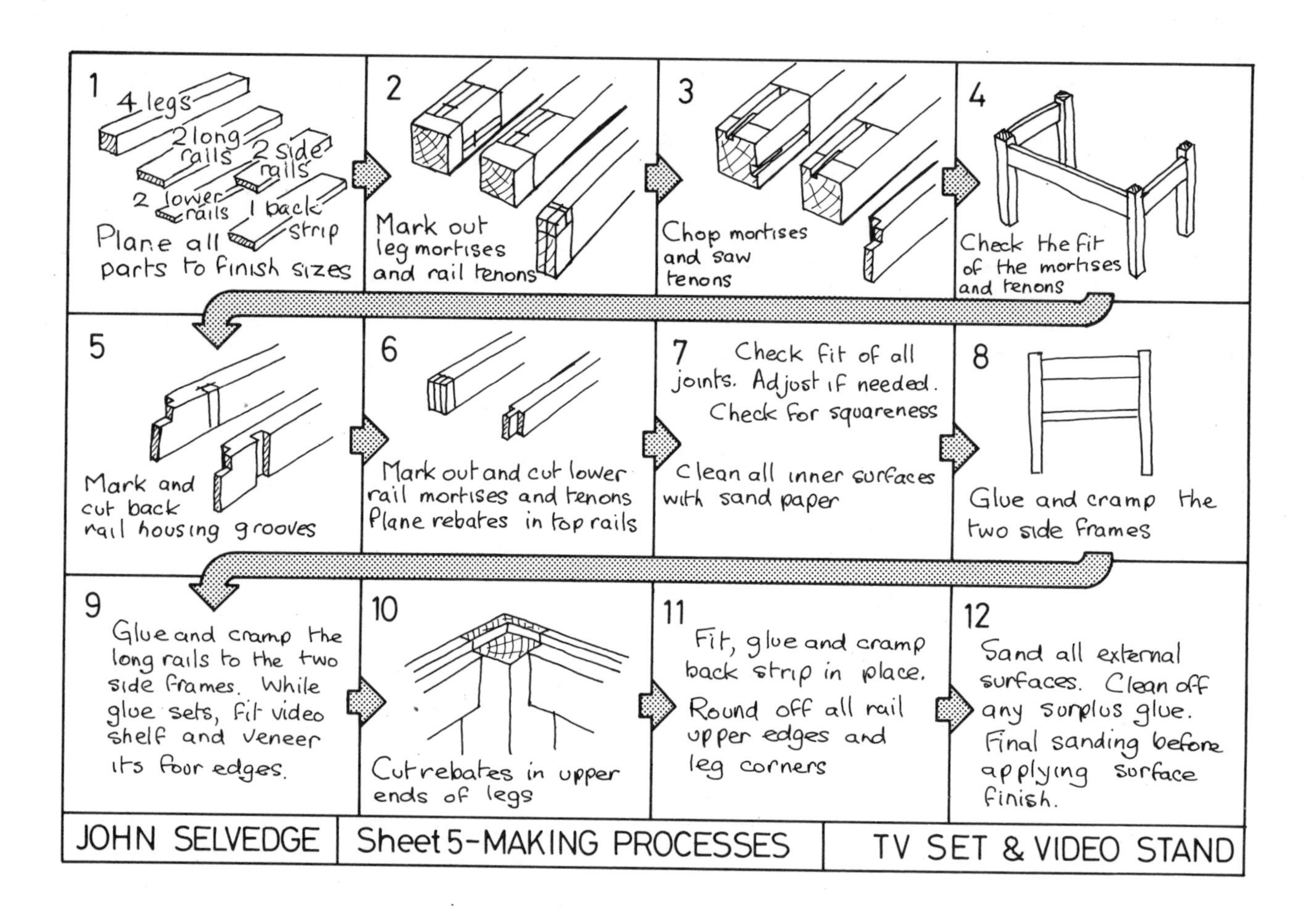

Note: Dimensions in mm. Finished sizes.

No.	Part	Dimensions	Remarks	No. off
1	Legs	450 x 40 x 40	Afrormosia	4
2	End rails	360 x 60 x 18	,,	2
3	Front rail	650 x 60 x 18	,,	1
4	Rear rail	650 x 51 x 18	,,	1
5	Lower rails	360 x 30 x 18	,,	2
6	Back strip	652 x 70 x 18	,,	1
7	Shelf	670 x 380 x 15	Veneered blockboard	1
8	Veneer strips	670 x 15 x veneer	Teak veneer - pre-glued	2
9	,, ,,	380 x 15 x veneer	,, ,, ,,	2
10	Castors			4
11	Screws	40 x gauge 8	Brass	6
JOHN SELVEDGE	Sheet 6 - PARTS LIST		TV SET & VIDEO STAND	

Energy and power

Some of the major sources of energy and forms of power are described in drawings on pages 131 to 132. Examine the drawings. Can you name other sources of energy and forms of power not illustrated by the drawings? Note that in this book, power supplies are regarded as having been derived from original sources of energy. Thus the electrical power which most of us have supplied to our houses, schools and places of work and which is used to light our streets at night, has been originally derived from sources of energy such as oil, coal or nuclear energy. Can you name other sources of energy which are converted by machines to supply our electrical power?

It is said that all energy has been derived originally from the sun. Can you explain this? How is it, for example, that coal and oil can be regarded as having been derived from the sun? Can it be said that nuclear energy has been originally derived from the sun? How can water movement as a source of energy be regarded as having been originally derived from the sun?

Gas

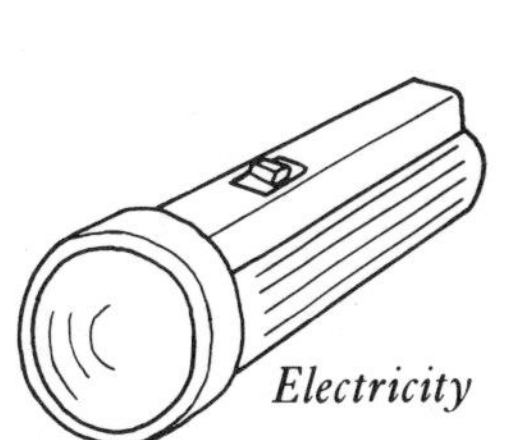

Electricity

Forms of power

Internal combustion

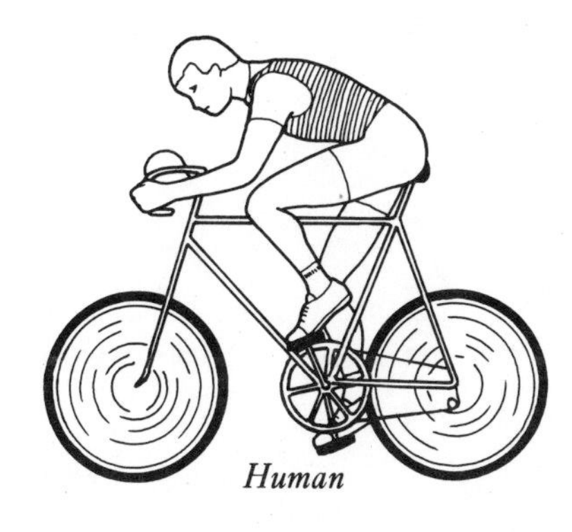

Human

Animal

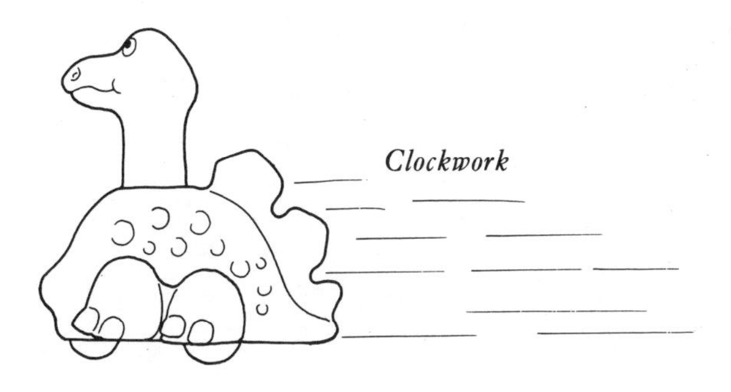

Clockwork

Press to wind up

Sources of energy

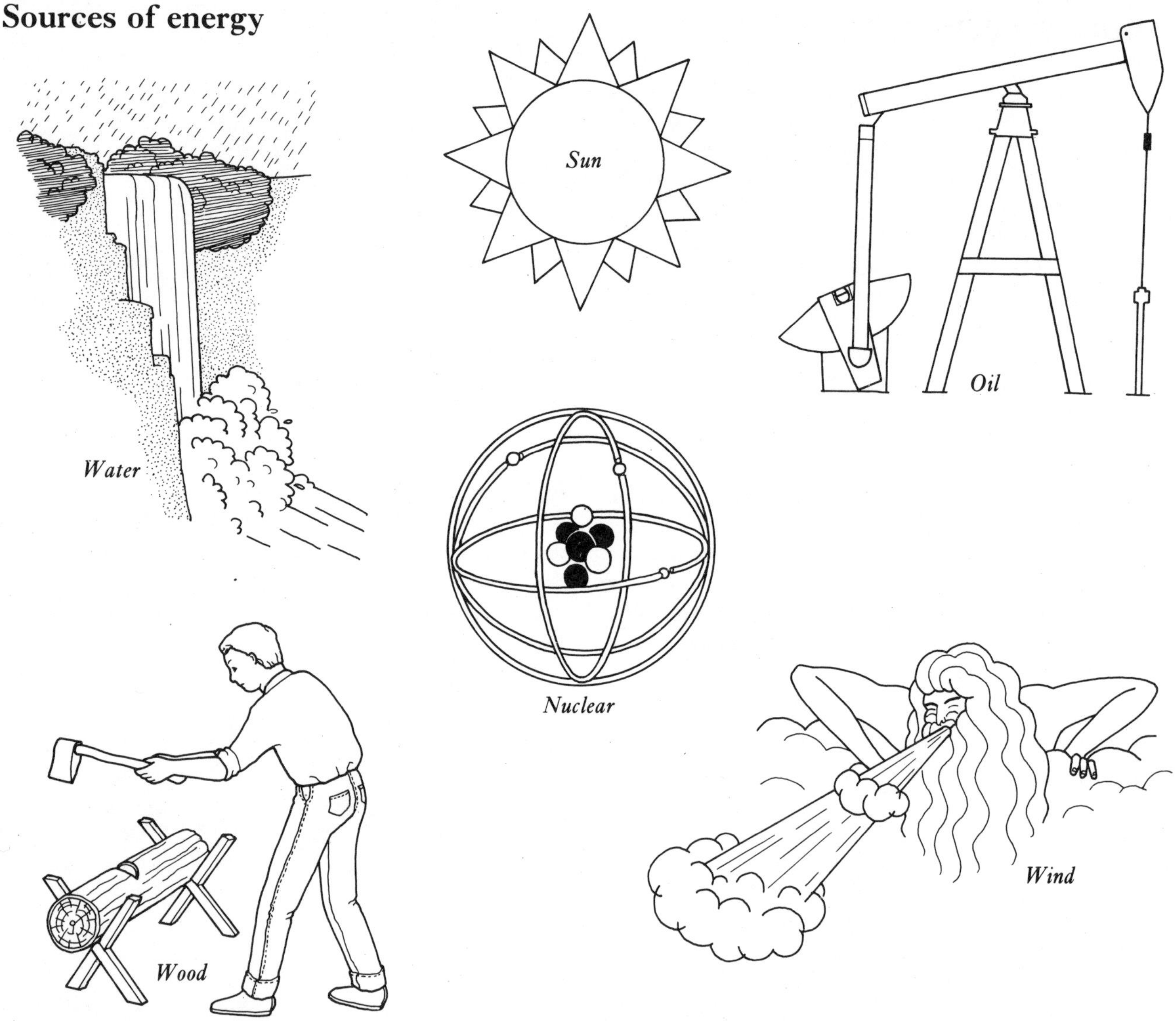

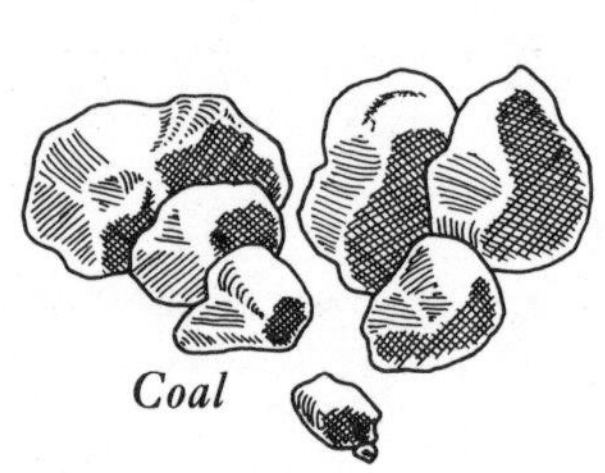

Sun

Oil

Water

Nuclear

Wind

Wood

Coal

Gas

Renewable and non-renewable sources of energy

All sources of energy can be regarded as belonging to one of two groups – those which are renewable and those which are not renewable. Those sources of energy which we can go on converting into power indefinitely are classed as renewable. As an example, power can be obtained from moving water so long as moving water can be found. The water is not used up. Other sources of energy must be regarded as being finite, in that they cannot be renewed once they have been completely used up. As an example, once all the oil in the earth's crust has been extracted, oil as a source of energy will have become exhausted. The non-renewable sources of energy are the *fossil* fuels – mainly coal, oil and gas.

Make two lists, one headed *non-renewable sources of energy*, the second headed *renewable sources of energy*. Add notes explaining why you have included each source in each list.

Energy and design

Many of the forms of power at present available have been derived from non-renewable sources of energy. Because of this, if the design you are working on requires some form of power, you should consider the need to economise in the consumption of that form of power. The aim should be to obtain the maximum benefit from the minimum power consumption. In other words, power and hence energy *conservation* should be regarded as an essential part of the design process in such technology projects.

Energy, materials and design

All the materials we employ in our design projects have been derived from the earth's crust or from its atmosphere. The raw materials from which all our design materials are made need energy to convert them into design materials. In some cases, very large amounts of energy and power are necessary, such as the heat needed to convert metal ores from the earth into the metals for design work. In other cases, power is needed to drive the machinery which, for example, saws logs from trees into timber of sizes suitable for our designs. Remember that energy and power are required to mine the ores from the earth and cut down the trees in the forest.

Can you name and explain other examples of the links between energy, materials and design?

EXERCISES

1. Forms of power are derived from sources of energy. Explain this statement.
2. Apart from natural gas, what other sources of energy are converted into gas for domestic use?
3. Name a source of energy other than crude oil which can be converted into petrol and oil to drive the internal combustion engine of a motor car.
4. Can wood be regarded as a source of energy derived from the sun? Explain your answer.
5. Explain the need for conservation of energy when planning a design project which needs power to drive a mechanism.
6. Explain the link between energy, power, design and design materials.

Materials

School and college craft designs are commonly made from metals, woods or plastics or combinations of these materials. Other materials are occasionally involved, such as glass, ceramics, leathers etc.

Metals

Metals can be divided into two main groups:
Ferrous – in which iron is the major ingredient.
Non-ferrous – in which iron is rarely present.
Another group of metals is formed by *alloys*.
Alloys are mixtures of different metals or metals and chemicals other than metals.

Ferrous metals

Carbon tool steel – An iron alloy with about 1 per cent carbon. For cutting tools.
Mild steel – From 0.1 to 0.6 per cent carbon. For constructional work in steel. Many fittings are made from mild steel. Also nails, screws, rivets, etc.
Stainless steel – An alloy of iron with at least 12 per cent chromium.
Cast iron – Iron with about 3.5 per cent carbon. Brittle. Used for large castings such as for machines.
Malleable cast iron – Heating and baking cast iron makes it malleable rather than brittle.

Non-ferrous metals

Aluminium – Lightweight; does not corrode; easily bent to shape. Many uses in modern designs.
Copper – Gold colour; high resistance to corrosion in wet conditions.
Zinc – Frequently used as a coating for other metals because of its high resistance to corrosion.
Brass – An alloy of copper and zinc. A metal which machines well. Good resistance to corrosion.

Woods

Woods are generally placed in two groups – *hardwoods*, which are obtained from broad leaved trees and *softwoods* which are obtained from coniferous trees. Most hardwoods are harder than most softwoods.

Hardwoods

Oak – Tough and durable. Fairly hard to work. Straw colour with pronounced grain. Some boards have 'flash' figure.
Beech – A tough and fairly hard wood. Light straw colour with not very pronounced grain.

Chestnut – A medium-hard hardwood. Straw colour with very pronounced and decorative grain. Easy working.
Elm – Medium hardness. Brown to nearly purple. Unseasoned elm distorts badly. Decorative grain.
Sycamore – Medium hardness. White to light brown colour. Some boards display a fine flash figure.
Mahoganies – Honduras mahogany, the only true mahogany, is easy to work; rich pink brown colour; a very stable material. Other 'mahoganies' include the so-called 'African mahoganies' which are generally harder and of a deeper colour than Honduras mahogany.
Afrormosia – A hard, but reasonably easily worked African timber with a deep brown colour and good grain.
Teak – Brown colour. Some boards have a good grain figure. Very resistant to decay.

Softwoods

Redwood – The common 'red deal' or 'yellow deal' derived from the Scots Pine tree. Widely used in building construction and in the making of 'pine' furniture.
Parana pine – From the monkey puzzle tree. Even texture; from light brown to purple. Splits fairly easily.
Spruce – A lightweight but strong timber (for its weight).
Western Red Cedar – Pink to brown colour. Soft and easily worked. Very resistant to decay, so widely used for outdoor work such as for sheds, greenhouse frames etc.
Balsa and *Jelulong* – Strictly speaking these are hardwoods, because they come from trees with broad leaves and open pores, yet they are soft and easy to work. Both are woods which are very suitable for the making of models because they can be worked so easily.
Yew – a common European tree. The wood is very hard, with a pleasing, variably coloured grain. Suitable for high class furniture and for turning.
British Columbian Pine – U.S.A. A good building and construction timber. Very variable pronounced grain.
Larch – a common European tree. The larch grown commercially is usually Japanese larch. It is light in weight, reddish brown in colour. Not often used in woodworking, its main value is for telegraph poles.

Notes

1. In school and college work, you may find you will be working with softwoods more often than with hardwoods. This is because hardwoods are more expensive than softwoods and also, as their name implies, much harder to work with. This does not apply to the making of models, which are often made in balsa or jelulong, which as stated above are really hardwoods.
2. The most common softwood is Redwood.

Plastics

Plastic materials can be grouped into three classes.
Thermoplastics – When heated become plastic and malleable and can thus be formed to new shapes.
Thermosetting – Once formed to shape cannot be reformed after heating.
Elastomers – Rubbers and rubber-like materials.

Thermoplastics

Acrylic – An example is Perspex. Mainly used in craft in sheet form, but available in rods, squares and extrusions.
PVC – Polyvinylchloride. Rigid or flexible.
Nylon – Many fittings are made from nylon. A useful bearing material allowing for almost friction-free running of shafts.
Polystyrene foam – Useful for model making.
Polyurethane foam – For upholstery.

Thermosetting plastics

Urea formaldehyde – A light coloured plastic, suitable for making electrical fittings and other coloured fittings, requiring electrical insulation. Also for fittings such as door and drawer handles. A useful wood adhesive – Cascamite glues being an example.
Phenol formaldehyde – 'Bakelite' is a phenol formaldehyde. Similar to ureas, except that phenols are brown. Used in the manufacture of items such as kettle handles, hot iron handles. Phenol formaldehyde glues are waterpoof and are therefore of value for waterproof plywood. The brown backing of 'Formica is based on phenol formaldehydes.
Melamine formaldehyde – The top layers of sheets such as 'Formica' are made from melamine formaldehyde.
Epoxy resin – Valuable as an adhesive which will form strong joints between practically any materials.
Polyester resins – The main plastic component of GRP (glass reinforced plastic) mouldings.

Elastomers

Neoprene – Chemically similar to natural rubber. High resistance to oils and chemicals. Used for belts and motor mountings.
Butyl – High resistance to tearing, flexing and abrasion. Hoses, cable insulations, seals for foods and chemical containers.
Urethane – Strongest and hardest of the elastomers in common use.
Silicones – Elastomers based on silicone and oxygen. Excellent resistance to oils and to chemicals at either high or low temperatures. Used for insulation and for the encapsulation of electronic components.
Isoprene – A synthetic plastic elastomer with properties simiolar to natural rubber. Items made from Isoprene have greater extensibility than natural rubber.

Constructions

Constructions and forming of metals

A small selection from the numerous methods by which metals are constructed and formed to shape is given in the illustrations on this page. The methods shown are those which are most commonly employed in schools and colleges.

Bolts, screws, nuts and washers

These are intended for making temporary joints. By unscrewing a nut from its bolt a joint can be taken apart. Note that when a washer is included it should be placed under that part of the bolt, nut and washer assembly which is turned to tighten the assembly. Note also that a bolt threaded along the whole length of its screwed diameter is commonly known as a screw.

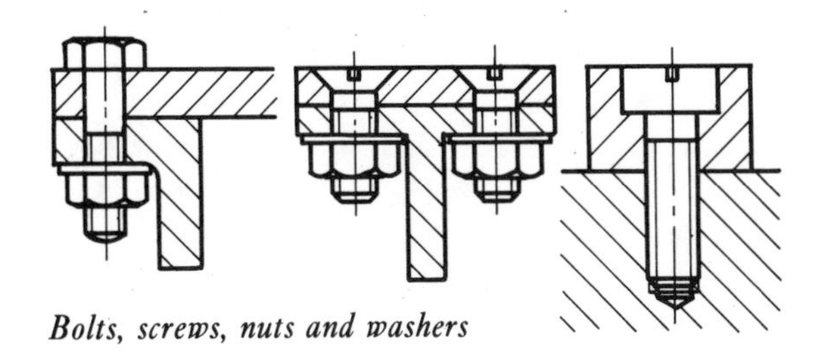

Bolts, screws, nuts and washers

Rivets

These are for making permanent joints, in that the parts of the joint can only be taken apart if the rivets are drilled out of their places. Rivets are made from a variety of metals, chiefly mild steel, but aluminium and copper rivets are fairly common.

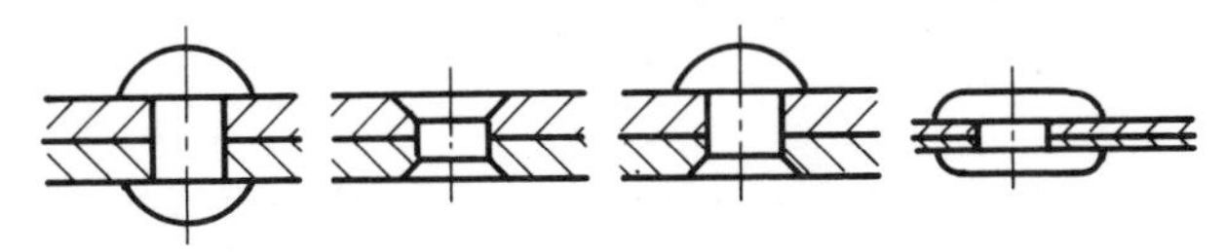

Round head, countersunk and flat head rivets

Soft solder joints

These are most frequently made with the aid of a soldering iron using tinman's solder (an alloy of lead and tin). Soft soldering is also employed for the fixing of electrical wiring in its final position in electric and electronics circuits.

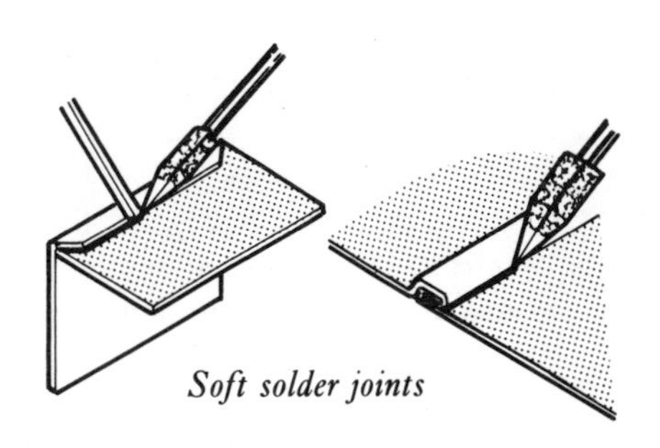

Soft solder joints

Silver solder

Silver soldered joints are considerably stronger than soft soldered joints. However, because silver solder (an alloy of silver and copper) requires a higher melting temperature, heat for silver soldering is usually applied with some form of blowlamp torch.

Brazing

The operations of brazing are very similar to those of silver soldering. Generally, higher temperatures are needed to melt the brazing spelter (an alloy of copper and zinc) into the join. A well made brazed joint is almost as strong as the metals which have been joined together by the method.

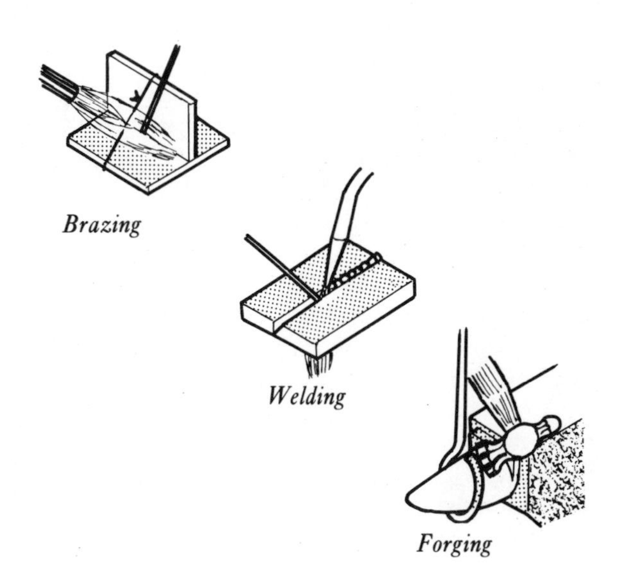

Brazing

Welding

Forging

Welding
Welding requires very high temperatures and either oxy-acetylene welding equipment or electric arc welding equipment. If, say, mild steel is being welded, then the welding stick will be a form of mild steel of a melting point close to that of the metals being joined. Welding in schools should only be carried out under the personal supervision of a teacher who has had training in the use of welding equipment.

Forging
This is for shaping metals which have been heated in a forge.

Raising, sinking and planishing
These are three operations involved in shaping sheet metals such as copper, gilding metals and silver. Depending upon the form which is being shaped, the sheet metal is first either raised or sunk, then polished and planished.

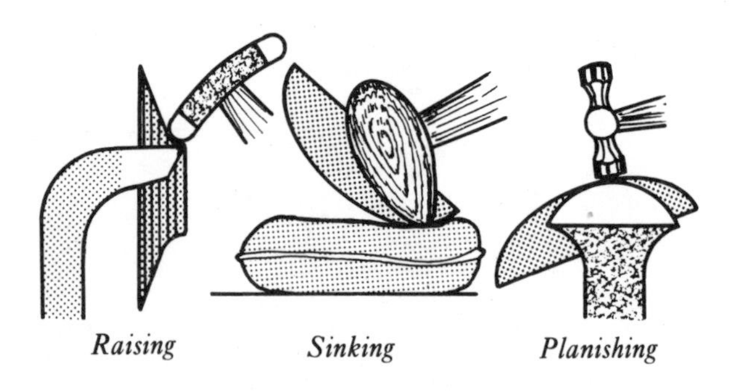

Raising *Sinking* *Planishing*

Casting
When metals are heated to temperatures at which they melt, they can be cast into moulds. In school work, metal casting is usually restricted to aluminium because other metals can only be smelted at temperatures which are too high for safety using the equipment available in schools.

Casting in aluminium

Adhesives for metal joints

Strong joints between metal parts can be made with an epoxy resin adhesive such as 'Araldite'. Not only will an epoxy resin adhere to the surfaces of all metals, but because the resin does not shrink as it hardens, there is no danger of the joint cracking as the adhesive sets. In addition, properly made clean joints made with epoxy resin are extremely strong.

Constructions for woods

The more commonly used methods of forming joints between wooden parts are shown in twenty isometric drawings on this page.
1. Nailed corner joint – When nailing corners of a box, such as that shown, a stronger joint will result if the nails are driven in a 'dovetail' fashion as shown.

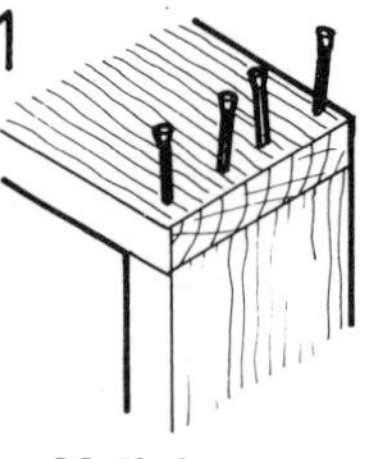

Nailed

2. Screwed corner joint – When screwing into end grain with wood screws, it is advisable to drill and plug holes in the end grain to prevent the screws pulling out easily.

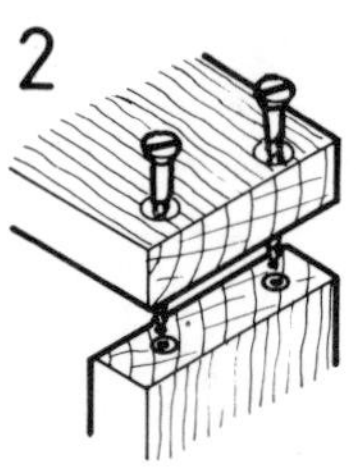

Screwed

3. Corner rebate – A common, easily made box corner joint, which must also be glued, nailed or screwed.

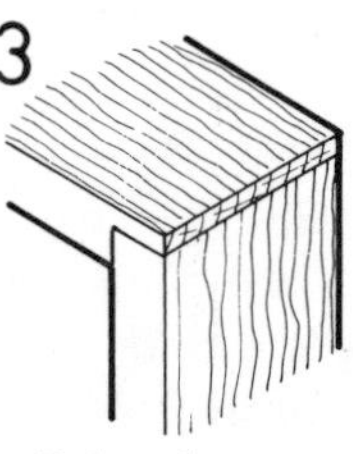

Rebated

4. Nailed partition joint – Note again the dovetailing of the nails.

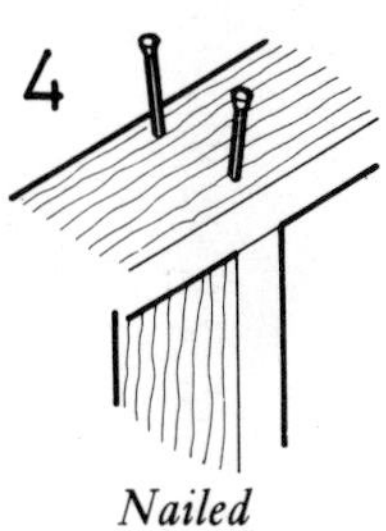

Nailed

5. Housed partition joint – When making a joint for a box partition or for a shelf, the partition can be fitted into a groove.

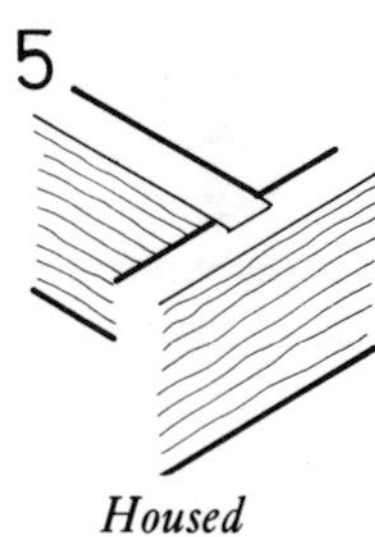

Housed

6. Stopped housed partition – The advantage of this method over the previous housed joint is that the joint cannot be seen from one of the edges.

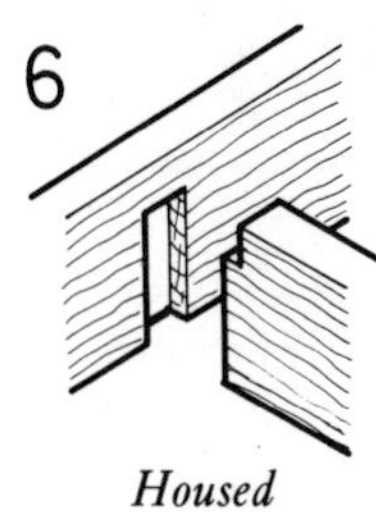

Housed

7. Grooved corner joint – When carefully fitted and glued this forms a strong corner joint.

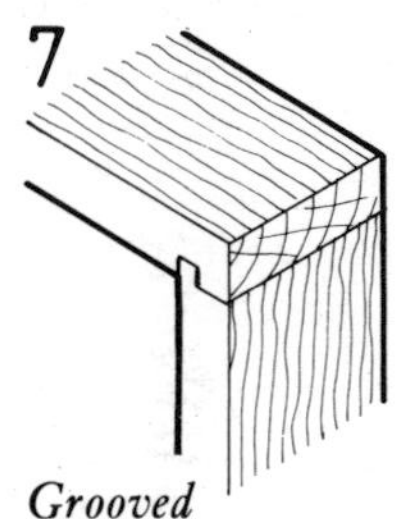

Grooved

8. Dovetailed corner joint – Possibly the strongest of all box joints.

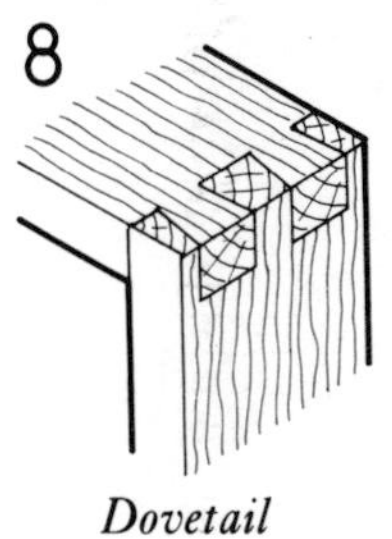

Dovetail

9. Lap dovetailed corner – When the dovetailed parts should not be seen from one side, a lap dovetail is a suitable joint.

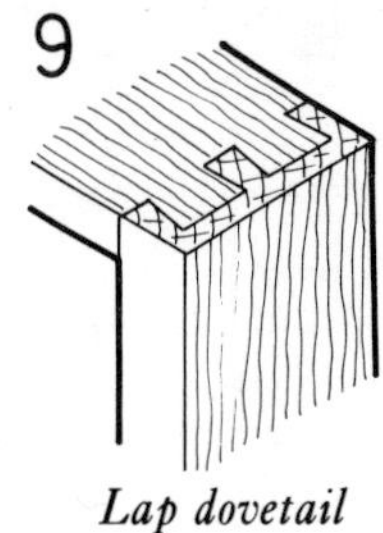

Lap dovetail

10. Halving corner joint – Half of each part of the frame is cut away. This joint must be glued and screwed (or nailed) otherwise it will not hold together.

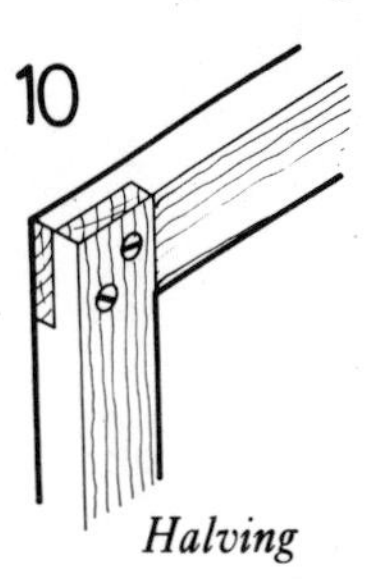

Halving

11. Bridle corner joint – A strong frame joint when well and accurately made.

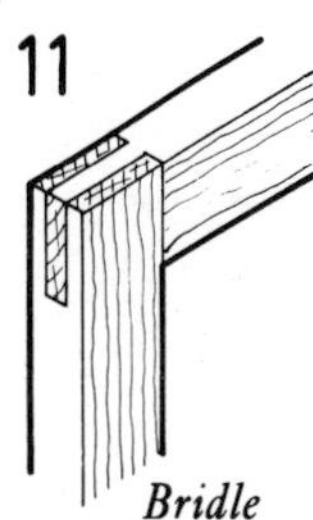

Bridle

12. Mortise and tenon joint – The most common of all framing joints. A large range of different types of mortise and tenon joints is in common use.

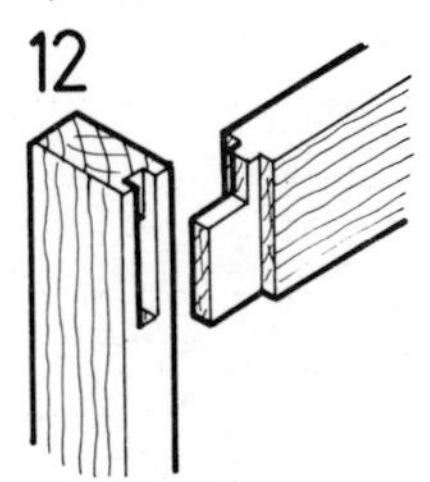

Mortise and tenon

13. Tee bridle joint – Another strong frame joint.

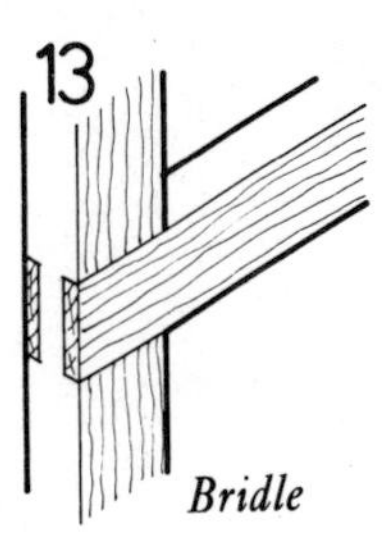

Bridle

14. Tee mortise and tenon – A second example of a mortise and tenon.

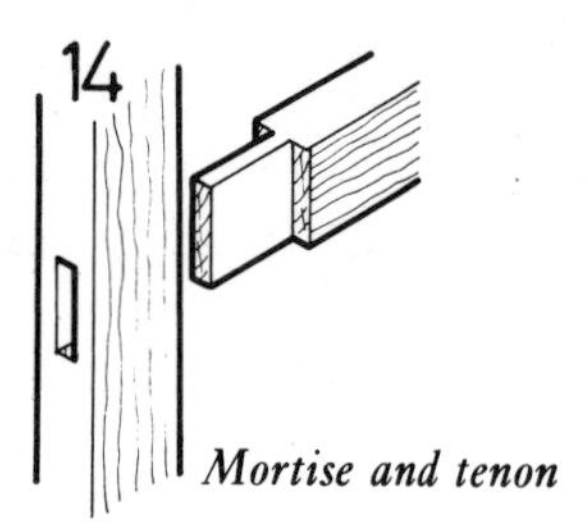

Mortise and tenon

15. Dowelled joint – Dowel joints can be made more quickly than can mortise and tenon joints but do not form so strong a construction.

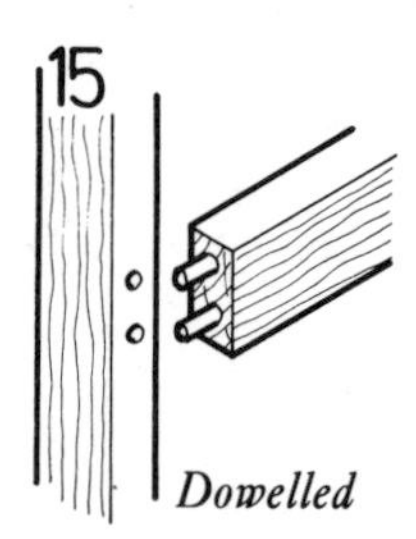

Dowelled

16. Halving joint – A cross halving. Half of each part being joined is cut away to form the cross.

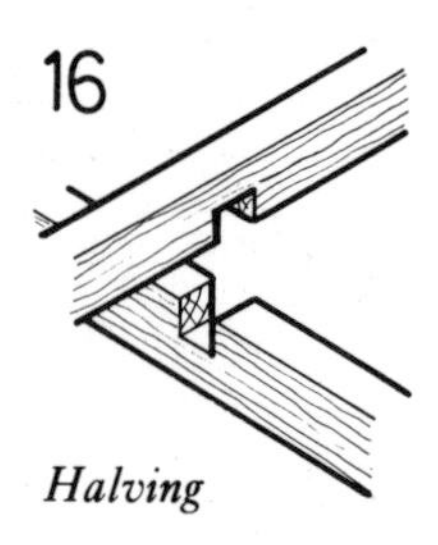

Halving

17. Mortise and tenon corner joining – A third mortise and tenon example.

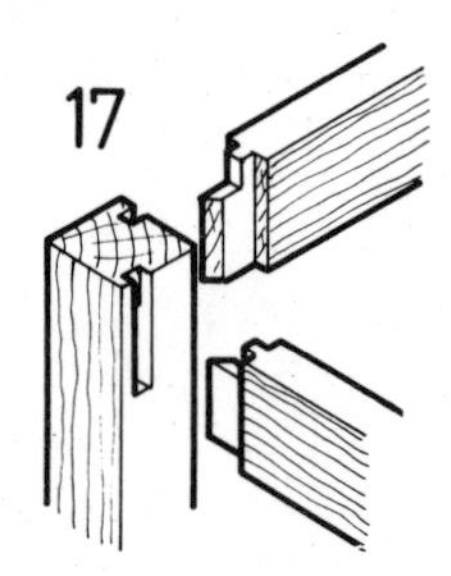

18. Dowelled corner jointing – A second dowelled joint.

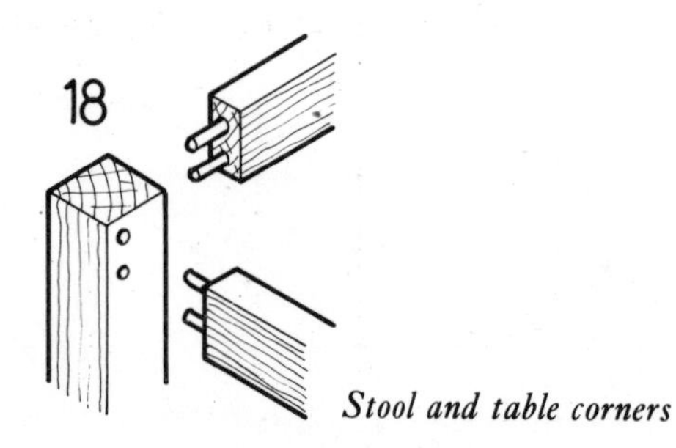

Stool and table corners

19 and 20. Laminating – Very strong shaped parts can be made by gluing thin strips to each other in some form of mould.

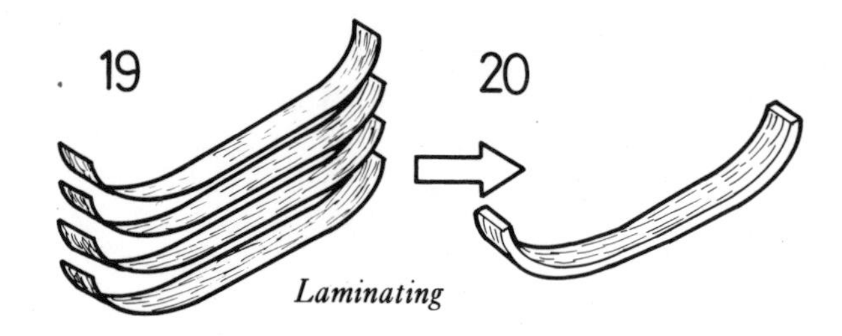

Laminating

Adhesives for wood jointing

A number of different glues are available for joining wood. These are mostly plastic glues – PVA wood glue; urea formaldehyde glues such as Cascamite; hot melt glues. Impact glues, which are rubber based, can be employed for joining sheet materials – fabrics, leather, thin plywood, veneers – to wood.

Forming plastics to shape

The diagrammatic illustrations on this page show industrial methods of forming plastics into the huge range of shapes which modern design demands. The plastic materials are most often fed into the machines which perform the operations in a powder or granular state. Heat is required to turn the plastic powder or granules into a sufficiently pliable and plastic melt to enable it to be easily shaped.

Calendering

PVC and polythene sheet are produced by this method. A ribbon of heated, pliable plastic is fed from an extruder over heated, polished rollers. The sheet is successively made thinner by being passed between suitably spaced rollers before being fed away on cooled rollers.

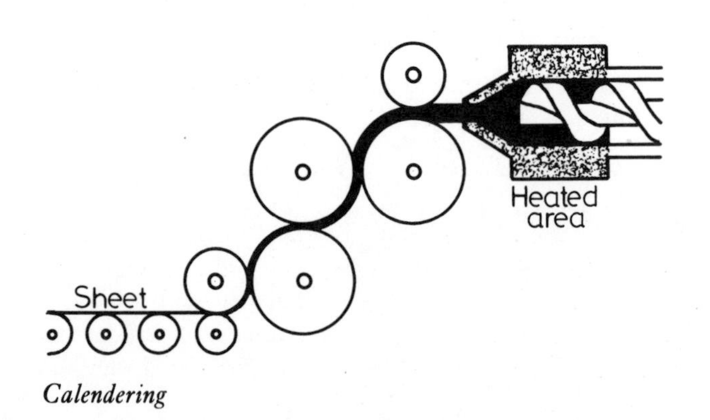

Calendering

Injection moulding

This is probably the most widely used method of producing articles from thermoplastics. Plastic powder is fed from a hopper into a cylinder in which a ram works. The ram forces the powder through a heated area where, under the action of the heat and pressure, the plastic melts. Each stroke of the ram forces a quantity of molten plastic into the spaces between dies where the plastic cools and sets.

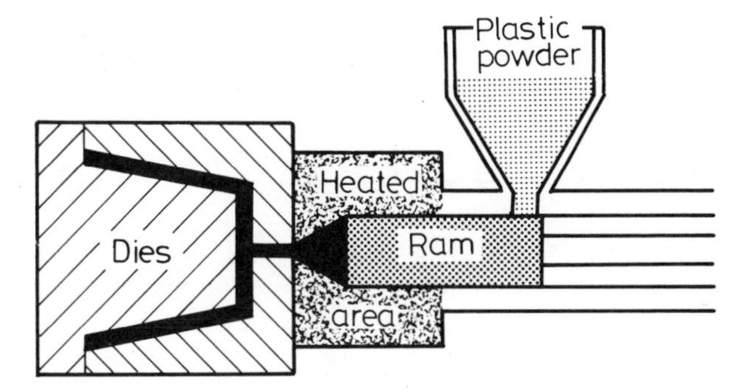

Injection moulding

Extrusion

Extrusion is used for producing pipes, rods and filaments from thermoplastics. A revolving screw drives thermoplastic granules through a heated area where the plastic melts. It is then passed through suitably shaped dies which determine the profile of the extruded material.

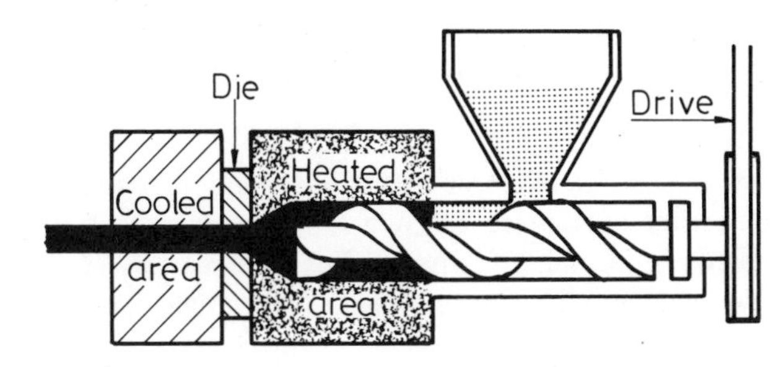

Extrusion

Compression moulding

This is mainly used for producing mouldings from thermosetting plastics. Uncured thermosetting powder or granules is placed between suitably shaped parts of a split mould, which is closed under heat and pressure. The heat and pressure set the uncured powder or granules. The setting of plastic material is known as polymerisation.

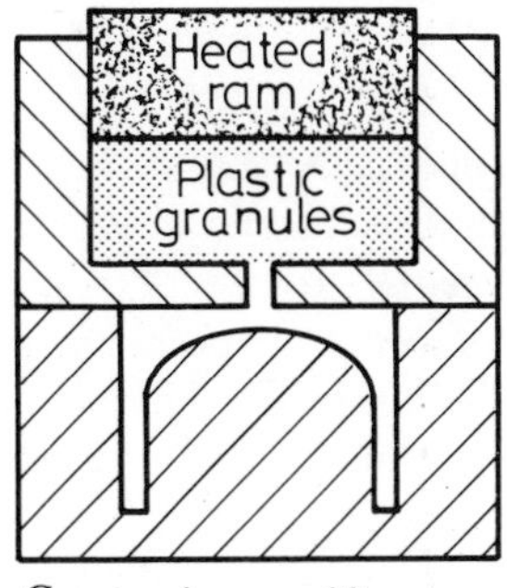

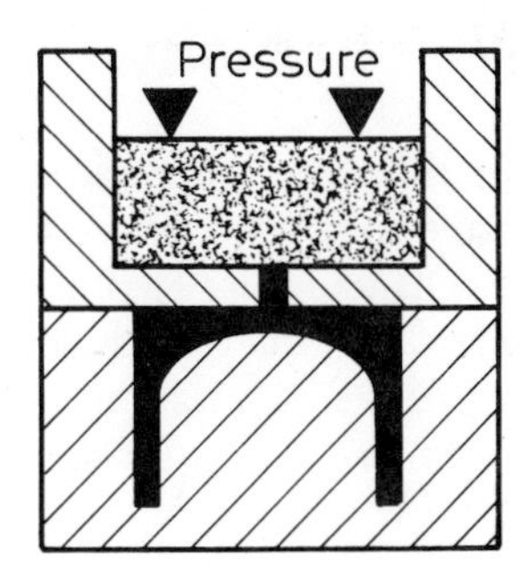

Compresion moulding

Vacuum moulding

This is a method of shaping thermoplastic sheet. The sheet is firmly clamped between the two parts of a holder, then heated until it is in a rubber-like and pliable state. The heaters are lifted or swung away and the warm, malleable sheet is then brought into contact with a mould by forming a vacuum between the sheet and the mould.

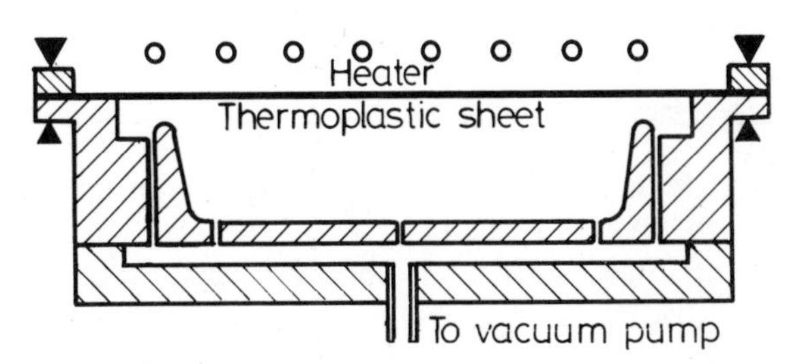

Vacuum moulding

GRP – glass reinforced plastic

This is a method of forming plastic shapes on moulds which can be carried out in schools. A mould made from wood, fibre glass or other suitable material has its surface polished and then coated with some form of release agent. Polyester gel coat resin is then painted onto the mould surface. When this has partly polymerised, several layers of fibre glass mat are laid on the mould and worked onto its surface with polyester 'lay-up' resins with brushes. To obtain a reasonably clean surface after laying-up has been completed, a thin surface tissue fibre glass is laid on with a brush and polyester resin. After the moulding has been allowed to set, it can be removed from the mould.

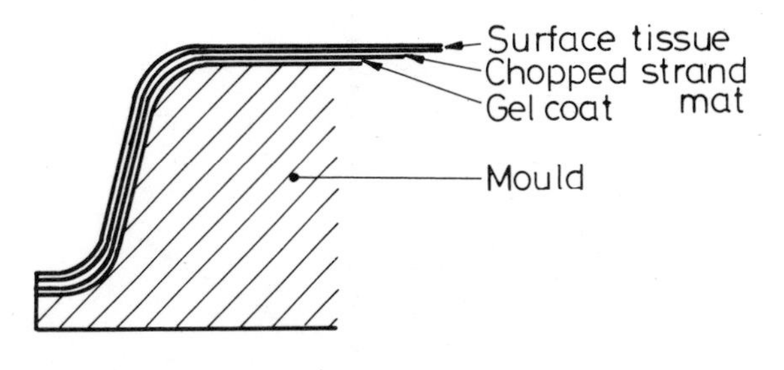

GRP—glass reinforced plastic

Mechanisms

In the next few pages we will briefly discuss levers, pulleys, gears, cams, link systems and screws. All these mechanisms are elementary *machines* for converting an *effort* into the *work* which allows us to shape, form and move the materials and goods essential to our environment. Machines make use of the power derived from energy in order to perform that work.

Models of the elementary machines described here can be made from wood, metal or plastics, mounted on boards with pins. Such models can also be made from kits such as LEGO®, Meccano and Fischertechnic. Some examples of models of this type are included here. You are advised to undertake the following two exercises if you wish to understand fully how important these elementary machines are in our everyday life.

1. Make models of the machines described.
2. Look about you to find examples of the machines and try to understand the functioning of the examples you see.

Units

As a general guide, the weight of 1 kg is approximately equal to 9.81 newtons (1 kg = 9.81 N). As the precise value of a newton depends upon gravitational force, its value varies according to the position in the world at which its value is taken. The figure 9.81 is so close to 10 that, for practical purposes, a weight of 1 kg is taken here as being equal to 10 N (1 kg = 10 N).

Levers

In general there are three types of lever as shown in drawings 1, 2 and 3. Examples of applications of each type are: 1 lifting a paving stone with a crowbar; 2 lifting a loaded wheelbarrow; 3 lifting a spadeful of concete. Can you give further examples of each type of lever?

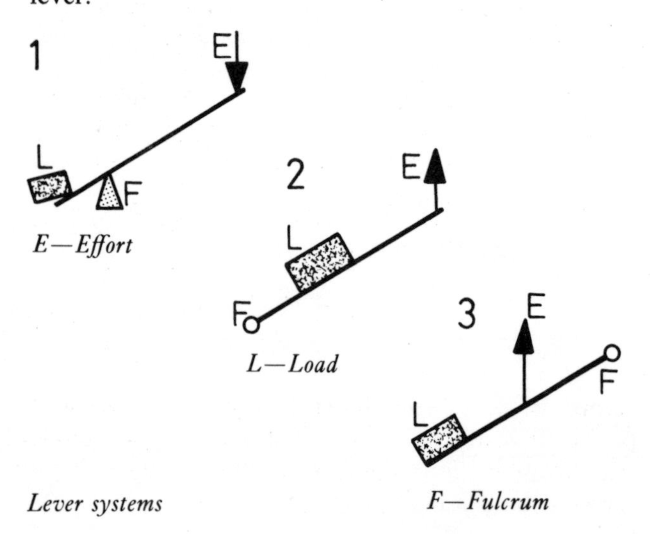

E—Effort *L—Load* *F—Fulcrum*

Lever systems

The relationship between the effort (E) required to lift the load (L) about the lever fulcrum (F) is shown in drawings 4, 5 and 6. The effort (E) multiplied by its distance (A) to the fulcrum is equal to the load (L) multiplied by its distance (a) to the fulcrum. Thus in each example:

$$E \times A = L \times a;$$

$$\text{and thus } E = \frac{La}{A} \text{ and } L = \frac{EA}{a}$$

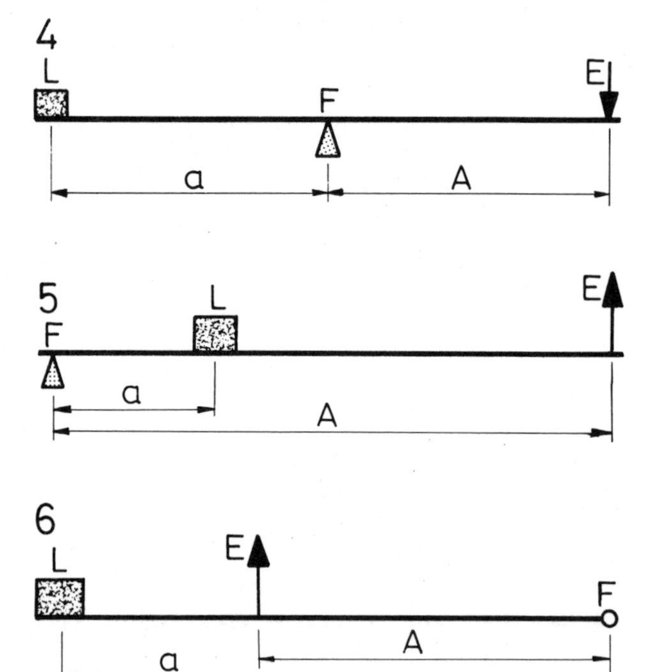

Drawing 7 is an example of a simple model illustrating the action of a lever mechanism. The levers of this model can be made from wood or thick card. A more permanent model could be made from metal.

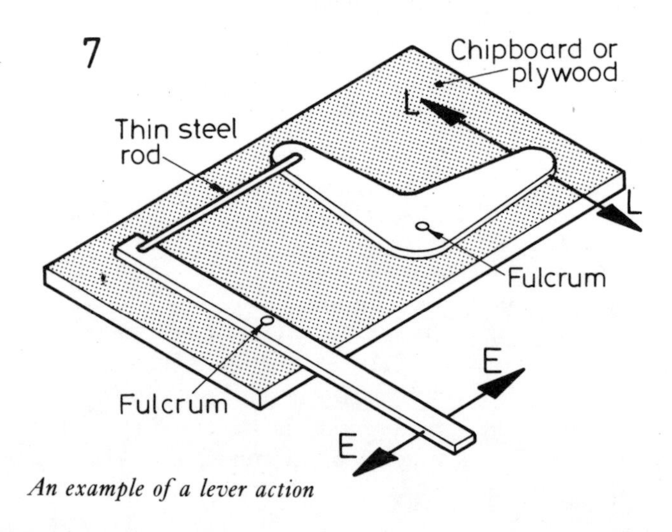

An example of a lever action

Can you state which is the fulcrum, where the effort is applied and how the resulting load is made to do work in the example of a pipe-bending machine shown in the photograph?

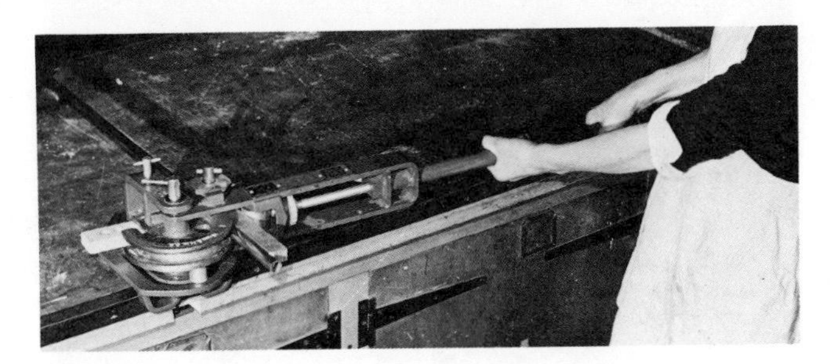

Pulley systems

Remember, in the following examples, that weight of 1 kg is taken as being equal to 10 N.

Five pulley systems are shown in drawings 1 to 5. In each example the effort (E) will lift a load (L). Two terms are used to describe the working of pulley systems. These are Mechanical Advantage (MA) and Velocity Ratio (VR). These two terms are also used in association with other machines.

$$\text{Mechanical Advantage (MA)} = \frac{\text{Load}}{\text{Effort}}$$

$$\text{Velocity Ratio (VR)} = \frac{\text{Distance moved by the Effort}}{\text{Distance moved by the Load}}$$

In theory MA should equal VR, but frictional losses in the pulley bearings prevent this being so in practice. Theoretically, however, the following apply.

Drawing 1 – An effort of 10 N will lift a load of 1 kg. The load will move upwards through the same distance that the effort moves downwards. MA = 1; VR = 1.

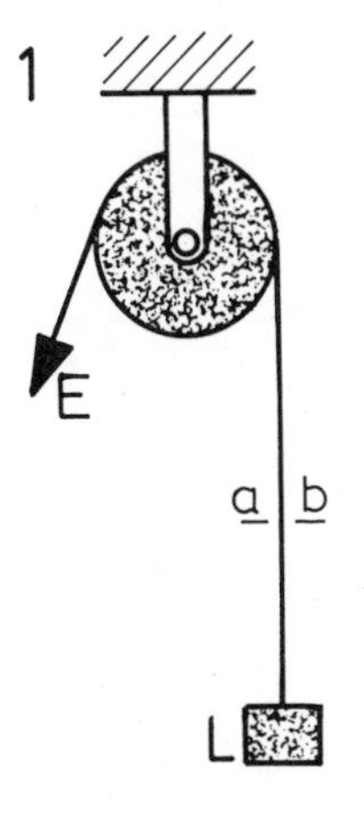

Drawings 2 and 3 – An effort of 10 N will lift a load of 2 kg. The load will move only half the distance through which the effort moves. MA = 2; VR = 2.

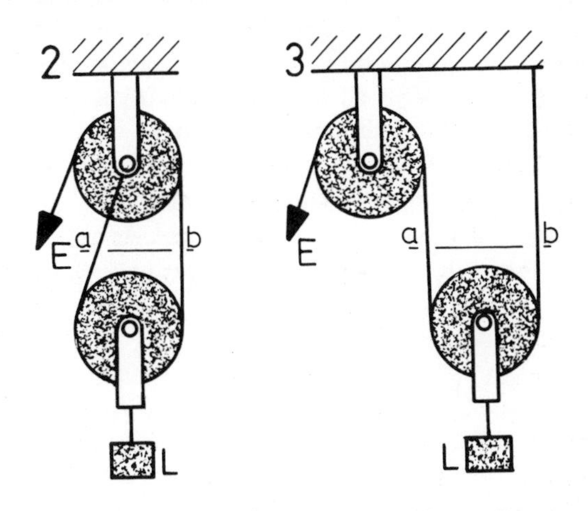

Drawing 4 – An effort of 10 N will lift a load of 4 kg. MA = 4 and thus theoretically VR = 4.
Drawing 5 – Effort of 10 N lifts a load of 6 kg. MA = 6; VR = 6.

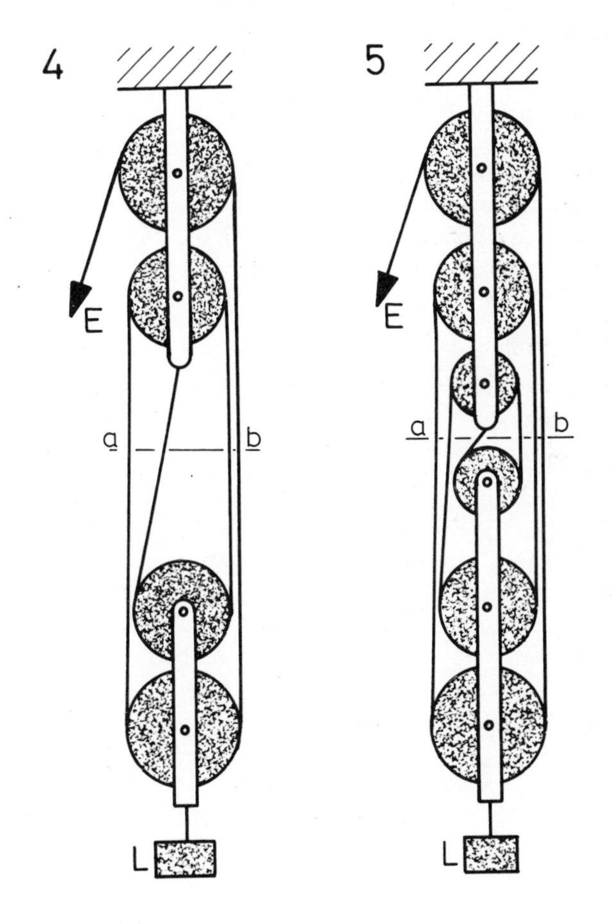

Note: In each of the drawings a line ab passes across the belts of the pulley systems. The lines ab indicate the MA and VR – count the number of times the line crosses the belt of each system.

Further pulley systems

Pulleys can also form systems for transmitting rotary movement from one place to another through the medium of belts passing around and between the pulleys. In such systems the relative speeds of rotation of the pulleys are in direct ratio to their diameters. In drawings 1 and 2 the rotational speed of the small pulley can be found by dividing the diameter $ØP_1$ by diameter $ØP_2$. Thus if the smaller pulley is of a diameter of say half that of the larger, its revolutions per minute (rpm) would be twice that of the larger.

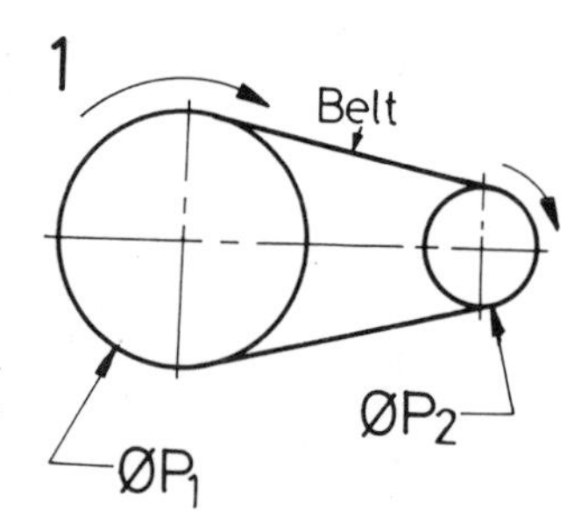

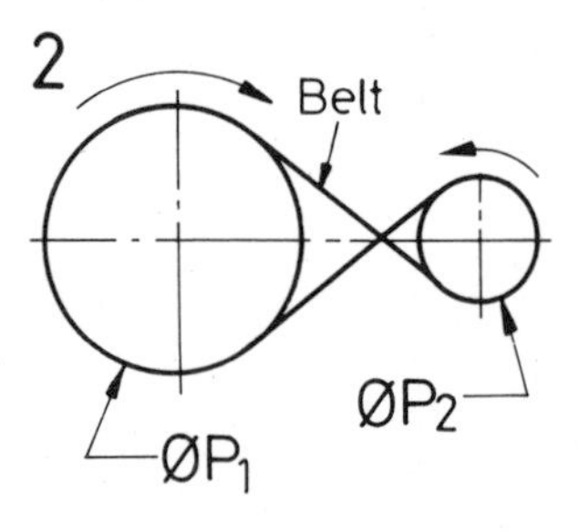

Note that to reverse the direction of rotation the belt between the pulleys needs to be crossed as in drawing 2.

Drawing 3 – If the diameter of each small pulley is 50 mm and the diameter of each large pulley is 100 mm and pulley A revolves at 100 rpm, what will the rotational speed in rpm of C be? If pulley C revolves at 100 rpm, what will the speed in rpm of pulley A be?

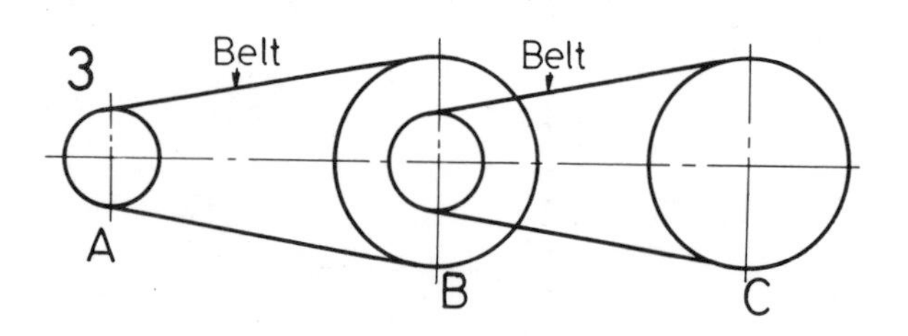

A pulley system

The photograph is of part of a film winding device.

Drawings 4, 5 and 6 – These are examples of some types of belts used between pulleys.

Drawing 7 – The driving device of the common bicycle illustrates a lever system applied to a pulley system. Can you see which is the lever and which is the pulley system?

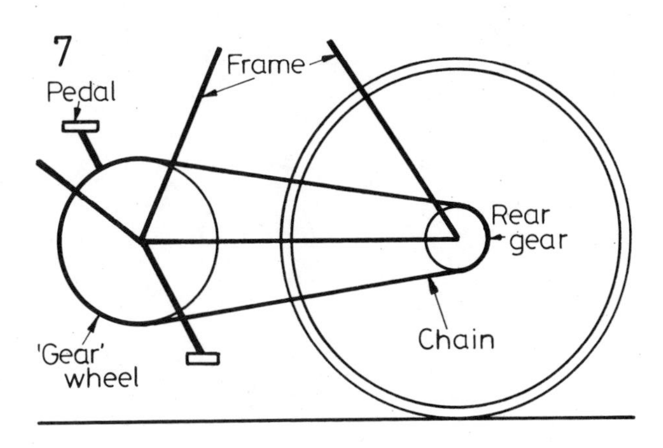

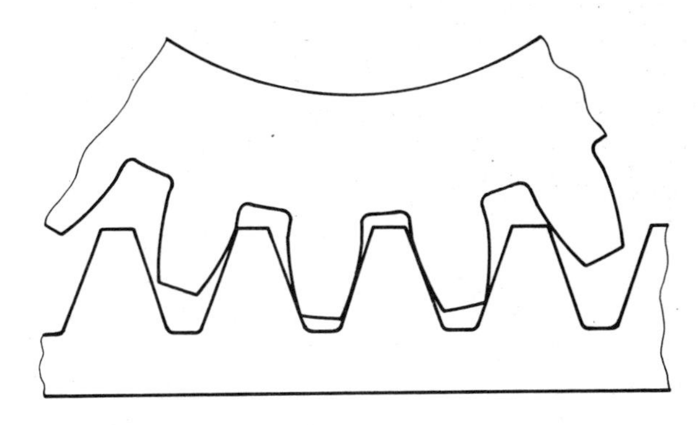

A gear rack

Gears

Applications of gear systems can be found in machinery of all types. Perhaps the best known is that of the 'gears' of cars and lorries. The most common form of gear is the *spur* gear wheel. A photograph of two spur gear wheels from a screw cutting lathe is shown. Such gears are called *involute* gears, the shape of the teeth profile being in the geometrical form of an involute.

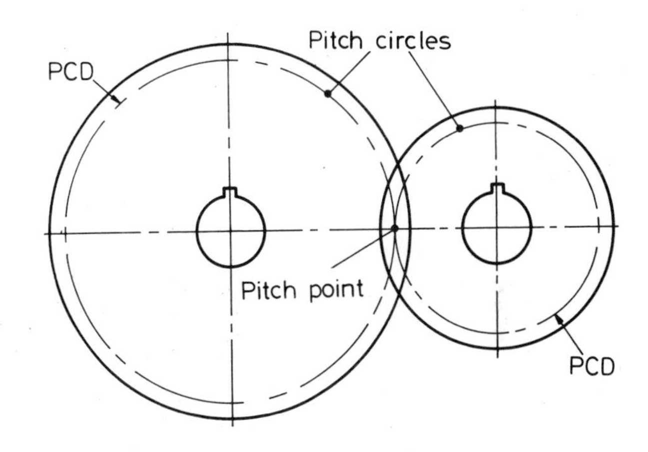

Terms associated with gears – British Standards conventions

Spur gear wheels from a screw cutting lathe

The geometry of a spur gear is based upon the *pitch circle* of the gear. Another form of gear system depends upon gear *racks* in which a moving rack causes a gear wheel to rotate, or in which a revolving gear wheel moves along a stationary rack.

A drawing shows spur gear wheels drawn to British Standard conventions. PCD stands for Pitch Circle Diameter.

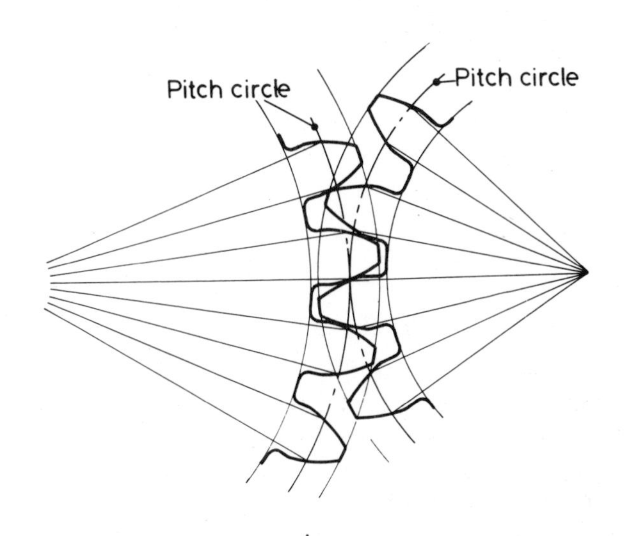

Meshing spur gears

Spur gears are most frequently included in gear systems to reduce or reverse direction of rotation or to increase rotational speeds between the spindles on which the gears are fitted. The relative speeds of rotation between meshing gears is in direct ratio to the number of teeth cut in the gears. The three drawings 1, 2 and 3 explain this.

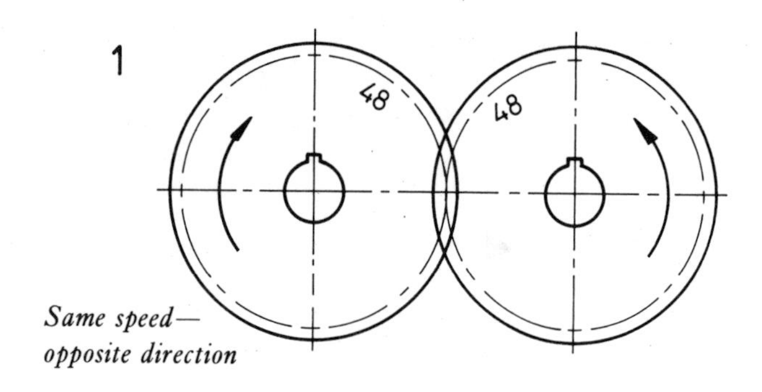

Same speed—
opposite direction

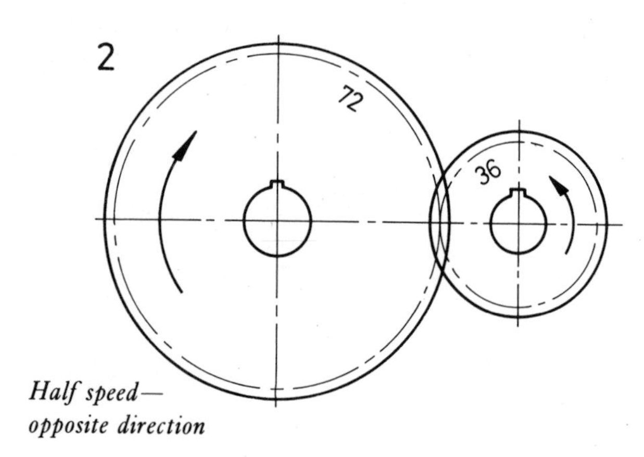

Half speed—
opposite direction

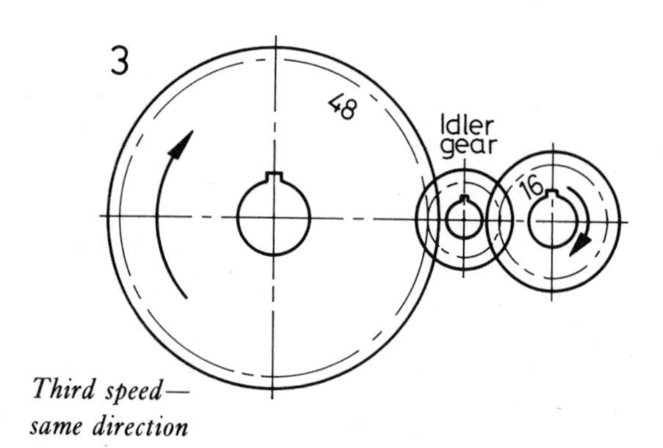

Third speed—
same direction

Bevel gear wheels

To change the angular direction of rotation of a pair of spindles in a gear system, bevel gear wheels are used. The first of the two given drawings of bevel gears is an orthographic projection of a ten-tooth bevel gear wheel. The second drawing is a sectional view through two bevel gears of unequal size. The second drawing is a conventional BS drawing.

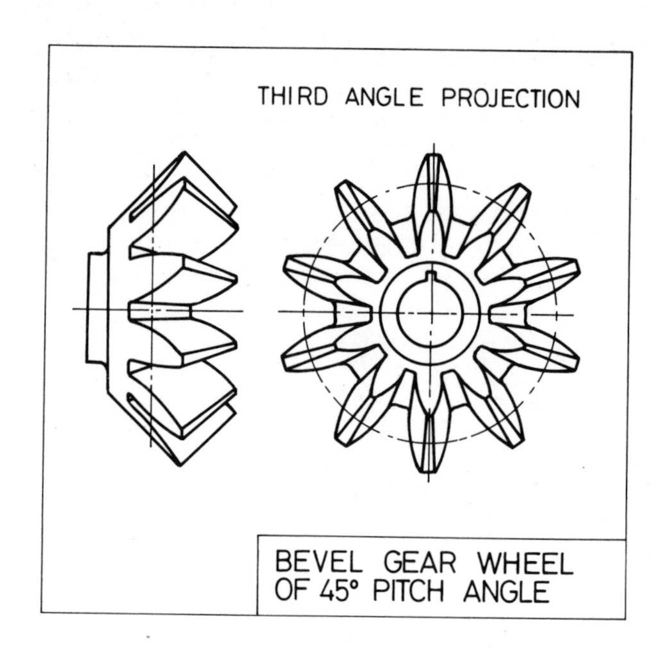

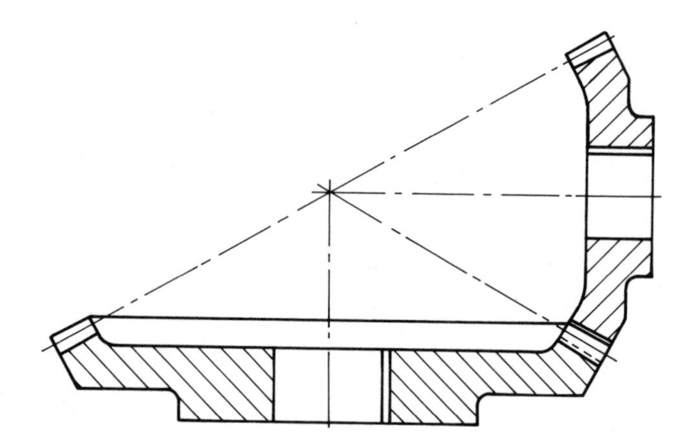

British Standards drawing of meshing bevel gears

Models

Models showing gear systems can be built up from Meccano parts.

Meccano gear wheels and pulley systems

Cams

Cams generally are designed to convert a rotary motion from a spindle (often called a cam shaft) into movement along a straight line. The design of the profile of a cam determines how the straight line movement occurs. In all cams, a *follower* rests upon the cam profile.

The seven drawings are sectional views of different types of cam mechanisms.

Drawing 1 – The follower is mounted *in-line* with the cam spindle centre. As the cam rotates, the roller follower follows its profile causing the part to which the roller is attached to move up and down.

Drawing 2 – Similar to 1 except that the end of the part resting on the cam is ground flat.

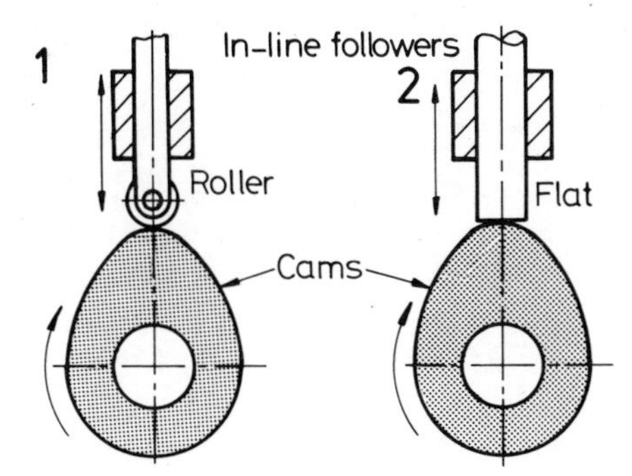

Radial cams

Drawing 3 – The rotation of the cam causes the bent arm to move up and down, which in turn imparts a straight line movement in the link to which it is attached.

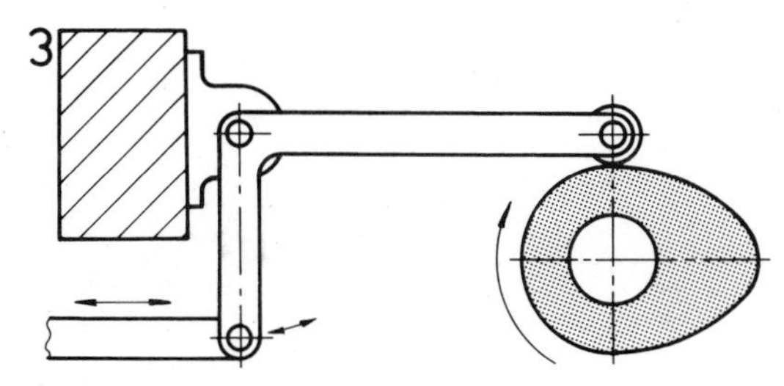

Follower on lever arm

Drawing 4 – As the cam slides sideways backwards and forwards, the roller follower moves the part to which it is attached up and down.

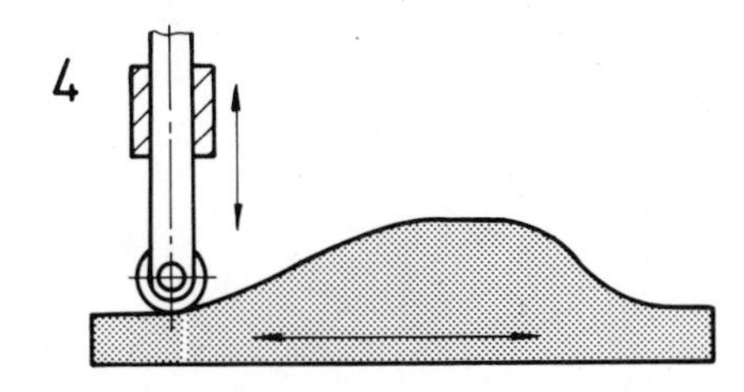

Slider cam

Drawing 5 – As the cam rotates, the roller following its profile moves up and down.

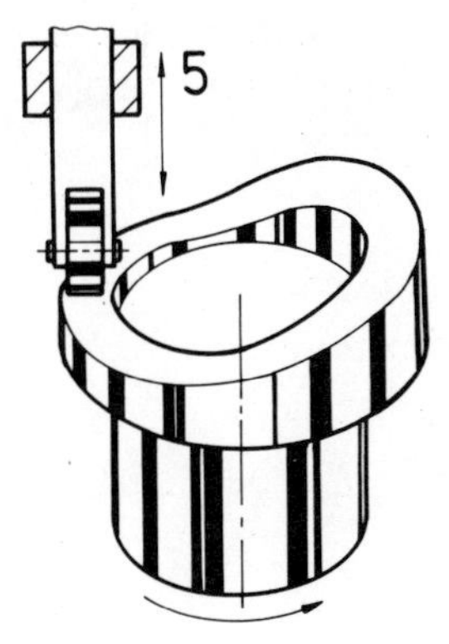

Cylindrical edge cam

Drawing 6 – A rotary cam movement causes the circular follower to move from side to side in its cam groove.

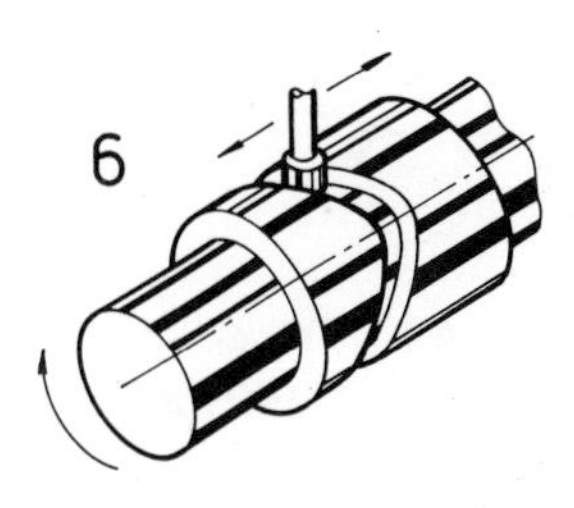

Cylindrical cam

Drawing 7 – A sectional view through a typical cam operation in an internal combustion engine.

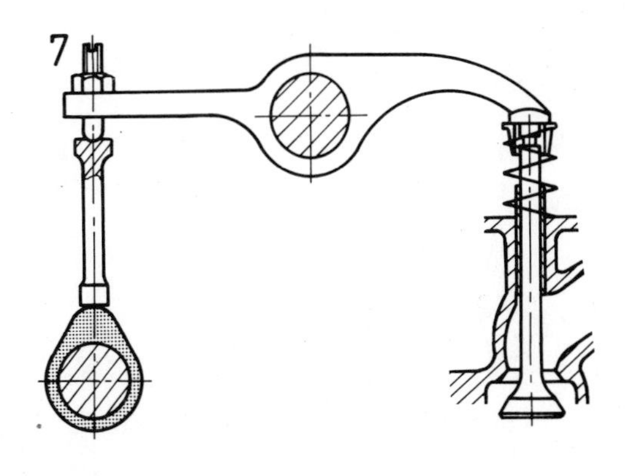

Valve cam from motor car engine

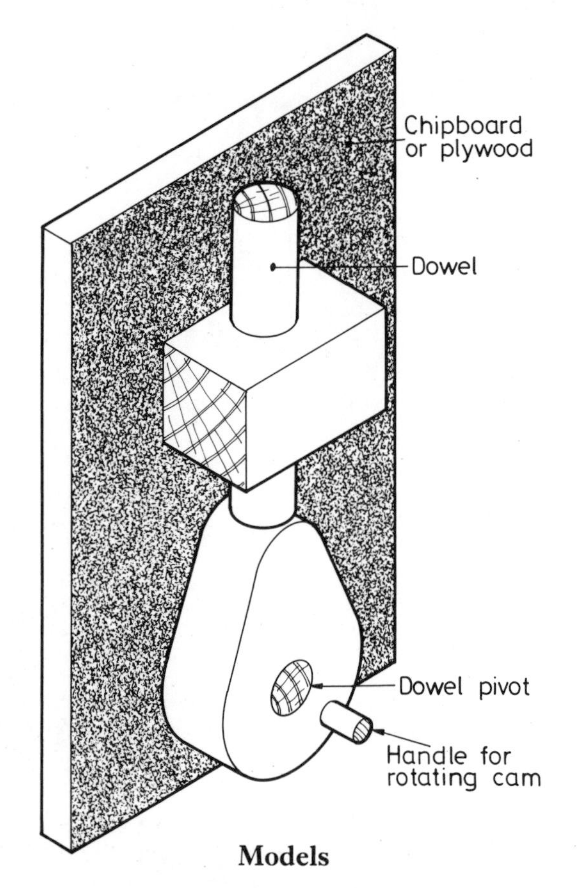

Models

Simple models such as that shown above will assist in understanding simple cam mechanisms. A number of different cams could be made for fitting onto the pivot of the model to experiment with cam outlines.

Link mechanisms

Link systems join parts of mechanisms in order to convert an input rotation or straight line movement into an output rotation or straight line movement.

Link mechanisms

Probably the most commonly used link mechanism is that in which the piston of an internal combustion engine is linked by a connecting rod to drive the engine's crank shaft. The up and down movement of the piston in its cylinder is converted to the rotational movement of the crank shaft. A sectional drawing of this link mechanism is given in *drawing 1*.
Drawing 2 shows the same system as a diagram.
Drawing 3 shows the geometry for finding the *locus* of a point on the connecting rod as the whole system moves.

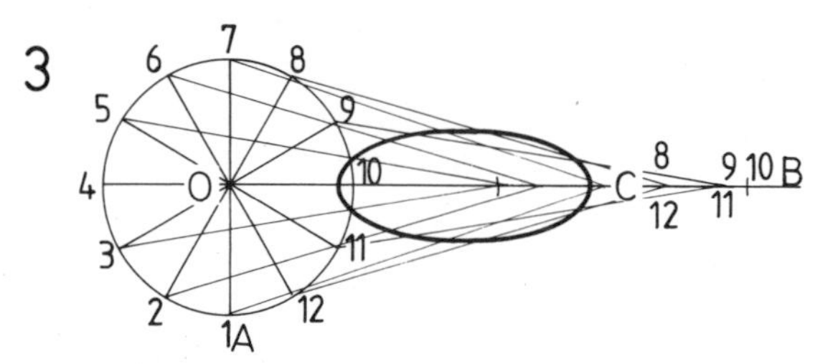

Drawing 4 is an enlarged view of part of the construction needed to plot points on the required locus curve.

The construction follows the sequence:

1. Draw the circle of rotation of the cam OA.
2. Divide this circle into 12 equal parts using a 30°, 60° set square.
3. Set a compass to the length AC and from the points 1 to 12 on the circle mark arcs along the line OB.
4. Join the points 1 to 12 on the circle to the marks made along OB with straight lines.
5. From the points 1 to 12 along OB (*drawing 4*) mark off the length of the point for which a locus is required.
6. Join the points which a neat freehand curve (*drawing 3*).

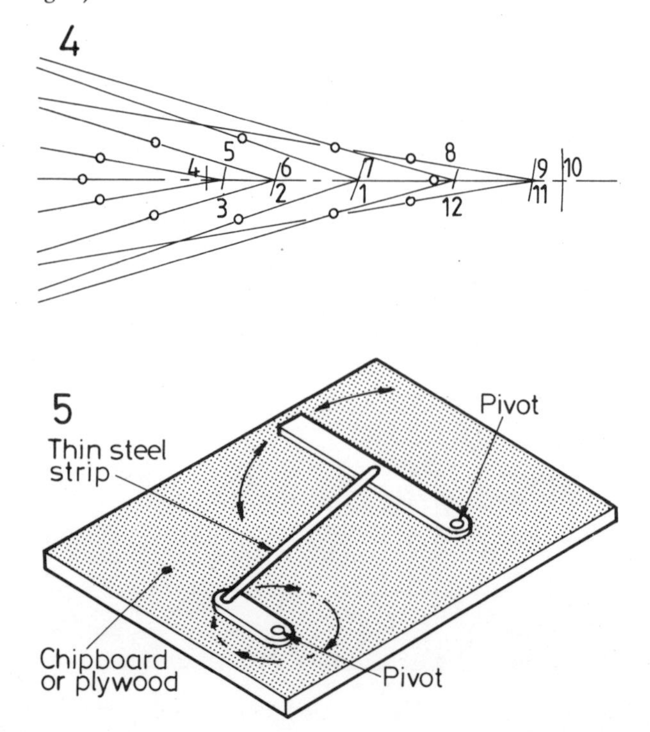

Drawing 5 is a model of one of the link mechanisms by which the windscreen wipers of a vehicle could be driven.

Drawing 6 is a diagrammatic drawing of a crank AC which rotates about a fixed point C. A slider S can only move along a slot LM. A link AP is connected to crank AC by a pivot at A and a pivot at S. As AC rotates find the locus of the end P of AP.

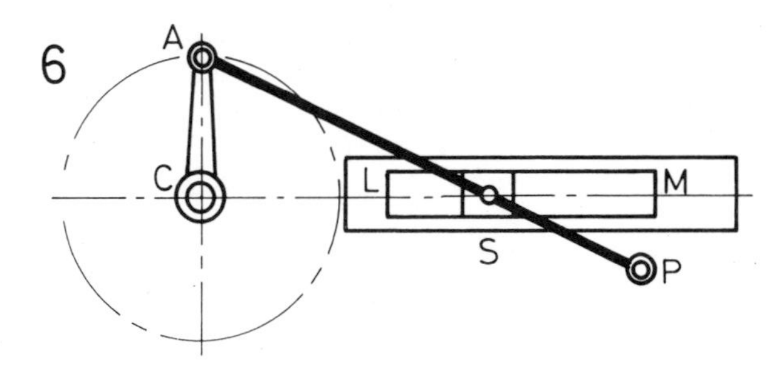

Drawing 7 shows the required locus and the construction for finding the points 9 and 11 of the locus.

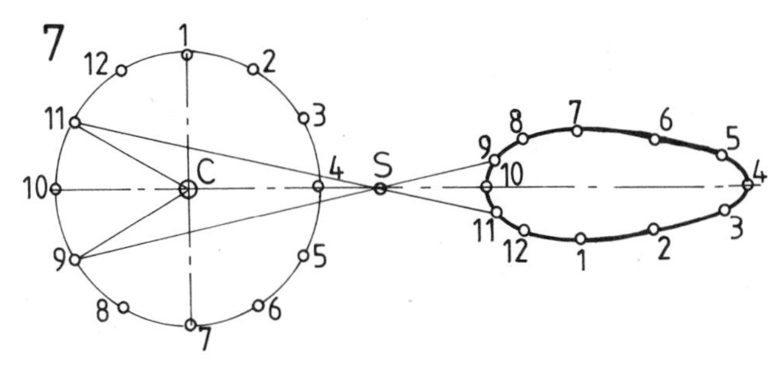

Locus of pivot P

A photograph of a link mechanism from a small draughting machine is shown. Name other link mechanisms you have seen and try to explain how they work.

The link system of a small draughting machine

Screws

The screw is the last elementary machine with which we shall deal. Screws are based on the *helix* (see page 46). Only two examples of screw forms are shown in photographs. The first is an aluminium casting and its wooden pattern, made as part of a pupil's technology project. The second is a connector from a hydraulics circuit. Using the hydraulics circuit connector as an example, its upper screw thread is of a diameter of 20 mm with a pitch of 2 mm. Thus for each complete turn of the screw that part of the connector will move forward 2 mm.

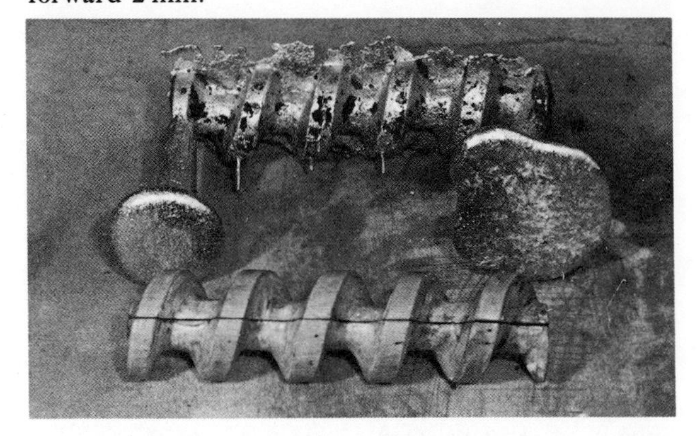

A casting in aluminium and its wooden pattern

A screwed connector from a hydraulics circuit

If a spanner of length 200 mm is used to tighten the connector the following forces are acting. If a force of 20 N is applied to the end of the spanner, the work applied to the spanner in making one complete revolution of the connector nut will be:

$$2\pi r \times 20 = 3.14 \times 2 \times 0.2\ \text{m} \times 20\ \text{N}$$
$$= \text{approx. } 25 \text{ Newton metres.}$$

Now, 1 Newton metre (Nm) is equal to 1 doule, which is the unit of work. Therefore the work done in turning the spanner through one complete revolution is 25 Joules.

$$25\ \text{Joules} \div 0.002\ \text{m} = 25\ \text{Nm} \div 0.002\ \text{m} = 12500\ \text{N}$$

Which makes the screw a very efficient machine!

Vectors

Triangle of forces

Drawing 1 – A simple shelf bracket must be strong enough to resist the forces acting on it by the mass of the articles placed on the shelf. These forces can be shown diagrammatically as in *drawing 2*. The shelf and its bracket are not moving and the forces acting in the system are said to be in *equilibrium*.

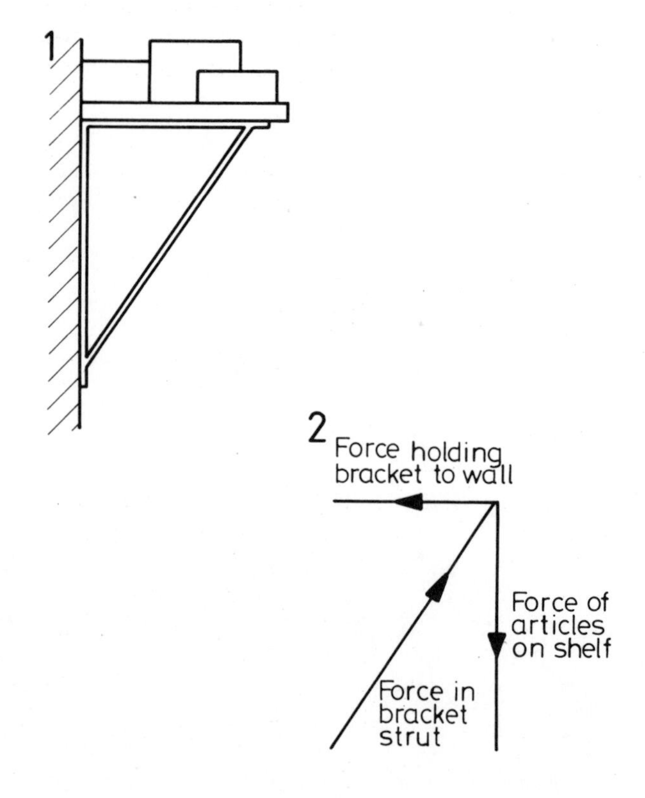

Drawing 3 – Three forces in equilibrium can be represented in a diagram by a triangle of forces in which the three forces A, B and C are drawn to a scale of length in newtons and the scaled lines are drawn at the angle in the direction in which the forces are acting.

Assuming that the mass of articles on the shelf exerts a force of 3 kg (30 N) vertically downwards, to a scale of 1 mm ≡ 1 N force A is represented by a line 30 mm long. Draw lines representing forces B and C parallel to the arm and strut of the bracket and a triangle of forces for the bracket is complete. Forces B and C can then be measured on the scale of 1 mm ≡ 1 N.

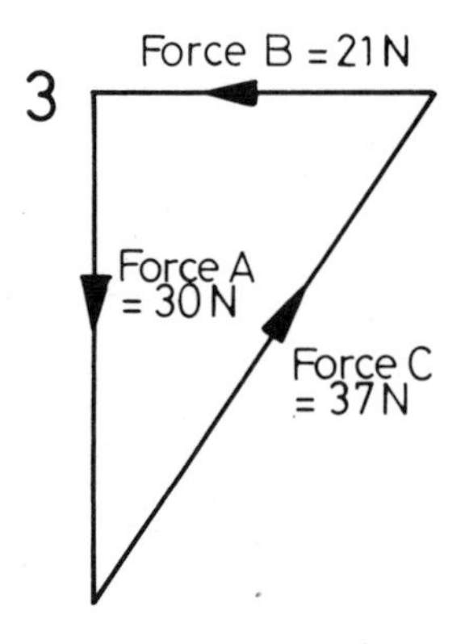

Triangle of forces

Note in drawing 3:

1. The three forces act in either a clockwise or an anti-clockwise direction around the triangle.
2. The triangle is complete – all three lines meet at the triangle vertices.
3. The lines in the diagram are *vectors* – they represent, to a scale, the magnitude *and* the direction in which the forces are acting.

EXERCISES

1. If the force A in the bracket is 50 N and the support strut is at an angle of 40° to the vertical, construct a triangle of forces for the bracket and from it find, by measurement, the other two forces acting along the members of the bracket.

2. Draw a triangle of forces diagram for the structure shown in drawing 4. From your drawing find the forces and their direction acting along the members of the structure. Measure the angles of drawing 4 with a protractor.

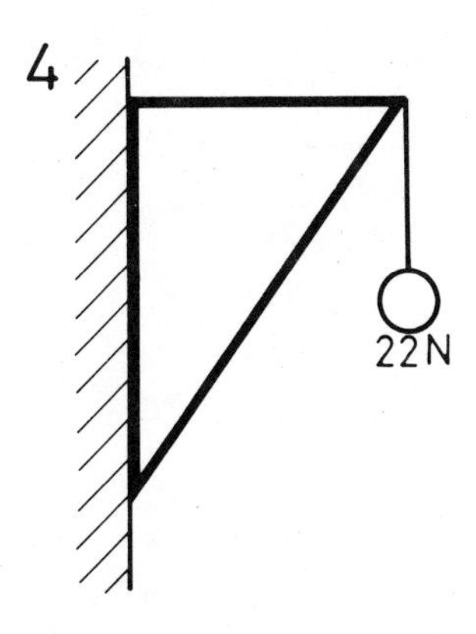

Forces in equilibrium

Drawing 1 – If a point P is held in equilibrium by three forces in the same plane acting on it, a vector diagram can be drawn representing the three forces as a triangle. To assist in the drawing of force vector diagrams a system of lettering known as Bow's notation is employed. In Bow's notation, the spaces between the lines of forces are given capital letters – A, B, C etc. The force between spaces A and B is then known as force AB and so on. In the vector diagram (drawing 2), the vertices are lettered with lower case letters – a, b, c etc. Thus, in a vector diagram, vector ab represents force AB.

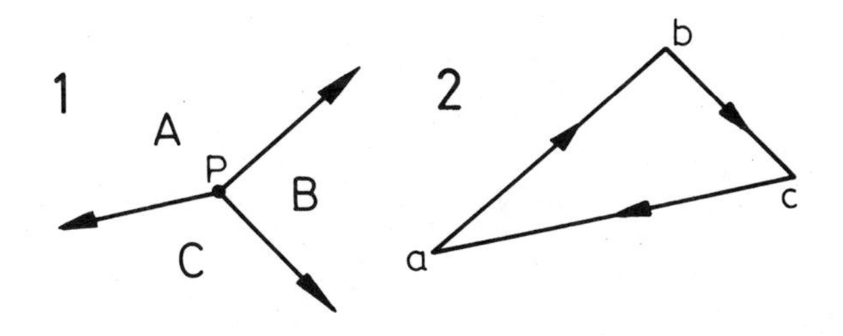

Example

Drawing 3 – A point P is held in equilibrium by a force of 20 N and two other forces acting in the directions shown. Draw the triangle of forces for this system and from your drawing calculate the size and direction of the other two forces.

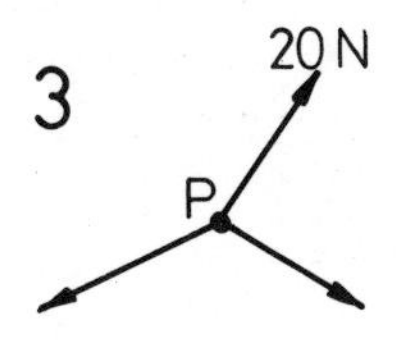

1. *Drawing 4* – Letter the drawing in Bow's notation.

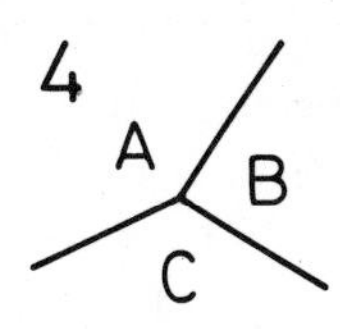

2. *Drawing 5* – Draw the triangle of forces. Letter the vertices.

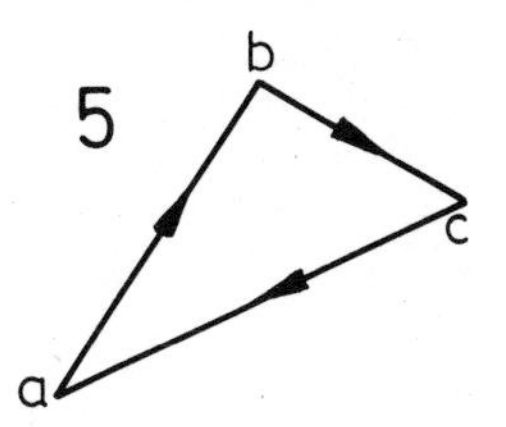

3. Measure the lengths of the two vectors bc and ca to find the values required. The directions of action of the two forces are given by the angles at which lines bc and ca are drawn.

EXERCISES

1. *Drawing 6* – A ring R is held stationary by three forces acting upon it as shown. Draw a vector diagram for the system and calculate the values for the other two forces.

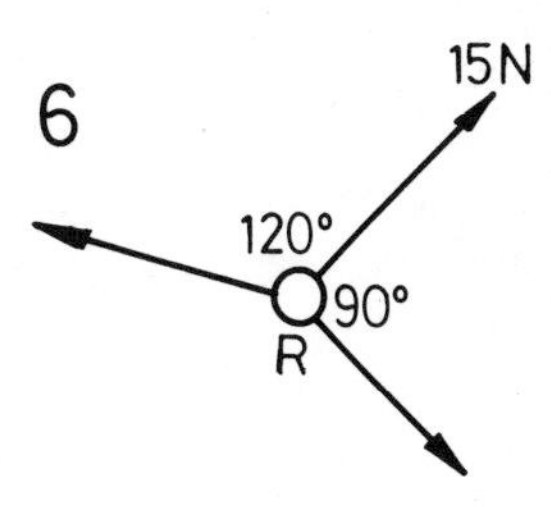

2. *Drawing 7* – If the point O is held in equilibrium, find the value of the other two forces.

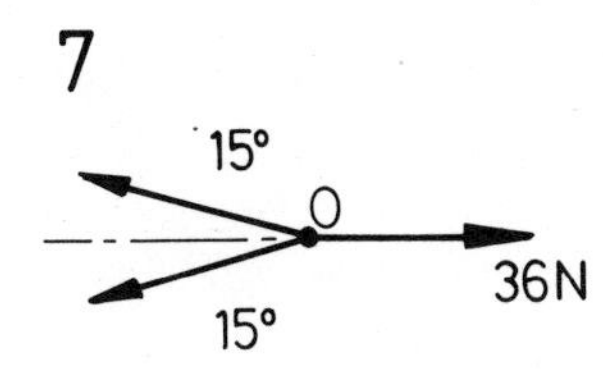

Resultants

If a point is acted upon by two forces, a single force can be employed to replace the two given forces.
Drawing 1 – The point P is acted upon by the two forces as shown. The two forces can be replaced by another single force R, which will have the same effect on P as the other two forces. This force R is the *resultant* of the other two. To calculate the value and direction of a resultant, a vector diagram can be drawn.

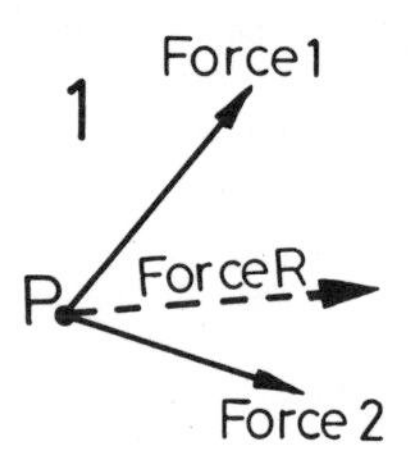

Example

Drawing 2 – The point P is acted upon by two forces of 20 N and 30 N as shown. Find the size and direction of the single force which could replace the two given.

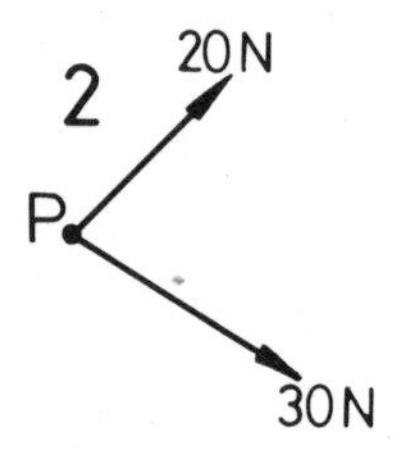

Drawing 3 – To a scale of 2 mm ≡ 1 N, draw ab parallel to the 20 N force and bc parallel to the 30 N force. Join ac and measure its length. The line ac is the resultant of the 20 N and 30 N forces.

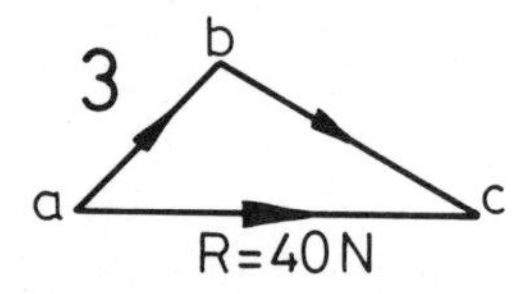

Note in drawing 3:
1. The two known forces are both acting in a clockwise or in an anti-clockwise direction around the vector diagram.
2. The resultant acts in the *opposite* direction to that of the other two forces.

EXERCISES

1. *Drawing 4* – A point P is acted upon by two forces as shown. Find the resultant of these two forces in both size and direction.

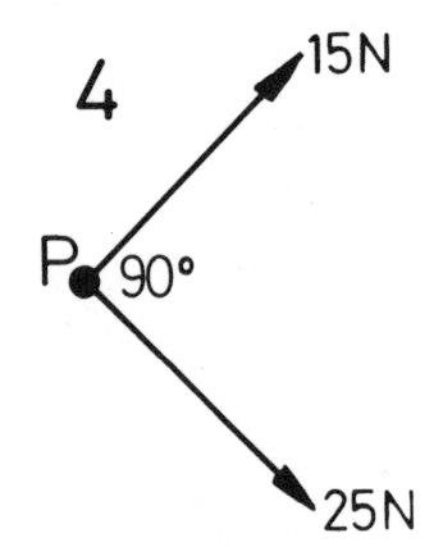

2. *Drawing 5* – A block B can just be moved along a surface by the two forces shown. Find a single force needed to move the block. In which direction will the block be moved under the action of the resultant force?

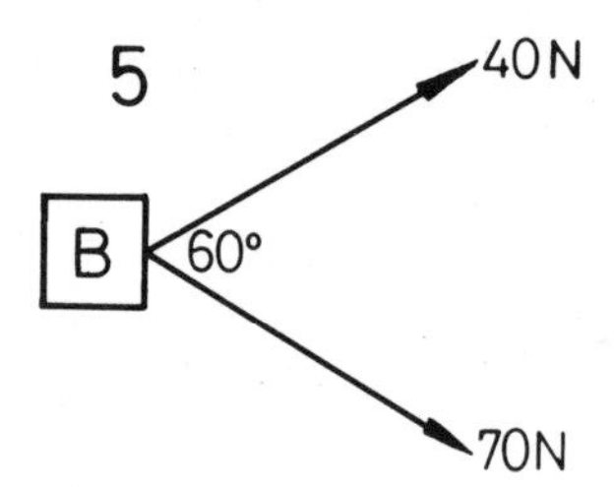

Polygon of forces

Any straight-sided polygon can be regarded as a series of triangles (*drawing 1*). Because of this, the theory of the triangle of forces can be applied to systems in which more than three forces are acting.

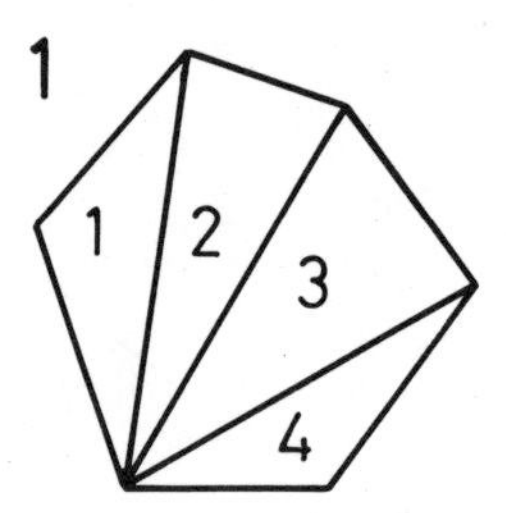

Hexagon = 4Δs

Example

Drawing 2 – The directions of action of four forces required to maintain P in equilibrium are given. Only two of the forces are known. Calculate the size of the other two forces.

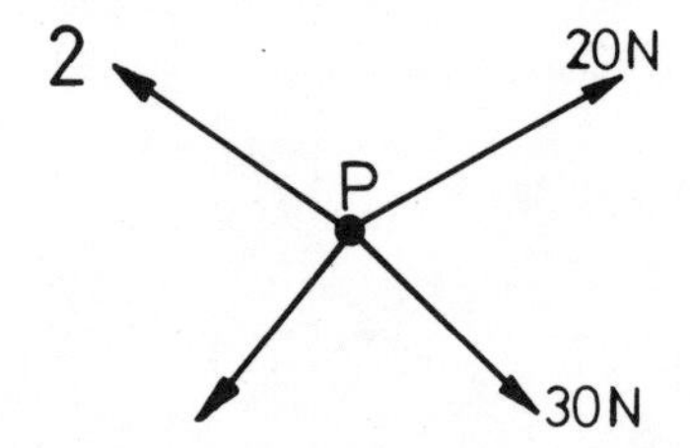

Drawing 3 – Draw the given system and add the Bow's notation lettering as shown.

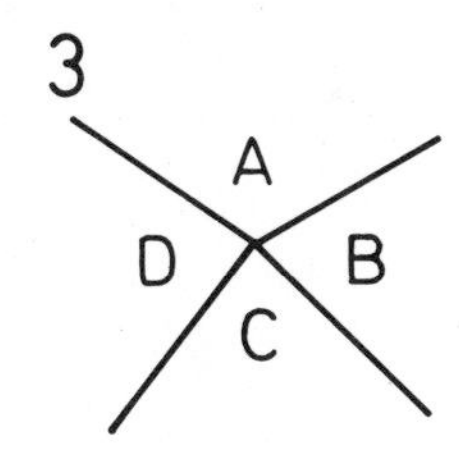

Drawing 4 – To a scale of 2 mm ≡ 1 N draw ab and bc parallel to the forces AB and BC. From c draw cd parallel to force CD. Join a and d. This completes the quadrilateral of forces abcd. Measure the two vectors cd and ad to find the required values.

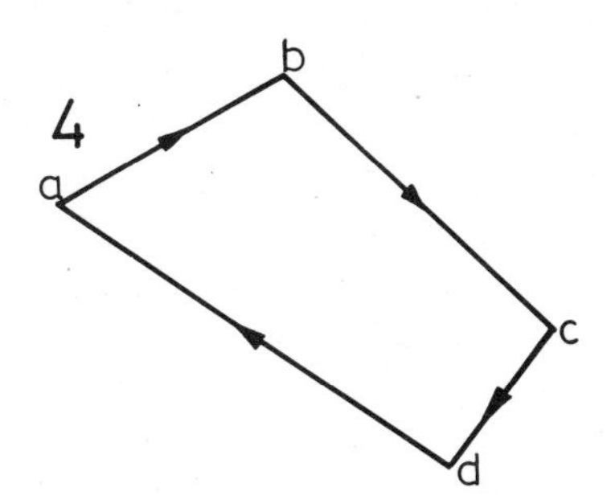

EXERCISES

1. *Drawing 5* – The point P is held in equilibrium by four forces of which three are given. By drawing a vector diagram find the size and direction of the fourth force.

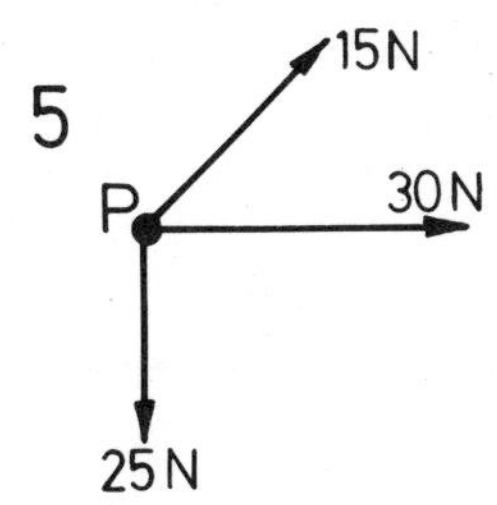

2. *Drawing 6* – Find the resultant of the three forces acting on the block B. In which direction will the system move if the resultant replaces the three given forces?

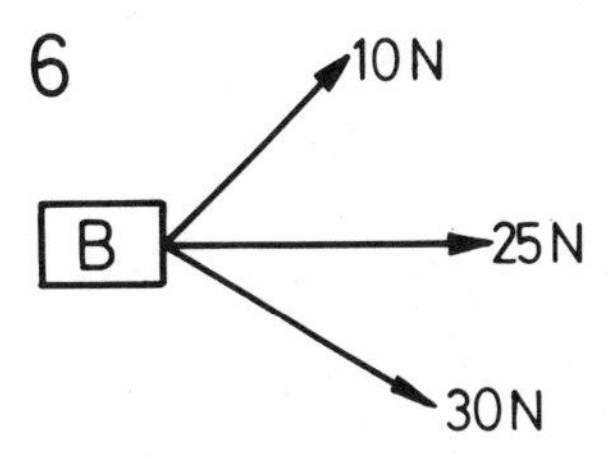

Circuit diagrams

When designing a technology project you may find it necessary to draw diagrams involving electrical, electronics, pneumatics or hydraulics circuits. The symbols representing components in electrical and electronics circuit diagrams are specified in BS 3939 (see page 33). Symbols for the components found in pneumatics and hydraulics circuits are specified in BS 2917 (some examples are given on page 154). Only very elementary types of circuits are shown in this book to demonstrate the methods of drawing such circuit diagrams.

Circuit diagram for a torch

One of the simplest of electrical circuits is that for a hand-held torch. The drawing shows the parts making up the electric circuit of a torch. Within an outer insulating casing made from plastics, the parts are: two 1.5 volt cells forming a 3 volt battery; a 2.5 volt light bulb; strips of metal and wire which act as conductors carrying the electric current between the battery and the light bulb.

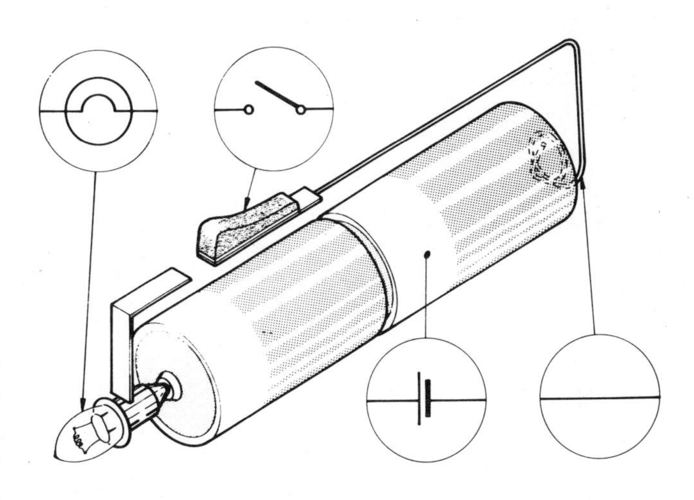

The various parts involved in the electrical circuit of a torch

A separate diagram shows the circuit for the torch using the British Standard symbols for electrical circuits.

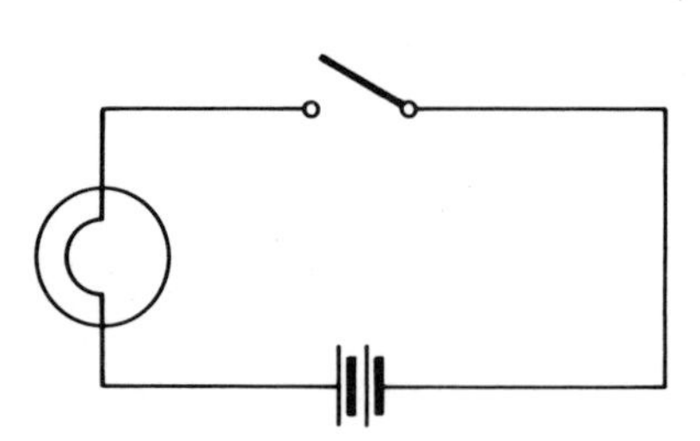

Circuit diagram of the torch

Note the following in this circuit

1. The four parts of the circuit are drawn as symbols – battery, conductors, light bulb and switch.
2. The parts of the circuit are connected by the conductors, which are drawn as neat straight lines.
3. The switch is in its 'open' position. When closed, the circuit conductors form a continuous line, making the circuit 'live'.

The parts of a simple bell push circuit and its circuit diagram are also shown. In this circuit, when the push switch is pushed down, the bell rings and the bulb lights up. When the push switch is released, the light goes out and the bell stops ringing.

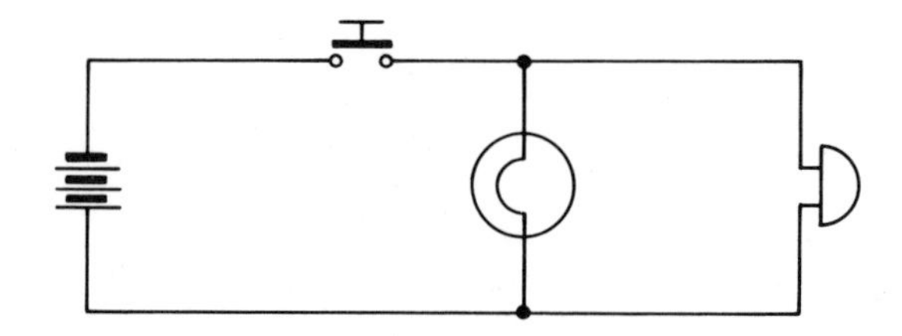

The circuit for the bell system

Drawing electrical circuits

The two simple circuits already given were drawn with the aid of drawing instruments – ruler, T-square, set square and compasses. When drawing more complex circuits containing a large number of components, you may find it necessary first to plan the drawing. This can be carried out on graph paper or on square grid paper, or you may find it easier to plan your circuit by drawing freehand on plain paper. Circle drawing aids will speed up the production of a circuit diagram – for drawing lamps, motors, sockets and the like.

Two circuit diagrams of a switching circuit for controlling an electric motor are shown. The first drawn on graph paper and the second drawn freehand on plain paper. In this circuit the power is provided from a 12 volt alternating current (a.c.) and the circuit is in three parts. A signal light will indicate that the circuit control switch has been made. A push switch controls the running of the motor. If a light is needed by the person operating the machine, a third switch controls a light which will illuminate the work being machined.

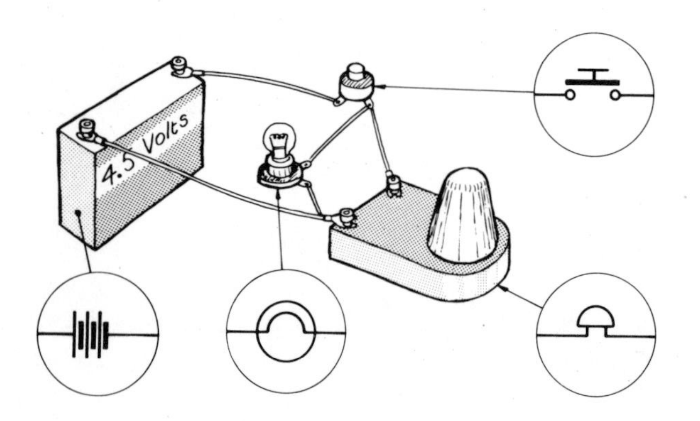

The parts of a simple bell system

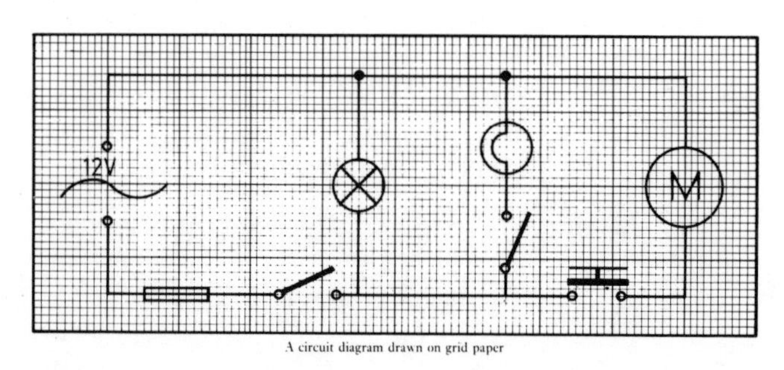

A circuit diagram drawing on grid paper

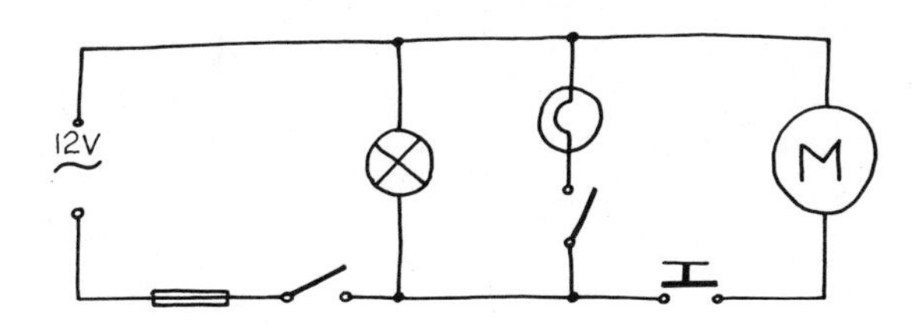

A circuit diagram drawn freehand

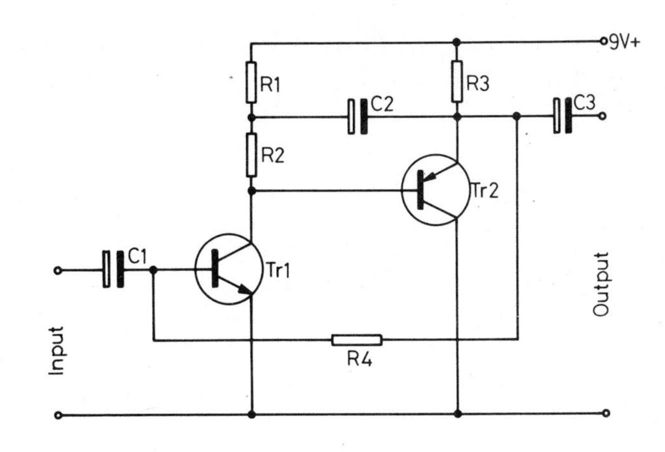

Circuit diagram of the transistorised amplifier

Electronics circuit diagrams

A circuit diagram of a type occasionally seen in books, magazines and instruction manuals is shown. This shows in diagrammatic form an amplifying circuit which includes two transistors. This type of drawing is of value for showing the actual appearance of a circuit when it has been assembled, although it is not often that you will see an actual circuit made up in such an exact rectangular shape. The drawing is a 30°, 60° planometric of the circuit showing the components, not as symbols but drawn approximately to the shapes as they would actually appear. The drawing could equally well have been made up as an isometric or perspective drawing. The British Standard symbols for each component have been added as labels in 'balloons' with leaders pointing to the components which they represent.

A circuit diagram for this transistorised amplifier is also shown. The components have been labelled – C for condenser, R for resistor and Tr for transistor. A list giving the values for each of the components could be added to the drawing if desired. This circuit could be used to amplify the sound of a person speaking into a microphone.

A third electronics circuit diagram shows a simple transistor radio circuit. This circuit is in four parts:

1. an aerial *tuned* to receive a radio frequency by a variable capacitor;
2. the audio frequency 'carried' by the radio wave is *detected* by a diode and capacitor system;
3. the resulting audio signals are *amplified* by the transistor circuit;
4. the resulting amplified audio frequency is fed to an *earphone*.

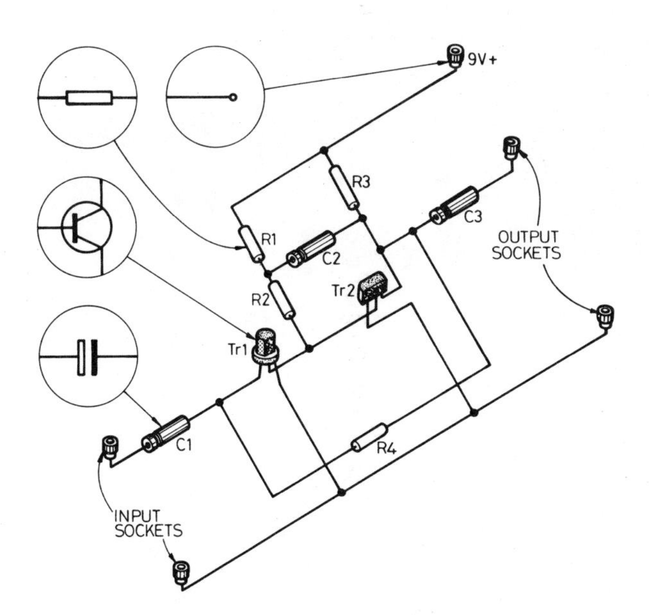

Pictorial diagram of a transistorised amplifier in a planometric drawing

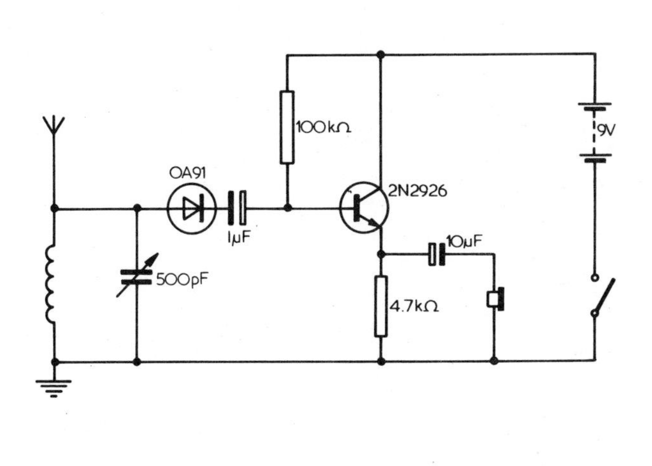

Circuit diagram of a simple transister radio

Symbols for pneumatics and hydraulics circuits

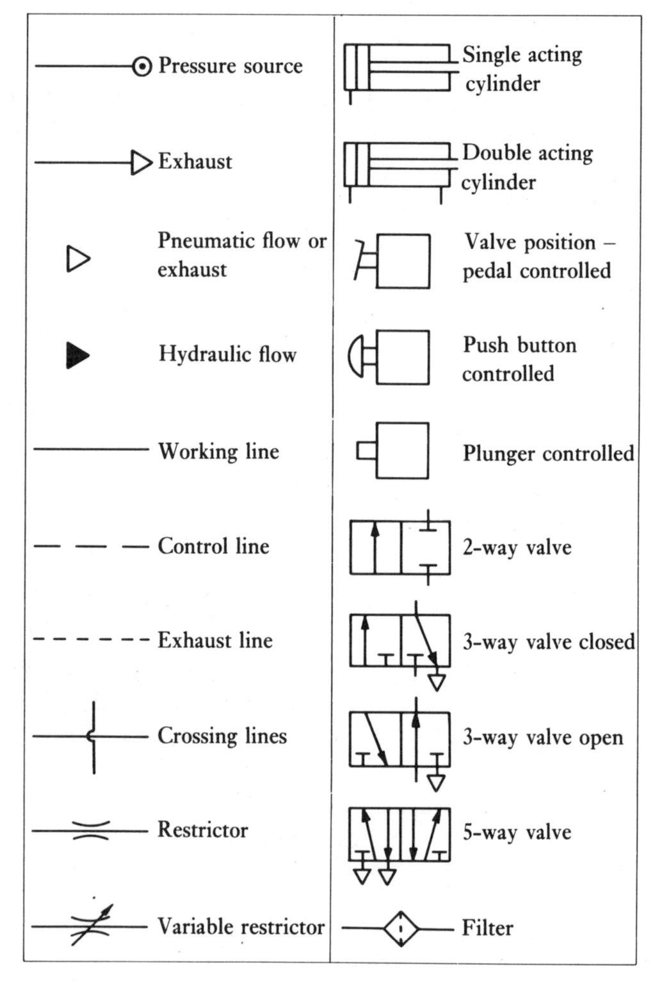

A selection of pneumatics and hydraulics circuit diagram symbols taken from BS 2917: *Specification for graphical symbols used on diagrams for fluid power systems and components* is given in a table. A simple pneumatics circuit diagram is also given in which three valves operate a double acting cylinder.

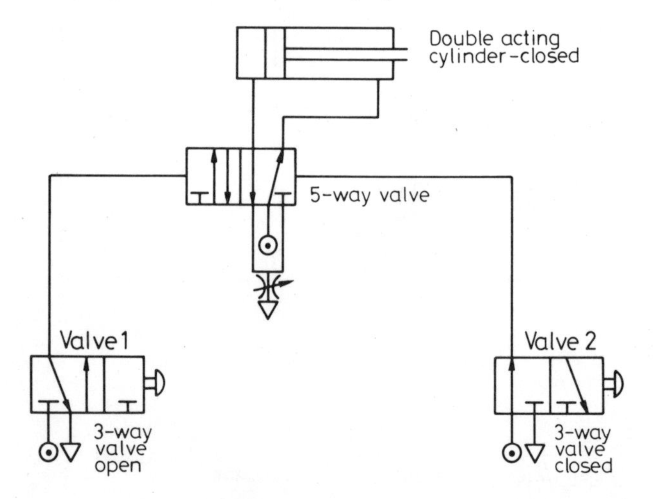

With Valve 1 in the open position the piston within the cylinder is as shown in the diagram. Press the push button of Valve 1 and the piston moves to the right. Then press the push button on Valve 2 and the piston moves back to its original position.

Although not shown in the given circuit diagram, each cylinder in a circuit would normally be given a letter – A, B, C etc. Against the diagram a list would usually be made with + and – signs denoting whether a piston is in a closed position (as in the diagram) or an open position. The list should be made up in an order which gives the sequence of operations in the circuit when valves are operated.

Part of the arm of a robot worked by a hydraulics circuit

Building drawing

Any construction projects, such as the building of a bungalow or house, a block of flats, a factory, a motorway, a school or any other type of building work, will have its own set of drawings, known as a project set. Each project set will include some or all of the following types of drawings – block plans, site location drawings, site plans, building or general arrangement drawings, contract drawings, fabrication drawings.

There is no need in this book to describe all these different types of construction drawings. From the list given above, four different types of drawing are included here. These are:

1. a site location drawing;
2. a site plan;
3. three drawings showing two types of general arrangement drawings;
4. a construction drawing.

In addition, two-point perspective drawings of a building and a room interior are included, together with a plan of a garden layout. These may prove of value to pupils and students who are designing a graphics project connected with building drawing.

Building drawing symbols

A number of symbols representing parts of buildings or components in buildings are given in a table. These have been taken from BS 1192: *Construction drawing practice.*

Cavity wall
Partition
Window
Door
North
Existing tree
New tree
Stair
B Basin
R Refrigerator
C Cooker
S Sink
Bath
RWP Rain water pipe
MH Man hole
Drainage lines
Building outline
Dimension
Dimension (modular)

Types of drawings

Site location drawing

This type of drawing is made to enable the exact location of a construction project to be defined in relation to its surroundings. A site location drawing will commonly be drawn to a scale of 1:1250 and will include details such as roads and streets in the area, together with their names; field boundaries; field numbers taken from a large scale Ordnance Survey map; the outlines of other buildings in the area; the direction of North. The main features of the drawing will be the outline plan of the proposed building set within the boundaries of the site. The given site location drawing is for Number 12, Queen's Close.

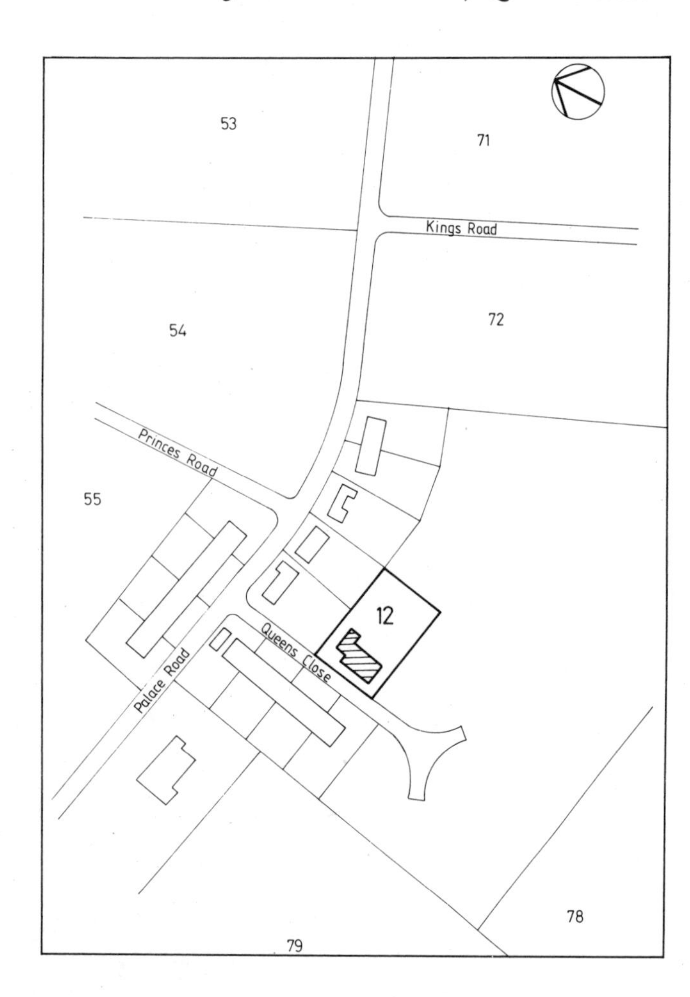

Note the following.

1. The outline plan of the house is drawn with thick lines.
2. The boundary of the house site is drawing with thick lines.
3. All other details have been drawn with thin lines.

Site plan

This type of drawing is made to show the exact position of a building within its site boundaries. A scale commonly used for site plans is 1:200. Note the following in the given site plan.

1. The outline plan of the building is drawn with thick lines.
2. All other lines in the drawing are thin.
3. The dimensions give the exact position the building is to occupy within its site.
4. Drainage systems are shown.
5. North is indicated.
6. The road adjoining the site is shown.
7. Paths and garage runway have been included.
8. The original drawing from which the given drawing was copied was made on an A4 sheet of paper.

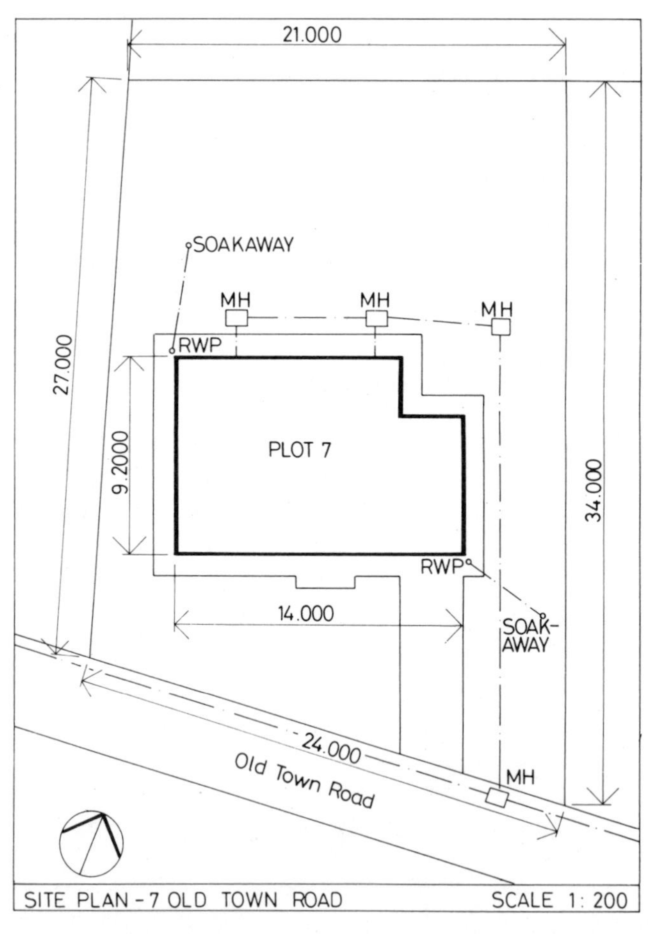

A building site plan

General arrangement drawings

Two types of general arrangement drawings are given. These show the general arrangement of the whole building as seen from the outside in orthographic projections and a plan of the building showing the general arrangement of the interior – rooms, stairs, cupboards, windows, doors etc.

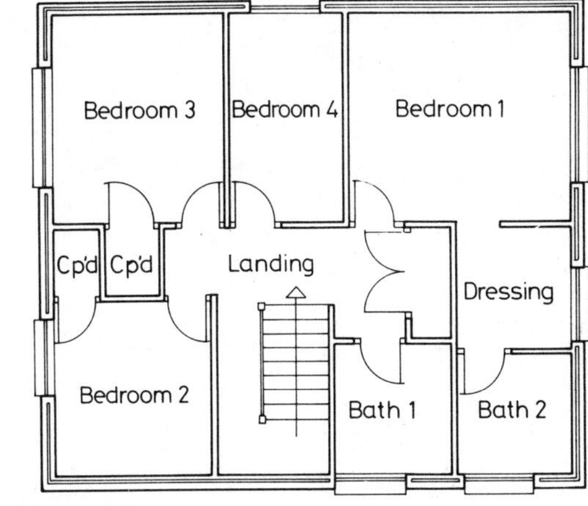

SECOND FLOOR

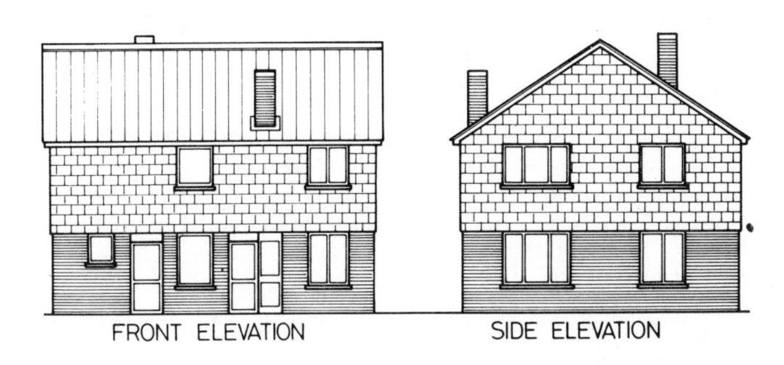

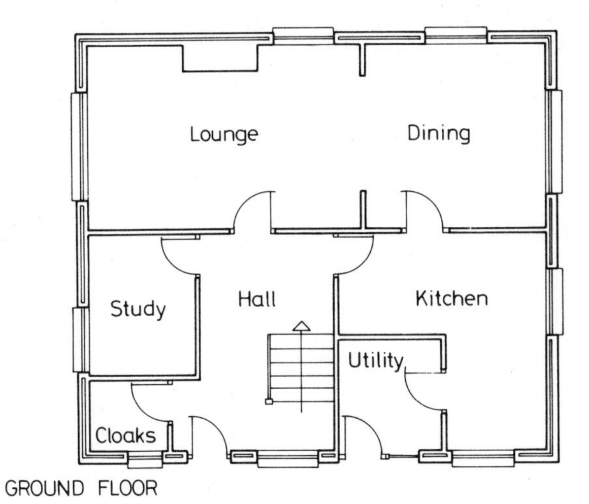

GROUND FLOOR

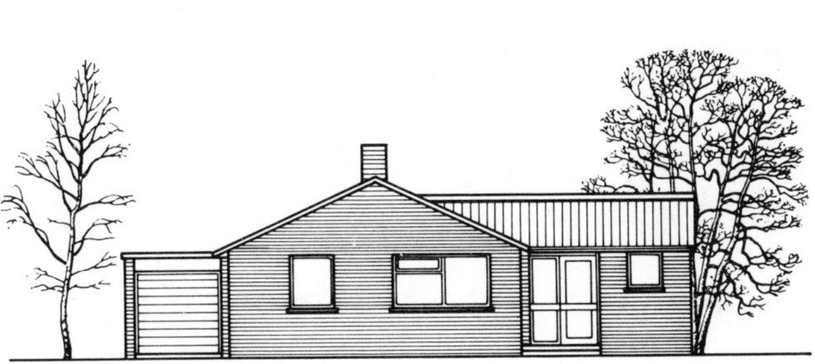
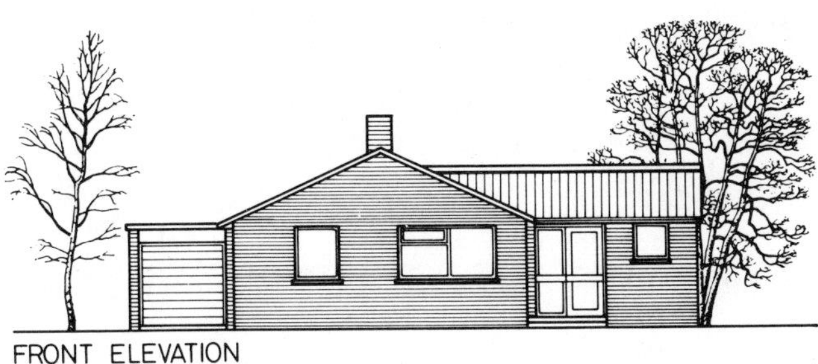
FRONT ELEVATION

The two drawings of the outside of buildings show front and side elevations of a two storey house and a front elevation of a bungalow. Note the following details in these two drawings.

1. Common practice in building drawings of this type is to draw an end view in Third Angle projection. If a plan is included it is drawn in First Angle projection.
2. Views are labelled – FRONT ELEVATION and SIDE ELEVATION.
3. Features such as tiling, brickwork, windows and doors are included.
4. The drawing of the FRONT ELEVATION of a bungalow includes trees, which have been added from Letraset dry transfer sheets (see page 36).
5. A scale commonly used for this type of drawing is 1:20.

The third general arrangement drawing shows the layout of spaces within a two storey house, Common practice is to label each room or space showing its intended use. Some drawings will include room sizes (e.g. Bedroom 1–4 m by 3.5 m).

Construction drawing

A sectional view through a bungalow is given as an example of a construction drawing. This includes some of the information necessary when the building is being constructed. Note the following in this drawing.

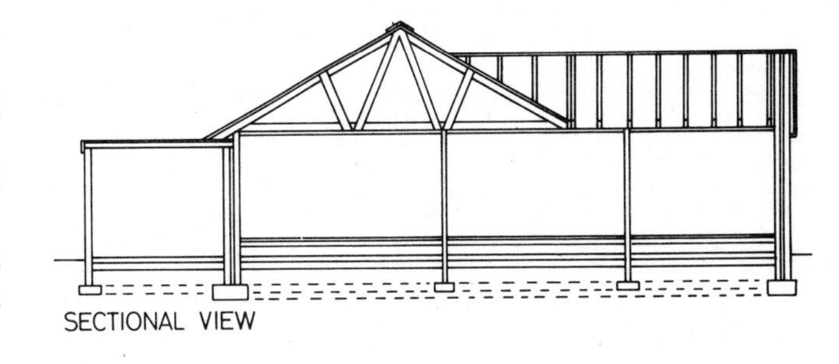
SECTIONAL VIEW

1. The position of 'footings'; the foundation 'raft' of concrete on which the bungalow will be built; walls and roof structures are clearly defined.
2. Notes would normally be added to a drawing of this type, of details such as joist sizes, dampproof course, type of roofing, type of flooring.

Many other construction drawings will be required when building such a bungalow. These will give construction details of features such as windows, doors, any cupboards, any fittings such as those needed when bathrooms and toilets are fitted out. Various scales are employed in construction drawings depending upon the size of the feature being defined. The given sectional view was drawn to a scale of 1:50.

Two-point estimated perspective drawings

The method of estimated perspective drawing is a good one for producing pictorial drawings of buildings or their interiors. The perspective may be based on two-point or single-point perspective. Both of the examples given here show two-point perspective, but single-point may well be suitable, particularly when drawing room interiors. The example given here show a large house which is intended for conversion into six flats and part of the interior of a room in one of the flats.

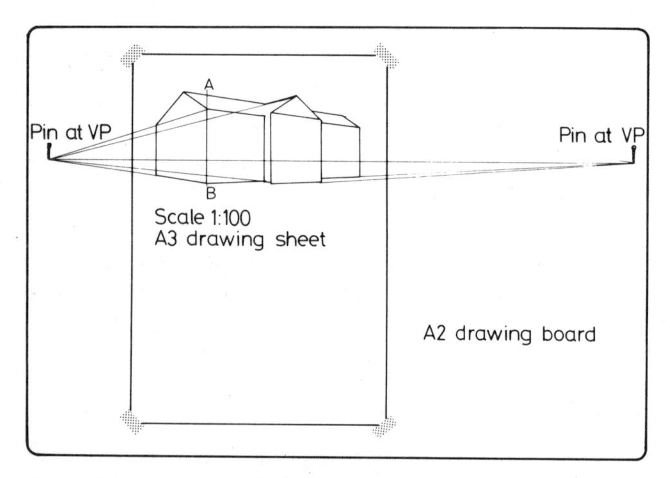

Preparation for estimated two-point perspective drawing of a building

Two-point estimated perspective drawing of a building

The drawing of the building was started as shown. A sheet of A3 drawing paper was taped in position on an A2 size drawing board. Two pins were inserted in the board in line with each other almost at the edges of the board. These gave two vanishing points. The vertical line representing the corner of the building nearest to the observer was then drawn – line AB. The drawing was made to an approximate scale of 1:100. The point B was taken at a scaled 1.7 metres below the line joining the two vanishing points. The drawing could then be constructed by taking all vertical measurements, scaled along AB, and drawing lines from the points so obtained to the two VPs. Sizes along the front and end of the building were approximated, remembering that the space on the drawing became narrower, the further they are away from the vertical line AB. Details of doors, windows, tilework, brickwork and a grassed area were added to the outline of the construction. The two pins at the VPs were used as pivots around which a straightedge could be manipulated when drawing lines to the VPs on the drawing.

Figures of people and the trees and shrubs were added to the finished sheet drawing from Letraset dry transfer sheets.

Estimated two-point perspective of a building

Two-point estimated perspective drawing of a room interior

This drawing was constructed in a manner similar to that for drawing the outside of the building. The major difference was that a larger scale was used – 1:20 instead of 1:100. It must be remembered, however, that scales used in perspective drawings of this kind can only be regarded as being approximate.

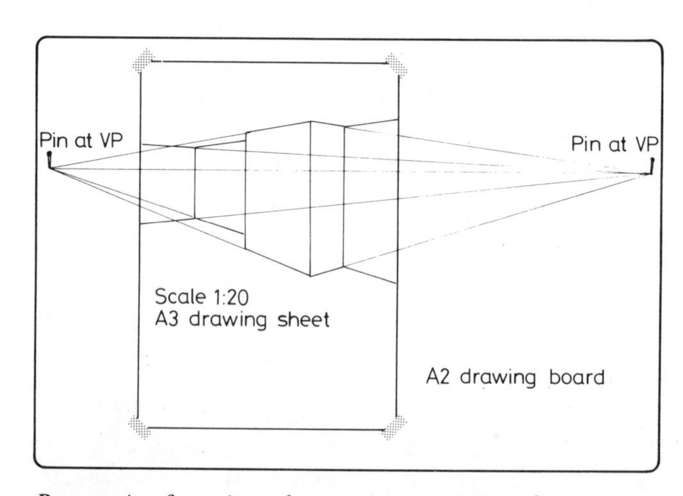

Preparation for estimated two-point perspective of a room interior

In the finished drawing, details include the stonework of the fireplace and appropriate shading, grain shading on the ledged and braced door, lines indicating the floor boards and a few pots of plants added to the drawing sheet from Letraset dry transfer sheets.

Estimated two-point perspective of a room interior

Garden layout plan

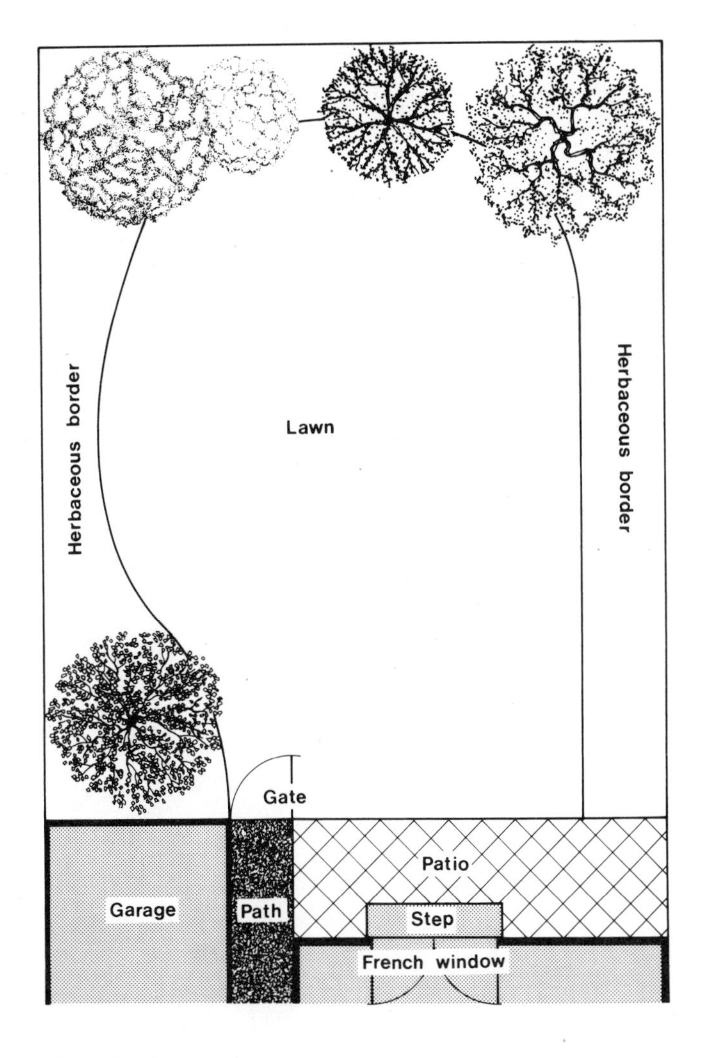

A garden layout plan

A garden layout plan can hardly be called a building drawing. This form of graphics is drawn by those who plan and design garden layouts. There is no laid down series of conventional methods of depicting features in a garden plan, although some suitable Letraset architectural dry transfer sheets can be obtained. The plans of trees and shrubs in the given drawing were obtained from dry transfer sheet. The layout for the garden was drawn to a scale of 1:20.

EXERCISES

1. An outline sketch of a small garden tool shed is given. The shed is to be made from wooden weather boards nailed to a framework. Construct the following drawings of the shed, working to any scale you consider suitable.

(a) Draw a front elevation, an end elevation and a plan which includes:

(i) details of a door;
(ii) details of a window;
(iii) the lines of the weather boarding;
(iv) details (in notes) about the roof.

(b) Construct a two-point estimated perspective drawing of the shed, with the details given in your answer to (a).

2. An outline sketch of a 'lean-to' garage to be built against the brick wall of a house is given. Choosing your own design and sizes:

(a) Draw a plan of the house and garage to include:

(i) a driveway to the garage door;
(ii) details of the garage doors and windows;
(iii) details of the garage roof.

(b) Construct a two-point estimated perspective drawing of the garage in position against the house wall.

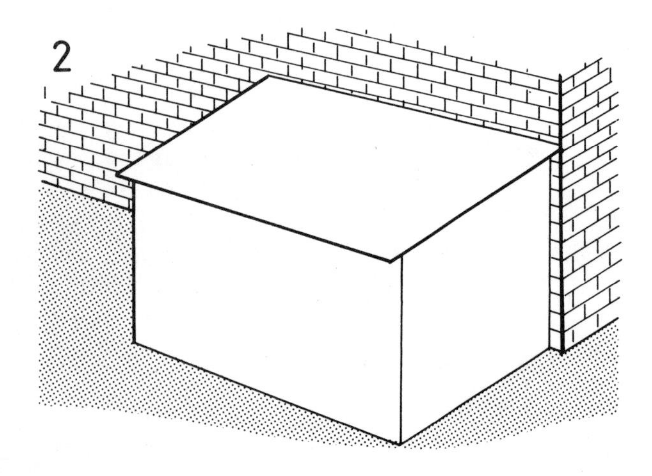

3. A plan of a bathroom on a 1 metre square grid is given. Re-draw the plan but with the bath, sink and W.C. in more suitable positions.

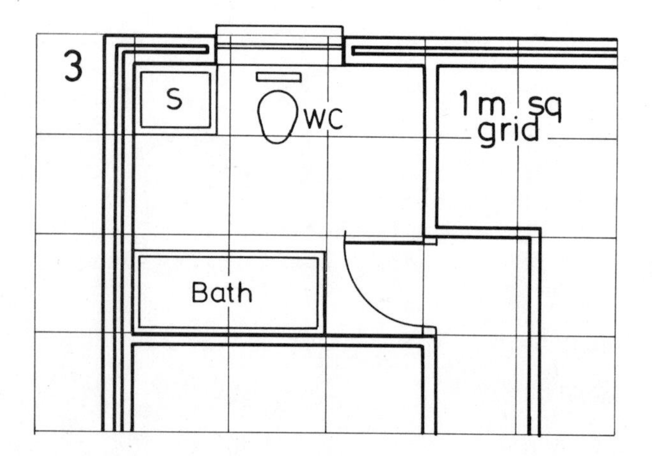

4. Draw a plan on a grid of 1 m squares (as in drawing 3) of your own bedroom; of the living room in your home; of any other room in your house.

5. Draw a plan of the school room you are in at the moment. Your drawing should be to a suitable scale and include the furniture and fittings, windows and doors of the room.

Draw a second plan in which you have re-designed the position of the fittings and furniture to allow more space to be available for each pupil.

6. Measure the site on which your own home is built and the outside dimensions of the building on the site. Construct an accurate scale 1:200 site plan of the building on its site.

7. Drawing 4 is a site plan for 4 Northwood Way. Copy the drawing to a scale of 1:200. Dimensions not given can be estimated. Within the rear garden area of your site plan, design and draw a garden layout. State which areas will be paved, grassed as lawns or made into flower borders and where trees or shrubs will be planted.

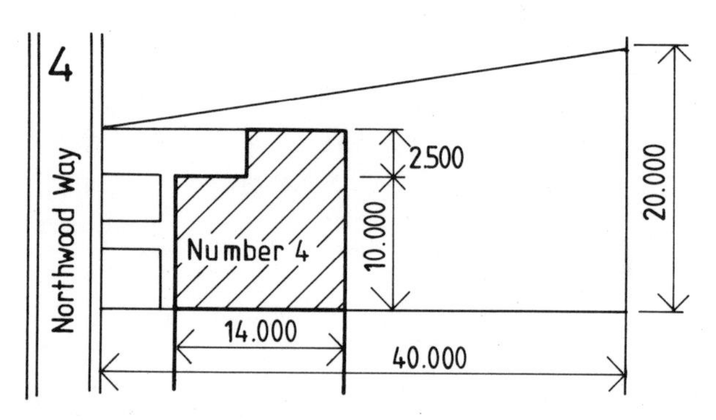

8. Drawing 5 is a plan of a corner of a bedroom drawn on a 1 metre square grid. Construct a two-point estimated perspective drawing of the bed, chests of drawers, wardrobe and corner fitment as if you were standing at about the spot S. Select your own dimensions for sizes which cannot be taken from the grid.

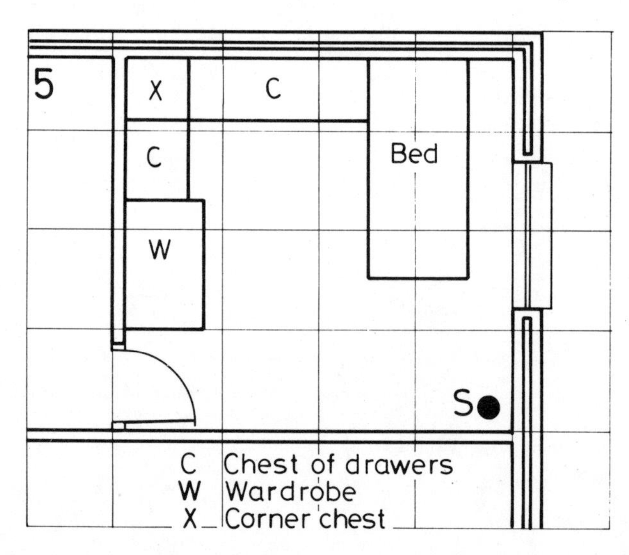

9. This exercise has been copied from a GCSE question as set in a CDT:Design and Communication specimen examination paper.

A single-storey extension is to be built on the rear of a house similar to that shown in drawing 6a. The extension is to comprise a room to be used as a studio/workshop and a porch.

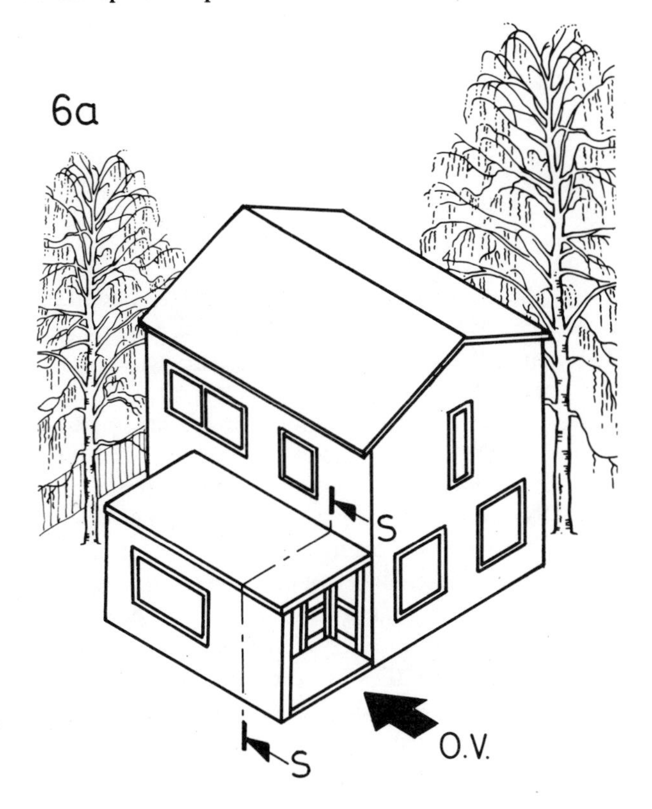

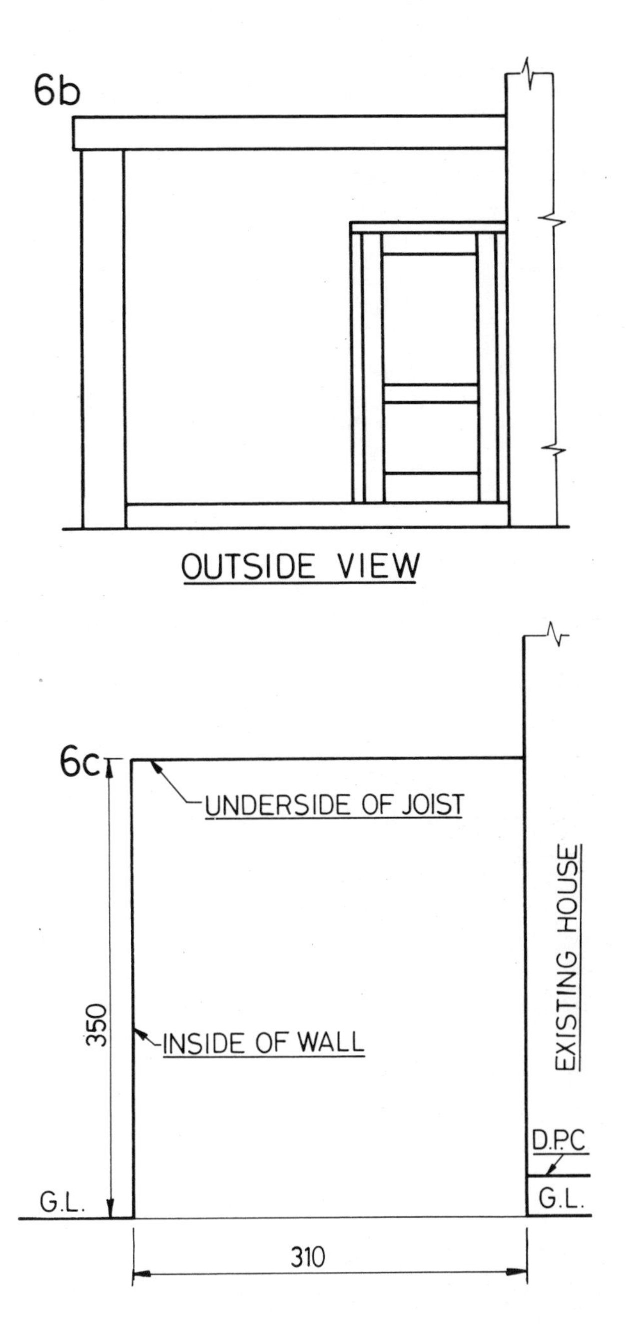

The general specification of the extension is to be as follows:

1. brick-built cavity walls;
2. flat roof;
3. solid floor;
4. panelled door to the pattern shown in drawing 6b.

Make a scale 1:1 copy of drawing 6c. Complete your drawing as a sectional elevation of the porch of the extension indicated by the section plane SS in drawing 6a. The position of the existing house, the ground line, the inside of the wall and the underside of the ceiling are given. Assume that the section line passes between two of the joints.

Your drawings must include details both above and below ground level.

Welsh Joint Education Committee

Computers

The Central Processing Unit (CPU) of a computer functions through the micro-electronics circuits contained in its silicon chips. These convert all input signals (drawing 1) into pulses of current ON or OFF

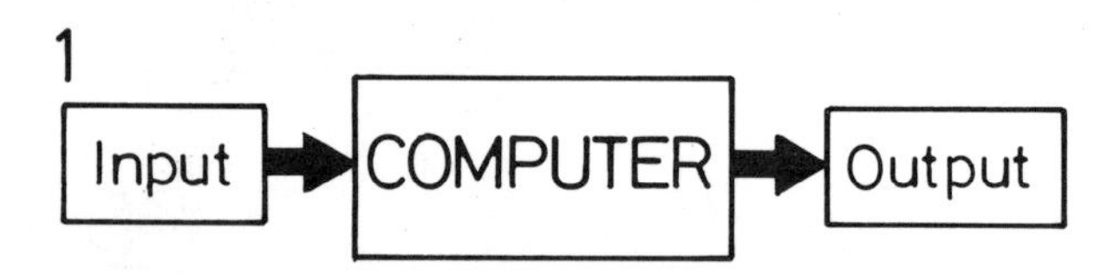

which can be represented by the mathematical system known as binary. The two digits (hence *digital* computer) of binary are 1 and 0. When a micro-electronics circuit passes a pulse of current this can be represented as in drawing 2. When no current is passed this can be represented as in drawing 3. A pulse of current can be

represented by the figure 1. No pulse can be represented by 0. A flow of ONs and OFFs can be represented diagrammatically and in binary as in drawing 4.

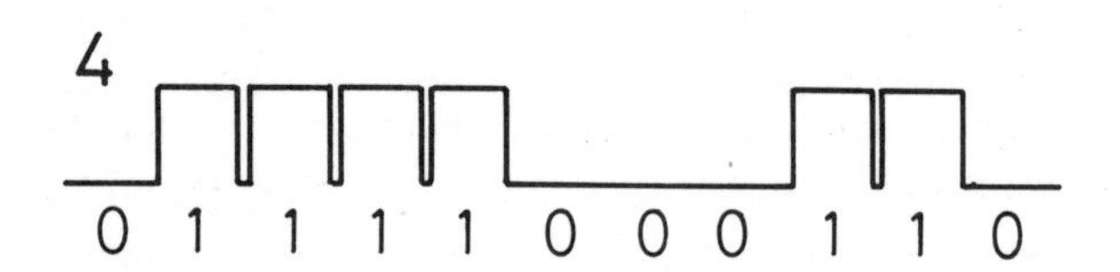

In a digital computer each pulse or non pulse of current represented by 1 or 0 is called a *bit*. A group of 8 bits is a *byte*. Computer storage and memory is measured in bytes, 1 kbyte usually being 1024 bytes (or $8 \times 1024 = 8192$ bits). It must be remembered that electric current flows at a speed of 300 000 kilometres per second. Thus a flow of thousands (indeed millions) of ONs and OFFs appears to be instantaneous.

A Visual Display Unit (VDU) is a television screen on which letters, figures, symbols and lines appear as and when commanded by the CPU. The VDU is an output unit. When the flow of current containing its bits of ONs and OFFs is passed from the CPU to the VDU, the screen is scanned (drawing 5). The scan starts at the top left corner of the screen, moves from left to right, flies to the left hand of the next line and so on down to the bottom right. From the bottom right of the screen the scan flies back to the top left of the screen. This all takes place in a period of time of about $\frac{1}{30}$ second and is repeated over and over again until the VDU screen is cleared by switching off.

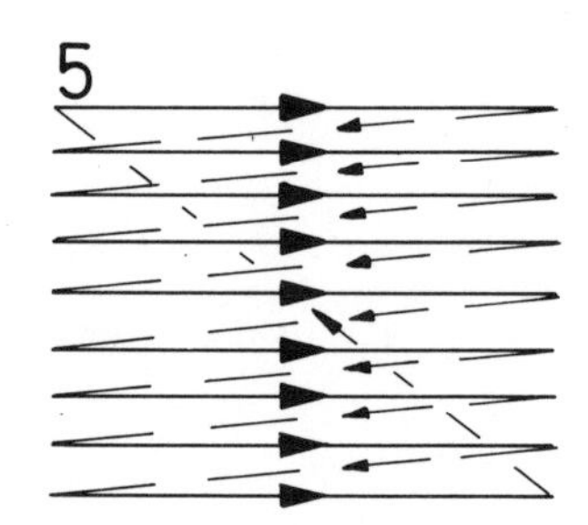

A typical VDU screen, such as the one for the BBC micro-computer, contains 256 lines each of 320 points, each of which can be lit up by an ON pulse or turned OFF by a non pulse. These points are known as *pixels*. 256 lines, each of 320 pixels, requires 81 920 bits to operate as a graphics display screen. 81 920 bits = 10 240 bytes = about 10 kbyte.

Drawings can be made to appear on the VDU screen by lighting up lines of pixels in response to the ONs (1s) of current. This is illustrated in drawings 6 and 7.

In drawing 6, nine lines each of nine pixels are shown with the scan switching ON some pixels in some lines.

6

0	0	0	0	0	0	0	0	0
0	1	1	1	1	1	1	1	1
0	1	1	1	1	1	1	1	1
0	1	1	0	0	0	0	0	0
0	1	1	0	0	0	0	0	0
0	1	1	0	0	0	0	0	0
0	1	1	0	0	0	0	0	0
0	1	1	0	0	0	0	0	0
0	1	1	0	0	0	0	0	0

Drawing 7 shows the lines at the corner of a rectangle as they would appear on that small part of the VDU screen controlled by the nine lines.

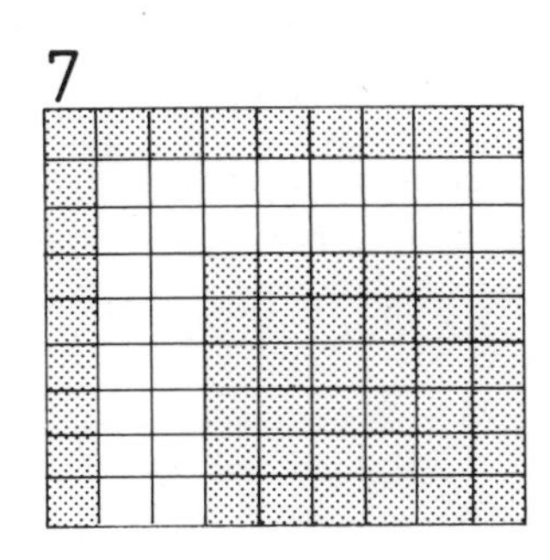

7

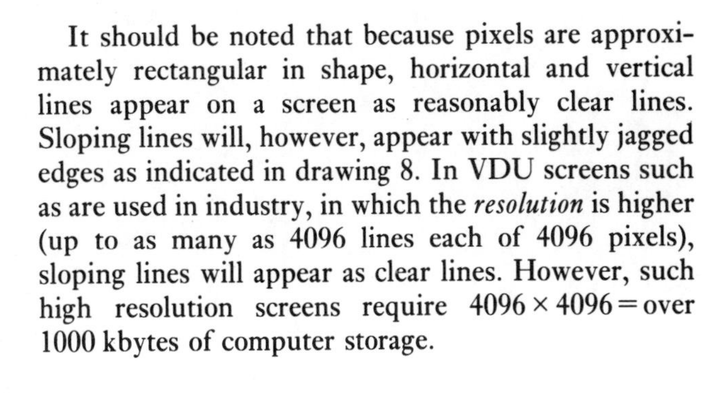

It should be noted that because pixels are approximately rectangular in shape, horizontal and vertical lines appear on a screen as reasonably clear lines. Sloping lines will, however, appear with slightly jagged edges as indicated in drawing 8. In VDU screens such as are used in industry, in which the *resolution* is higher (up to as many as 4096 lines each of 4096 pixels), sloping lines will appear as clear lines. However, such high resolution screens require $4096 \times 4096 =$ over 1000 kbytes of computer storage.

8

Graphics programs in BASIC

Simple forms of graphics on a VDU screen from a computer can be produced from programs such as those listed below. All the letters, figures and symbols of these programs are put into the computer via the keyboard. The programs given can be used only with a BBC computer. Different instructions would be required on other computers. These differences can be found by referring to the instruction booklets accompanying the particular computer being used. The instructions and programs given here are in BASIC language. BASIC stands for Beginners' All-purpose Symbolic Instruction Code.

Keying the instruction MODE 1 on a BBC computer keyboard results in a cleared screen which gives a *resolution* of 320 pixels by 256 pixels and makes four colours available – black, red, yellow and white.

Keying MODE 2 results in a lower resolution of 160 pixels by 256 pixels but allows sixteen colours.

Lines are drawn on the screen as between points in a co-ordinate geometry system. The points on the screen are given coordinates (x,y) from $x = 0$ to 1279 and $y = 0$ to 1023.

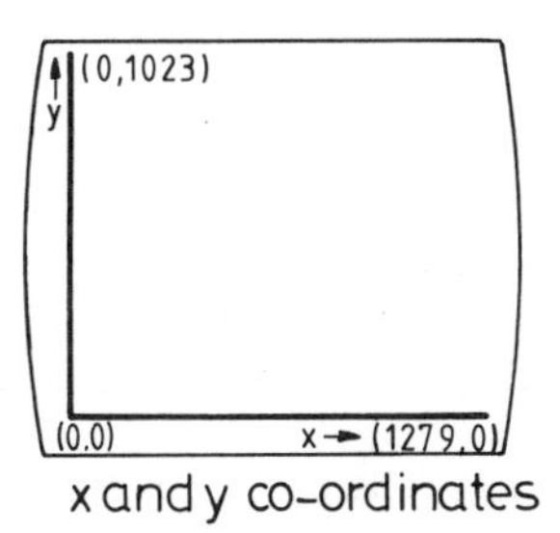

x and y co-ordinates

The command MOVE X,Y positions the beginning of a line at the point (x,y) in this coordinate system. The command DRAW X1,Y1 following the MOVE instruction will result in a line appearing on the screen between the coordinate points (x,y) and x_1y_1). This is the basis of the programs which follow. Note that other programs in BASIC will produce arcs, circles, ellipses and other geometrical shapes. These are not dealt with here.

In mode 1 the command GCOL0 instructs the computer to produce colours. GCOL0,1 (red); GCOL0,2 (yellow); GCOL0,3 (white); GCOL0,4 (black).

Program 1

```
  5 REM *** DRAW A SQUARE
 10 MODE 1
 20 MOVE 300,200
 30 DRAW 800,200
 40 DRAW 800,700
 50 DRAW 300,700
 60 DRAW 300,200
 70 END
RUN
```

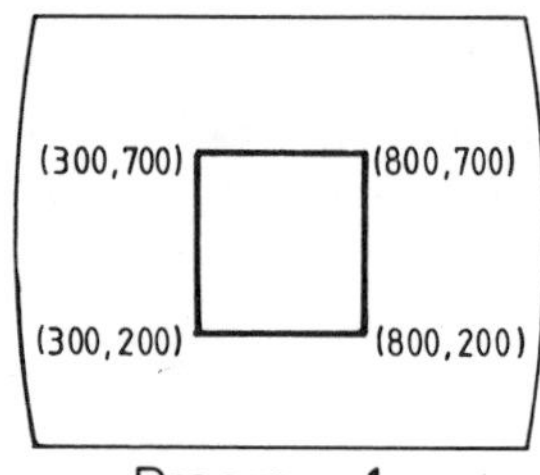

Program 1

Program 2

```
  5 REM *** COLOURED LINE SQUARE
 10 MODE 1
 20 GCOL0,1
 30 MOVE 300,200
 40 DRAW 800,200
 50 GCOL0,2
 60 DRAW 800,700
 70 GCOL0,3
 80 DRAW 300,700
 90 GCOL0,2
100 DRAW 300,200
110 END
RUN
```

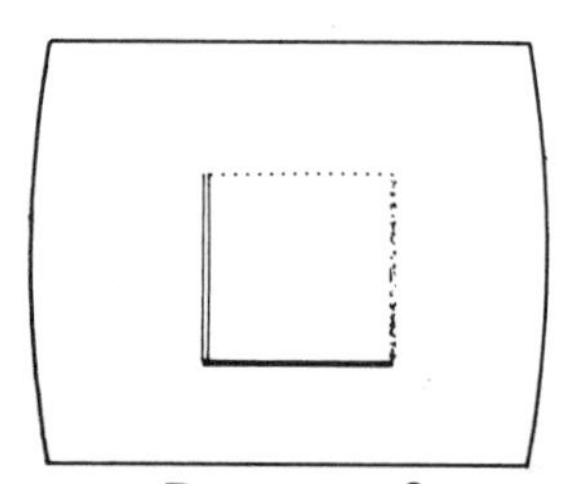
Program 2

Program 3

```
REM *** DRAW A TRIANGLE
10 MODE 1
20 MOVE 200,100
30 DRAW 600,100
40 DRAW 600,900
50 DRAW 200,100
60 END
RUN
```

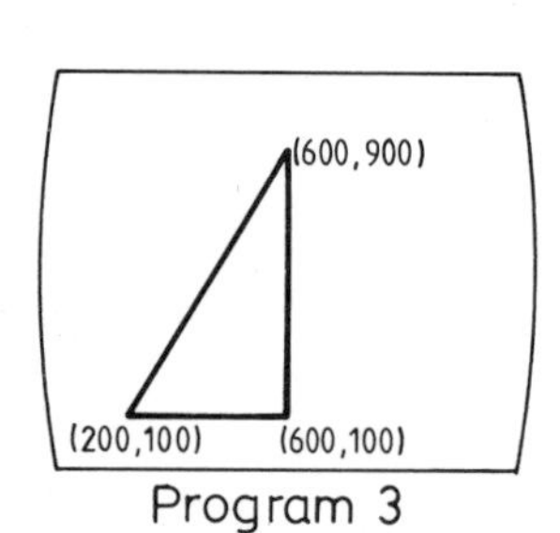

Program 3

The instruction PLOT 85 in a program results in a triangular area being 'painted' in colour on the BBC system. As the computer 'remembers' two previously given triangle vertices, the command PLOT X,Y results in the area between all three points being 'painted' in the colour given by the earlier command GCOL0 in the program.

Program 4

```
  5 REM *** COLOURED TRIANGLE
 10 MODE 1
 20 GCOL0,2
 30 MOVE 200,100
 40 DRAW 600,100
 50 DRAW 600,900
 60 PLOT 85,200,100
 70 END
```

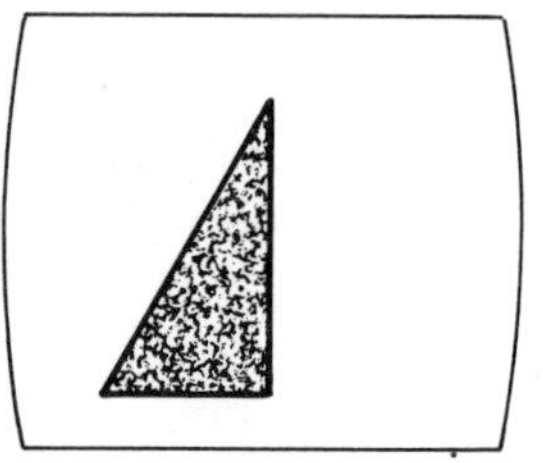
Program 4

EXERCISES

1. Write a program in BASIC for a BBC computer to draw the house outline shown in the given drawing. Use as much of the screen as possible with your drawing. Add to your program the items which will colour the roof red, the door yellow and the window black.
2. Write a program in BASIC for a BBC computer which will draw the rectangle and triangle within a square as shown. Select any suitable sizes. Colour the rectangle in red and the triangle in yellow.

Exercise 1 Exercise 2

3. Write a program in BBC BASIC for a BBC computer which will show three views in First Angle orthographic projection of a block 100 mm long by 60 mm high by 50 mm deep. Assume each millimetre is shown on the screen by 5 screen units.
4. Sketch the front view of a two-storey house, which has a pitched roof with overhanging eaves, 2 upper storey windows, 2 lower storey windows, 1 each side of a door. Now write a BASIC program which will display your sketch on the screen of a BBC computer.
5. Write a BBC BASIC program which will show the following geometrical shapes on the computer screen:
   a hexagon:
   an octagon;
   an octagon within a square;
   an equilateral triangle.
6. Re-write the program written in answer to Exercise 5, so each each figure is coloured in a different manner.
7. What is meant by the following terms:
   (a) VDU;
   (b) a bit;
   (c) a byte;
   (d) a kilobyte;
   (e) screen resolution?

Micro-computer systems for graphics

Programs such as those keyed into the computer using BASIC are not suitable for any degree of complication in drawings. This is partly because of the limitations of such programs and partly because they can become very involved and long if any degree of complexity is involved in the graphics.

The drawing on this page shows the various input and output systems which can be employed for the production of computer aided drawings. To avoid confusion here, the term CAD is taken nowadays to mean Computer Aided Design, embracing both drawing and design. Another term, CAM, which stands for Computer Aided Manufacture, is not referred to in this book, but in modern engineering computing CAD/CAM, embracing both engineering design and manufacture, is a frequently used abbreviation.

The Apple IIGS – 256K RAM and 128K ROM – high resolution colour

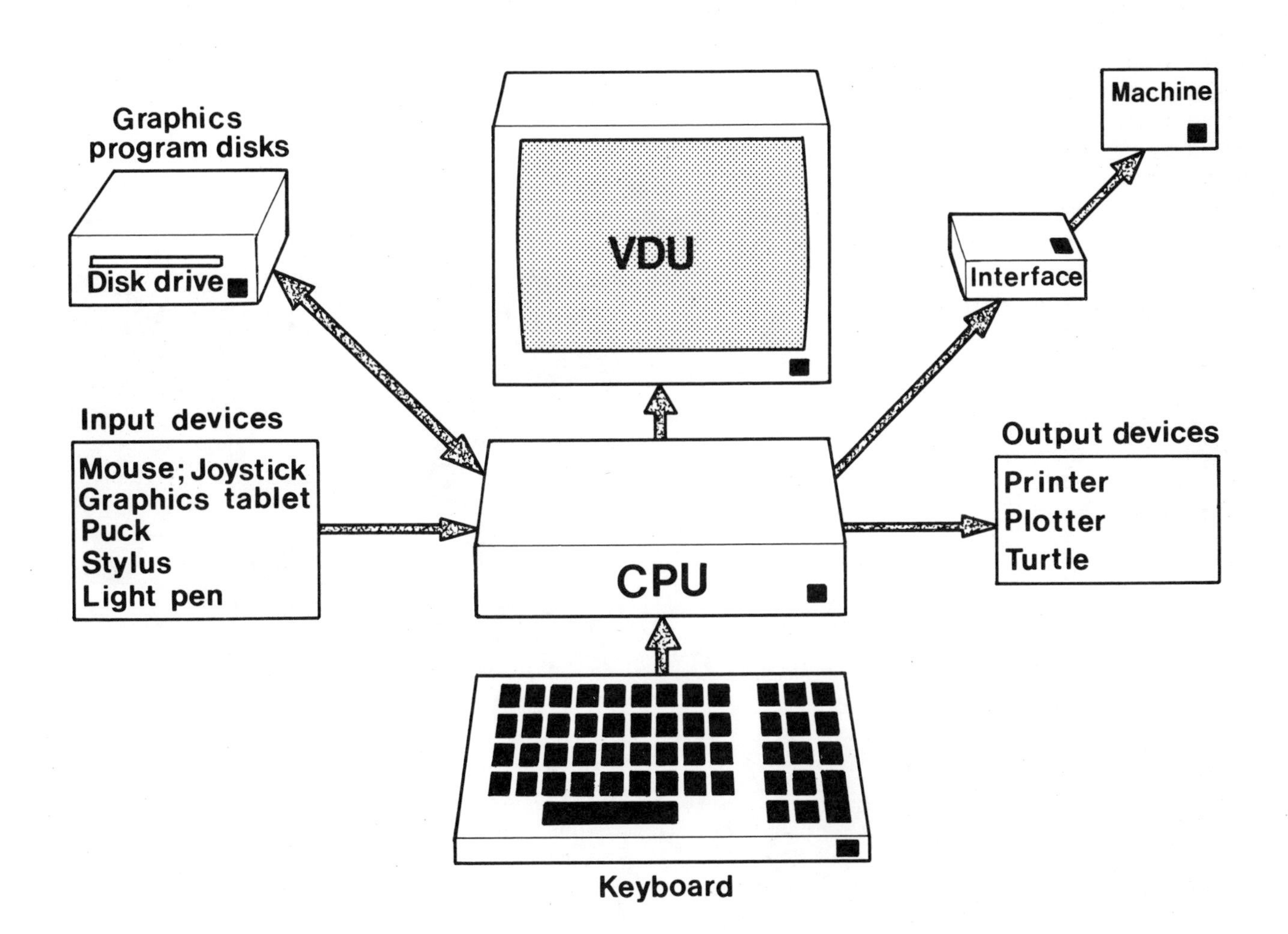

Input devices

A variety of input devices are available for feeding information into the Central Processing Unit (CPU) for conversion into digital signals which can then be passed on to a Visual Display Unit (VDU) or other output device. Cursors will frequently be presented on the VDU screen in connection with input devices. Cursors are intended to point to a small area on the screen or to draw attention to a detail on the screen. In graphics cursors are generally one of four types, as in the drawing below.

Different styles of cursor

The gaps in the cross and arrow cursors are present so that the cursor lines do not actually cover the particular pixel being pointed out. With some input devices the cursor will flash on and off on the screen so as to draw attention to its position.

Mouse

A mouse is a small, hand-held unit which contains a number of control buttons – typically between one and four. It is attached to the computer by a cable. As the mouse is moved over any hard surface, such as a table top, its movements are followed by a cursor on the VDU screen. The mouse shown in the photograph works on the principle of variation in voltage across a pair of crossing potentiometers (variable resistors) set in the body of the mouse and controlled by a ball set centrally in the underside of the mouse. As the mouse is moved the variations in voltage pass to the computer via the cable and are converted to digital signals to be shown on the VDU screen.

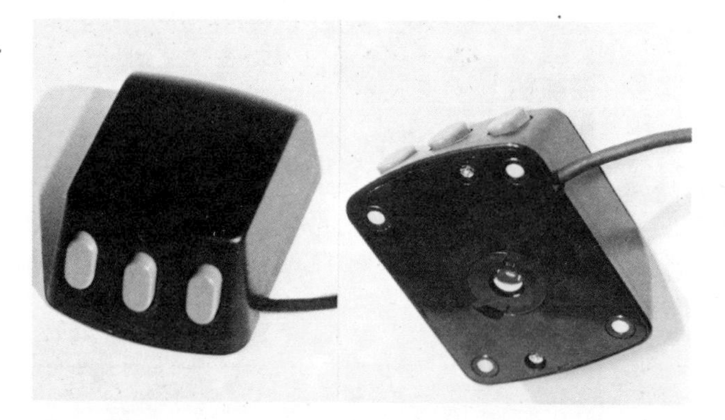

Views of a mouse from above and from below

Joystick

A photograph of a joystick is shown here. The photograph on page 166 shows a 'Bitstik' joystick attached to a BBC micro-computer, a visual display screen and a disk drive.

A joystick

The handle or 'paddle' of the joystick can be freely moved up, down, to left or right or at any angle in between. As the handle is moved a cursor can be positioned in any position on the VDU screen as required by the operator. By rotating the handle clockwise or anti-clockwise the scale of the picture can be enlarged or reduced. Operating the handle can produce

A BBC computer set up with disk drive and joystick

a 'zoom' effect by which any part of a drawing on the screen can be enlarged. The pushing of buttons on the face of the joystick container allows operations such as clearing the screen, selecting from a menu, 'freezing' a picture on the screen and so on to be performed. These operations allow a joystick to be used to produce a drawing on the screen. If a drawing is then required for future reference, it can be stored on a floppy disk, to be recalled when needed.

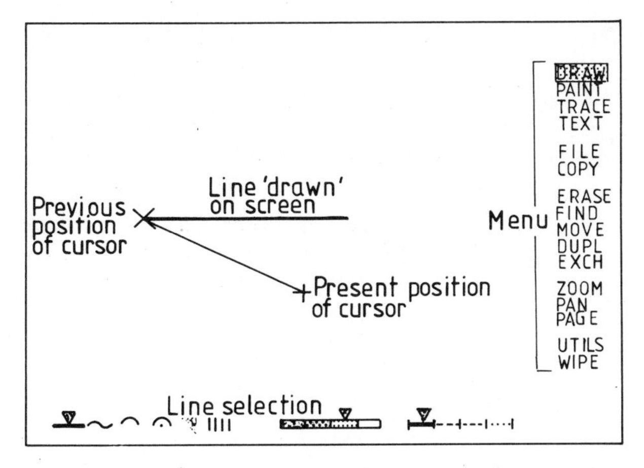

Lines drawn on a VDU screen with the aid of a joystick

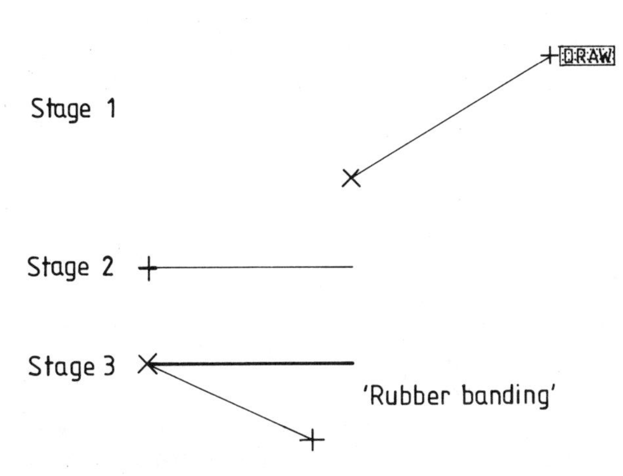

Stages in drawing with a joystick

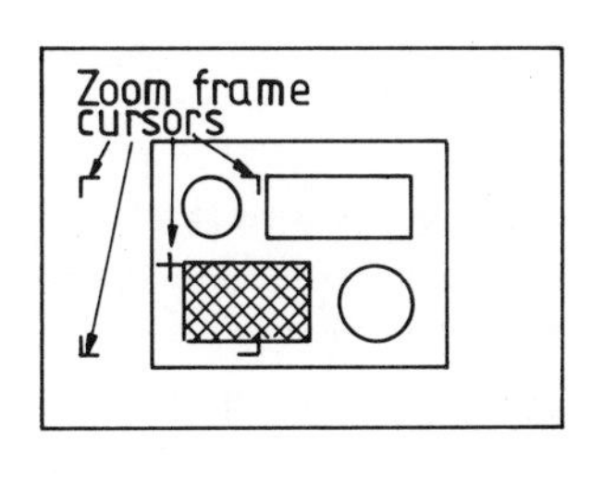

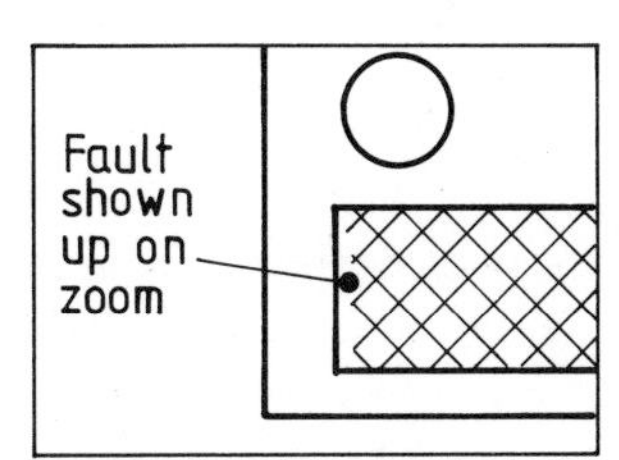

Joystick zoom facility showing up missing line joins in an enlargement

Most joysticks contain two crossing potentiometers (variable resistors) at right angles to each other (north to south and east to west) through which a small current is being passed. Movement of the joystick handle results in a variable voltage being passed to the computer which is read by the CPU as a digital signal which controls the picture on the screen. Simplified drawings of pictures on a screen produced by operating a joystick are given. Circles, arcs, shading, colouring and other features can be 'drawn' on to a VDU screen from details selected from a 'menu'. Other menus can be displayed from a variety of graphics programs.

Graphics tablet and input puck

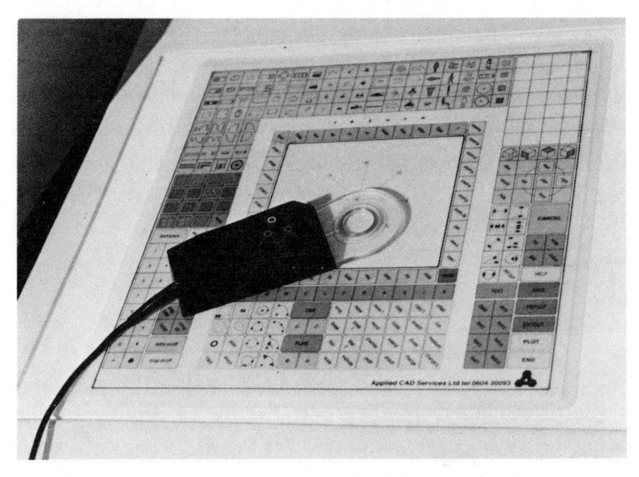

A graphics tablet designed by Applied CAD (Computer Aided Design) Ltd

A graphics tablet and a puck are shown in the photograph. Buried within the central surface area of the graphics tablet is a fine grid of wires. When the centre of the cross hairs of the puck is placed over any part of the central area of the tablet, its exact position is mimicked on the VDU screen in some way – either by a cursor or by crossing x,y lines. In general, graphics tablets work in one of two ways. Either the puck picks up signals passing through the grid of wires in the tablet, or the puck emits a signal which is picked up by the grid. Whichever method is used, the signal is passed to the CPU through the cable attaching the puck to the CPU.

When the puck is placed over those parts of the tablet showing symbols or words, a menu, a program or a symbol comes up on the screen when one of the buttons in the puck is pressed. The menu could be e.g. *drawing board*; the call could be to produce circles; a system for drawing in *orthographic projection* might be called up to the VDU screen. The various designs of graphics tablets allow different forms of graphics to be originated on the screen.

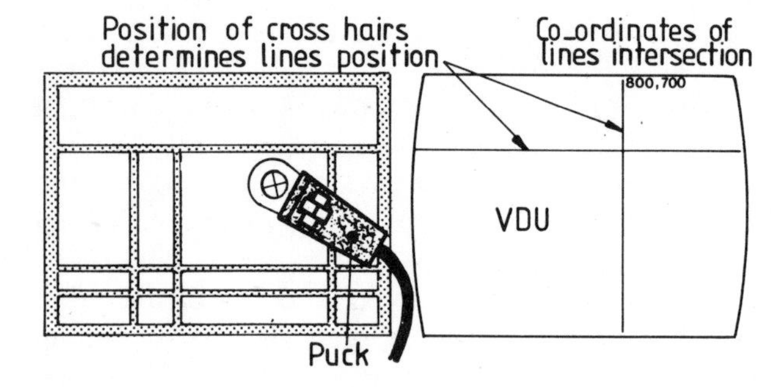

A puck and graphics tablet

The two diagrams show a graphics tablet and puck and the position of two lines on a VDU screen. The x and y coordinates of the position of the point of intersection of the two lines on the VDU are determined by the position of the intersection of the cross hairs of the puck window in the central area of the graphics tablet. The position of the x,y intersection shows up on the VDU as coordinate numbers as the puck is moved. By pressing buttons on the puck, the x,y coordinates can determine say, the end of a line. This is just one example of the action on a VDU screen of a graphics tablet and puck as used in the Autocad system (see page 170).

Stylus

The puck shown in the photograph with the graphics tablet is replaced in some systems by a pen-like stylus. The stylus is again attached to the CPU via a cable. Its purpose is the same as the cross hairs of the puck. Positions on the VDU screen, different menus or different graphics programs can be selected from the tablet by touching its surface with the end of the stylus.

Light pen

A light pen is a pencil-like tube in the end of which is a light sensitive diode set behind a lens. The light pen is usually connected to the CPU by a fibre optic cable down which a light signal is sent to the CPU. If the light pen is pointed at a cursor on the screen it can be used to re-position the cursor by 'dragging' it away from its position.

The central processing unit (CPU)

The essential part of a computer is its CPU. This contains the silicon chips and their associated circuitry. The chips and circuitry deal with the signals received from various devices or from the keyboard and process them into digital signals which can be passed to the various output devices, either as information on a VDU screen or as 'hard copy' from a printer or a plotter. The CPU will contain some information and some storage in its chips. Read Only Memory (ROM) chips contain the information by which the CPU processes input signals into output digital signals. Random Access Memory (RAM) chips contain storage for the programs being input into the CPU during the time an operator is working the computer system. When the computer is switched off all RAM is lost, but (of course) all ROM is retained. The storage in RAM can be transferred to a cassette tape, floppy disk or other backing storage device before the CPU is switched off if one wishes to retain what has been worked.

The Keyboard

The keyboard is essential to the working of a micro-computer system. Even when input devices such as a puck or a stylus or a graphics tablet are in use, some of the instructions to the CPU can only be keyed in from the keyboard. The keyboard functions as an accessory to all types of input device. A micro-computer keyboard is similar to a typewriter keyboard except that it contains more keys. The additional keys are used to perform functions not needed when using a typewriter. However, if one is familiar with the common typewriter keyboard, working at a computer keyboard will be that much speedier.

Software

So far we have been dealing with items such as the CPU, its keyboard, VDUs and input devices. Such items, which are for the implementation of computer programs are referred to in computer terms as *hardware*. The programs themselves are known as *software*. Programs need to be held, or stored. Apart from the ROM programs held permanently in the chips of CPUs and the temporary storage of programs in the RAM of CPUs, other items of software will be held in backing storage on cassette tapes, floppy disks and the like.

Backing storage

Audio cassette tapes
Programs can be stored on, and retrieved from, cassette tapes in an audio cassette player. As the tapes have to be wound and unwound, the finding of a program takes considerable time. However, the storage of the digital signals from a CPU on the magnetic tape of a cassette recorder is one of the cheapest methods of storing programs.

Floppy disks
These are thin discs of flexible plastic material, coated with magnetic oxide. They are available in two diameters – 89 mm and 133 mm. The disk is placed in a floppy disk drive machine which revolves it at high speed. Magnetic sensors are passed radially across the surface of the disk to read the digital magnetic signals recorded on it. Because of the speed of revolution of the

disk and the fact that the reading is radial, programs can be selected from the disk in very short times. The disk drive machines take either one or two disks and are attached to the CPU by cables. A 133 mm disk will store about 0.5 megabyte (Mbyte) of information. 1 Mbyte usually = 1024 kbytes.

Disk cartridge
This can carry up to 10 Mbyte, but is not generally used in schools.

Fixed head disks
These can carry up to 500 Mbyte. They are used only in large industrial and commercial applications.

Notes: When handling tapes and disks care must be taken to avoid touching their surfaces. If touched, the information contained on the surface may be lost or distorted. Tapes and disks should be kept in dust free containers. Dust can also be a source of loss or distortion of information. It is always advisable to make copies of tapes and disks which are required for future use. This ensures that if a disk is damaged, say by someone touching its surfaces, the programs it holds are still in the back-up copy.

Interfaces

The output of a CPU is a digital signal in binary of very low power (power-current × voltage). In order that a program from a CPU can be used to control a mechanism, the digital signal must be decoded and the power of the signal amplified. An interface is used for this purpose between the output sockets of the CPU and the mechanism. An interface must also be capable of blocking any feedback of power from the mechanism. Such feedback surges of power could ruin the chips of a CPU. Using properly designed interfaces, vehicles, robots, cutting machinery and other mechanisms can be controlled from programs passed via the CPU.

Robot model from LEGO® TECHNIC Control Set 1090 being run by a BBC-B Computer via a new LEGO® interface

Output devices

A VDU screen can be regarded as an output device in that it shows in a visual form a reading of the digital signal output from a CPU. If a copy of what can be seen on the screen is required on paper, it can be obtained from either a *printer* or a *plotter* or from a third output device known as a *turtle*. Copies on paper of programs from a CPU are called *hard* copy.

Printers

A dot matrix printer operates in much the same way as a keyboard operated typewriter, in that copies are obtained via a carbon ribbon on a sheet of paper. The difference is that letters and symbols are formed from a printer head controlled by the CPU to which the printer is attached by cable. Pins in the printer head respond to the digital signals by impacting independently of each other on the ribbon to form the required print on the paper. When a binary 1 is signalled to a particular pin a dot is printed. When a binary 0 is signalled the pin does not move. The speed at which a CPU controlled printer works is much faster than a typewriter can be operated by hand.

Plotters

An A3 size plotter is shown in the photograph. The pen held in the moving head of the plotter moves along x and y axes in response to the signals from a stored copy of a drawing program. When a line ends the pen moves up and off the paper surface; when a line begins, the pen moves down onto the paper surface. The movement of the pen follows all the lines on the drawing program held in the CPU (or on disk). The resulting *hard* copy is an accurate copy of what has been previously drawn on a VDU screen.

Turtle

This is a device on wheels which can move freely on commands given to it via a cable attached to the computer. A pen held in the turtle can be commanded to touch the paper surface or lift up from it. A program of commands such as:

> pen up; forward 50 mm; pen down; forward 200 mm; rotate right 45°; forward 200 mm; pen up

would make the turtle draw two lines each 200 mm long at 45° to each other on the paper.

Autocad

One of the photographs on this page shows a graphics tablet from an Autocad draughting system, with the aid of which an orthographic projection drawing has been programmed into the CPU of the system. The CPU has then been keyed to instruct an A3 plotter to print the drawing. The second photograph shows the plotter halfway through completing the drawing on an A3 sheet of paper.

Autocad is an advanced CAD (computer aided drawing) program which is in common use in technical colleges, running on computers which are more powerful than those available in schools.

There are a number of CAD software programs which can be purchased and which are more suitable for the types of computer found in schools, including the BBC range. Among these are *Novocad*, *Techsoft Designer* and *Bitstik*.

Such CAD programs are *MENU* driven. This means that when in use, the operations which can be selected for drawing on the screen are shown printed on the screen. The user selects the operation which he-she wishes to perform, eg draw a circle, and then a further set of operations are printed on the screen, eg position of centre, radius, etc. The required positions, sizes etc are keyed in by the user and the circle appears on the screen. A large variety of other operations make it possible for the user of such programs to quickly build up engineering, building, architectural, electrical and circuit drawings.

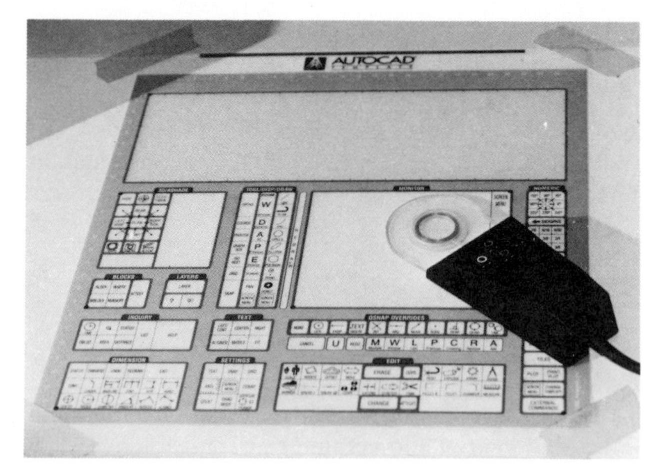

An Autocad graphics tablet

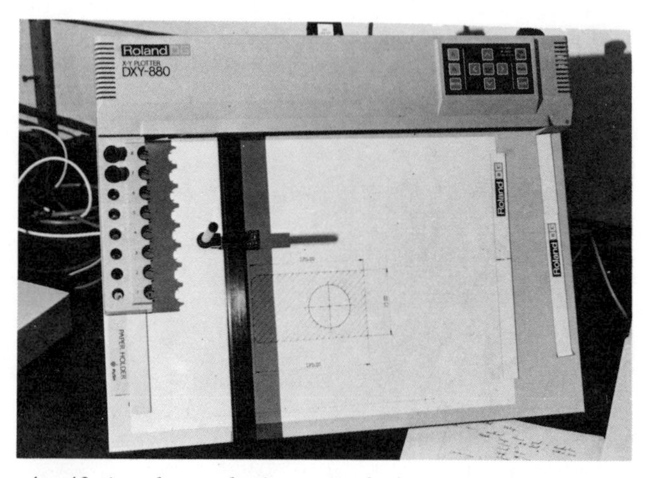

An A3 size plotter plotting a simple drawing

The computer as a design tool

When a drawing has been built up by a computer on a VDU screen, the program by which you have produced the drawing can be stored on a cassette tape or on a floppy disk. When you want to see it again, the drawing can be retrieved from its store and shown on the screen. If necessary the drawing can then be re-designed. Parts can be erased and amended. Proportions, dimensions, shapes can be changed by the computer. Either the re-designed drawing program, or the original drawing, or both, can then be again stored. In this way a complete design solution can be built up over a period of time. If required, each new drawing can be printed as hard copy.

Although at the present time schools will probably find that computer facilities offering 3D are too expensive, with the rapidly decreasing costs of computer graphics chips it may not be all that long before computers with 3D facilities are available to schools. With 3D programs, it is possible to see a design in pictorial forms and, in addition, to allow the design to be rotated three-dimensionally on the screen so that it can be viewed as if from a number of different directions. If dissatisfied with any part of a design, that part can then be amended by operating the computer and the amendment included then and there on the VDU screen. In addition, working parts of a design can be made to appear as if working so that fits and clearances can be tested and checked.

Remember, however, that the computer should be regarded as only one of the available graphics tools. It is very necessary to learn how to produce graphics with pen, pencil, instruments and colour. It is no good being able to call up a drawing on a computer VDU screen if you have not learned, through practice using hand methods, the meaning of the picture you see on the screen.

Revision

Geometrical constructions

The reader is advised to work through the constructions on pages 37 to 46, to any suitable dimensions unless sizes are already given. The following exercises are included here for revision practice. Please do *not* draw on the pages of this book. Copy the drawings onto a sheet of drawing paper, with the aid of tracing paper when necessary.

1. On the given base line AB construct:
(a) an equilateral triangle;
(b) an isosceles triangle with angles of 73° at A and B;
(c) an isosceles triangle with sides AB and AC of 90 mm;
(d) a right-angled triangle with a hypotenuse AC of 90 mm;
(e) a scalene triangle with the angle at B equal to 110° and with BC = 50 mm.

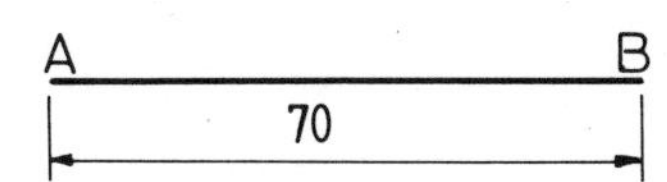

2. Name the given triangles A, B, C, D and E.

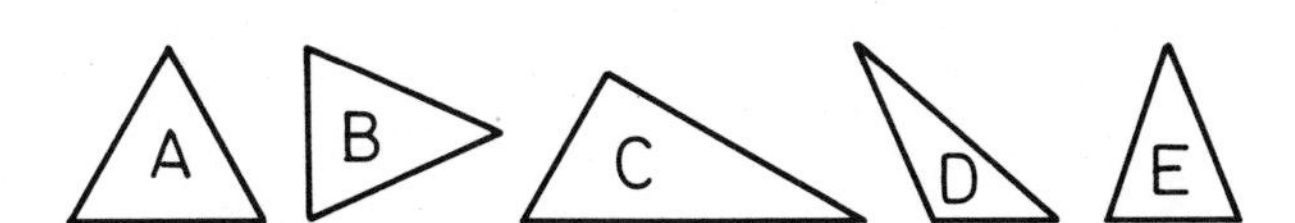

3. Divide CD geometrically into seven equal parts.

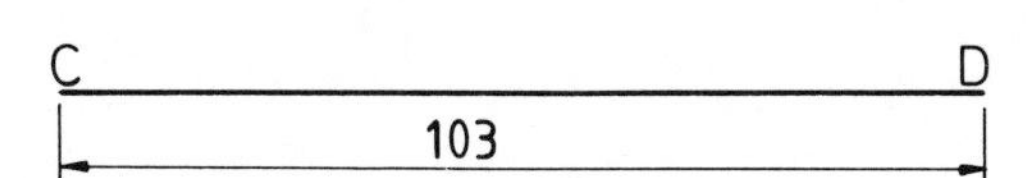

4. What size angles have been constructed at A, B, D, D and E?

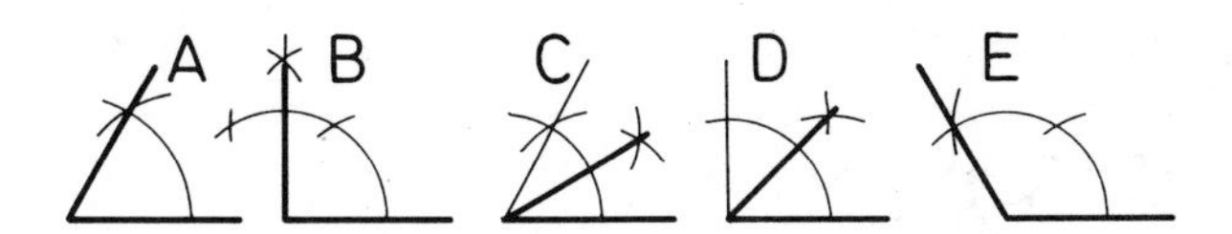

5. Name the two given circles A and B.

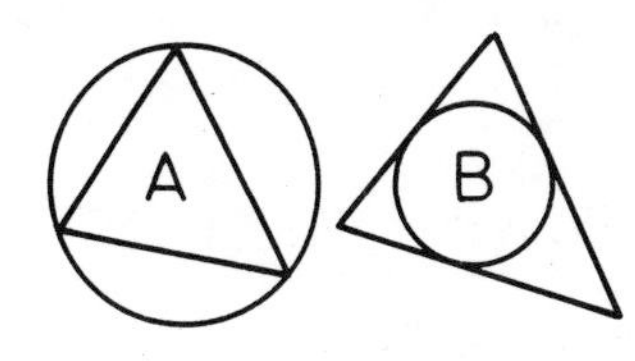

6. How many degrees in the angle A?

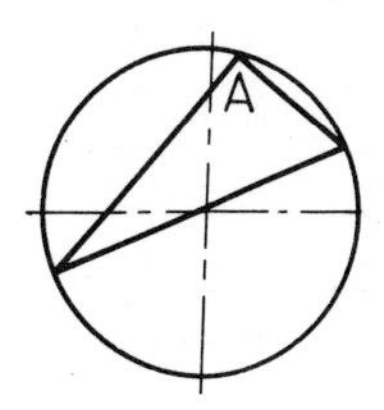

7. Construct a regular hexagon within the circle.

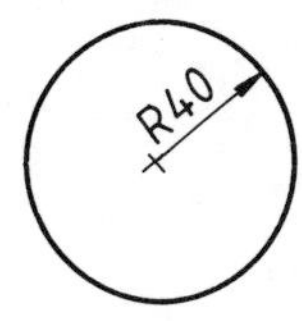

8. Construct a regular octagon touching the outside of the circle.

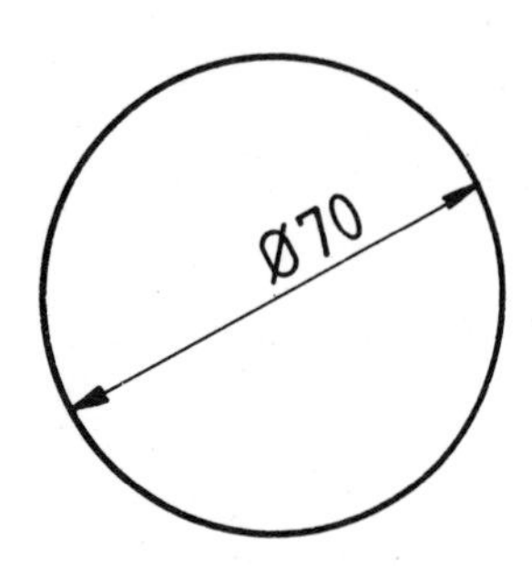

9. How many sides in a heptagon? In a pentagon?

10. On the base line AB construct a regular five sided regular polygon.

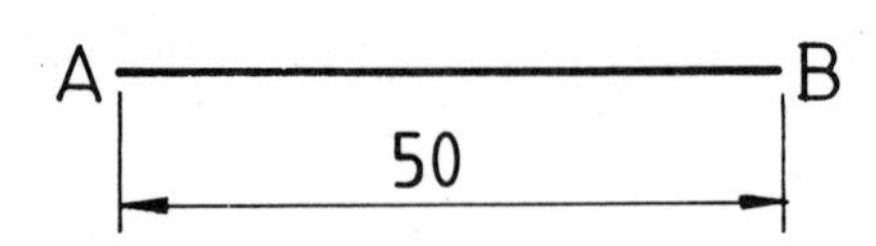

11. Name the four polygons A, B, C and D.

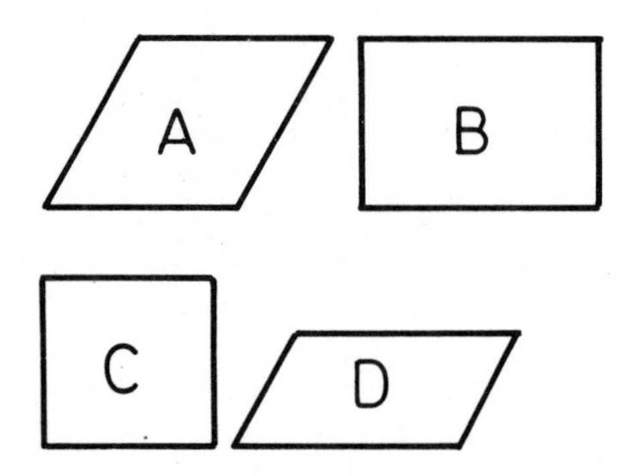

12. Complete the construction of the enlargement of the polygon so that its sides are twice as long as in the original.

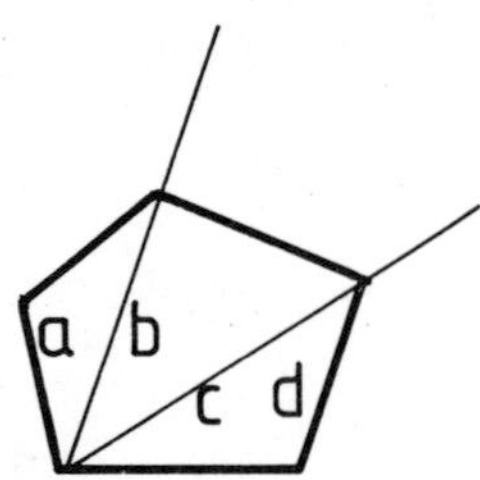

13. Name the curve.

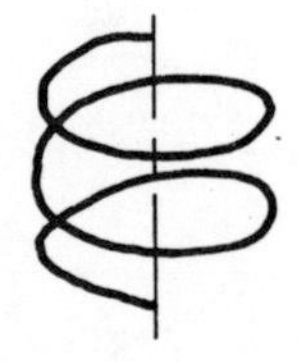

14. Complete the construction of the reduction of the polygon so that all sides are half the length of the original.

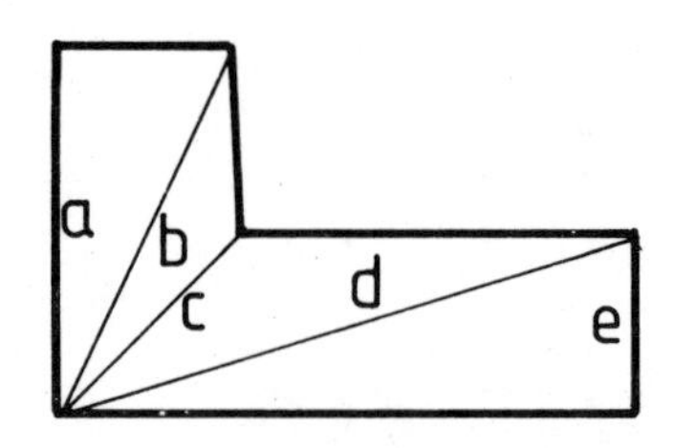

15. Name the lines A, B, C and D.

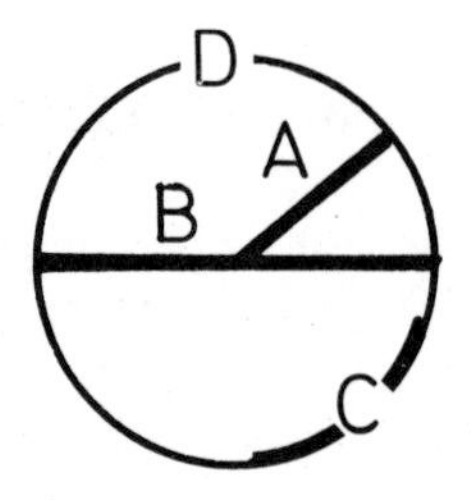

16. Name the areas A, B and C.

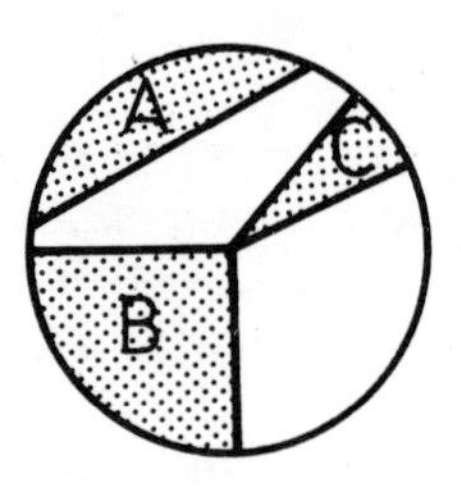

17. Construct a straight line tangent at T.

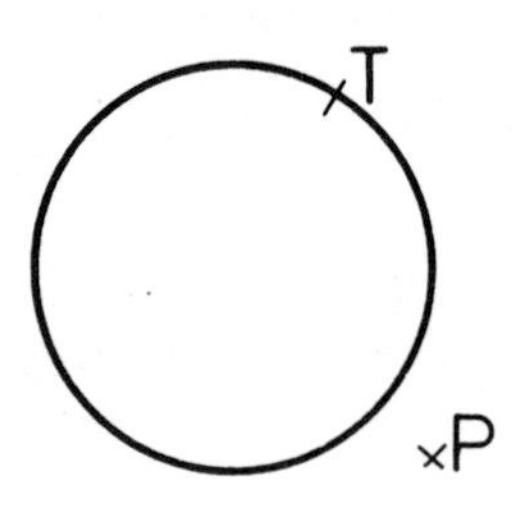

18. Construct a straight line tangent from P to the circle.

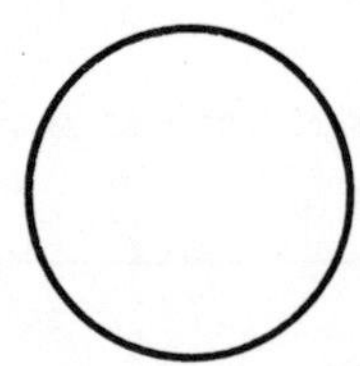

19. State the lengths between the pairs of radii AB, AC and AD.

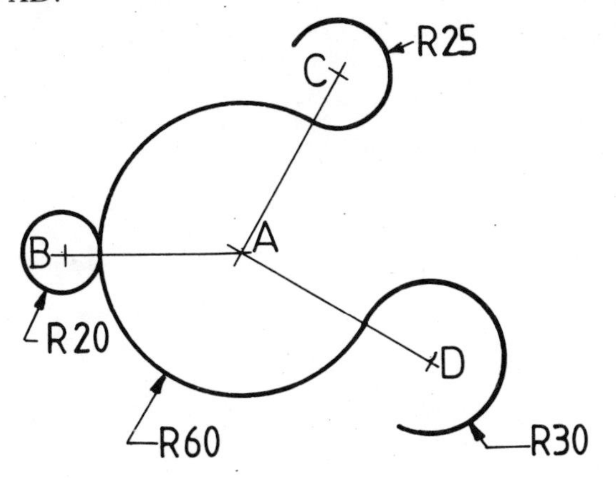

20. State the lengths between the two pairs of radii AB and AC.

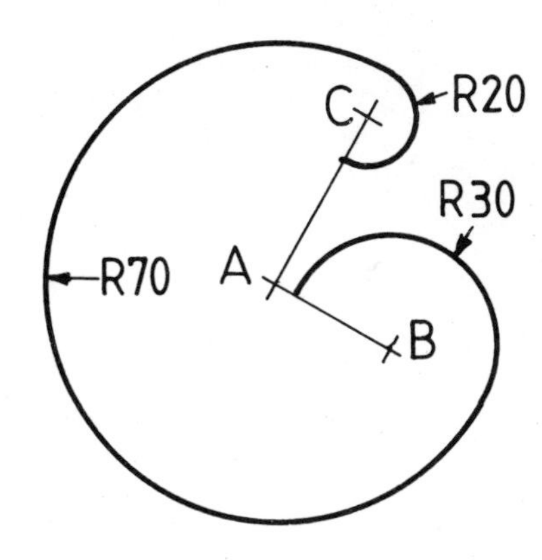

21. Construct radius corners at A (30 mm), B (40 mm) and C (35 mm).

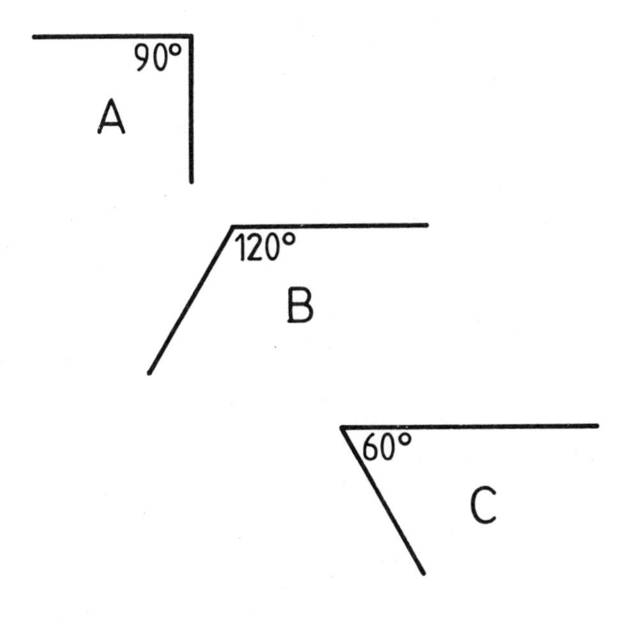

22. Calculate the area from the dimensions given in cm².

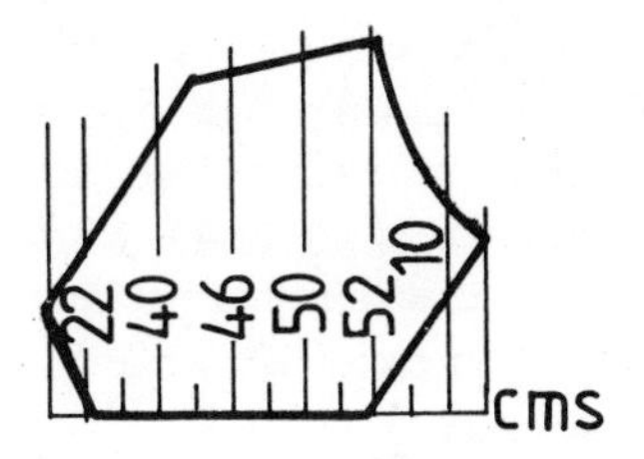

23. Calculate the area of the shaded figure in cm².

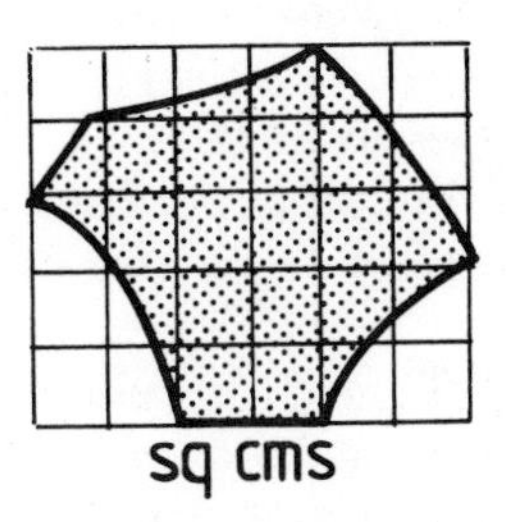

24. Name the parts of the ellipse a, b, c, d and e.

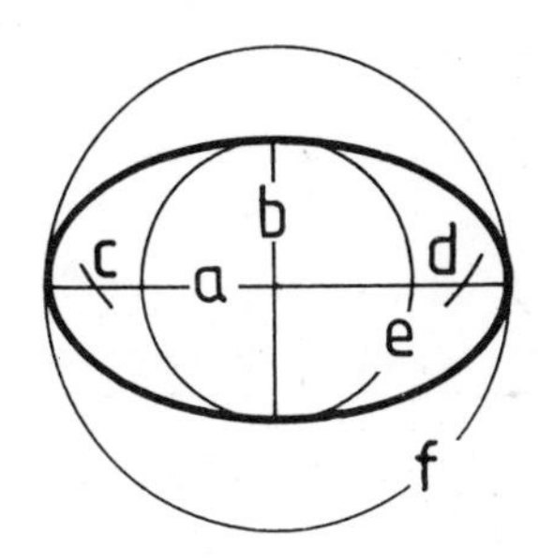

25. Construct an ellipse on the given centre lines.

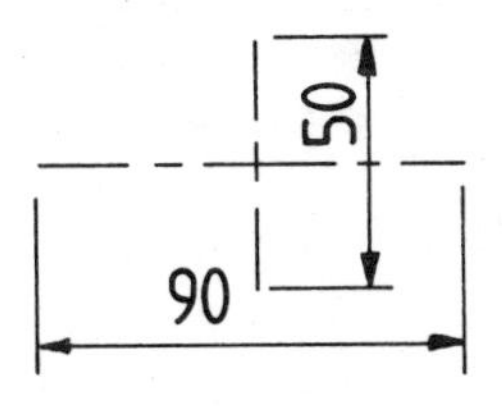

26. Construct a parabola with its vertex at V within the rectangle.

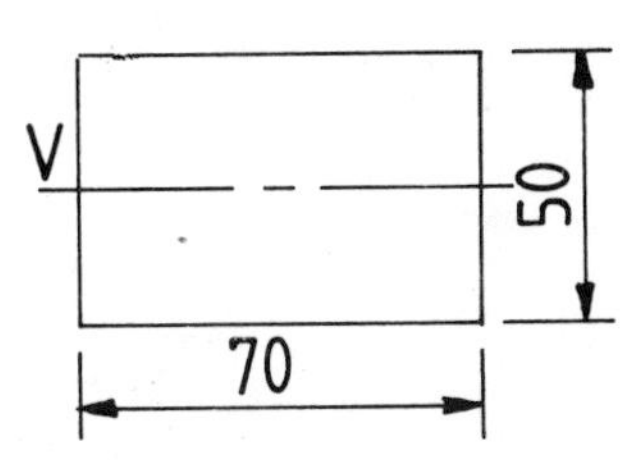

Exercises

1. Right is a sketch of a work top with a supporting end, fitted in a corner of a kitchen. The area under the work top is to be fitted with cupboards and drawers suitable for the storage of kitchen equipment.

On an A2 size sheet of drawing paper draw a vertical line 175 mm from the right-hand border. Make ALL the following drawings.

In the space to the right of the line you have drawn, make TWO FREEHAND sketches showing suitable designs for the cupboards and drawers to be fitted under the work top. Make a THIRD FREEHAND sketch a suitable design for a drawer or door handle for the fitment you have designed.

In the space to the left of the line you have drawn, and using your sketches as a guide, draw to a scale of 1:5:

(i) a front view of the work top and its supporting end and the cupboards and drawers you have designed;

(ii) a sectional end view taken on the line AA.

Add hidden detail and dimensions where necessary.

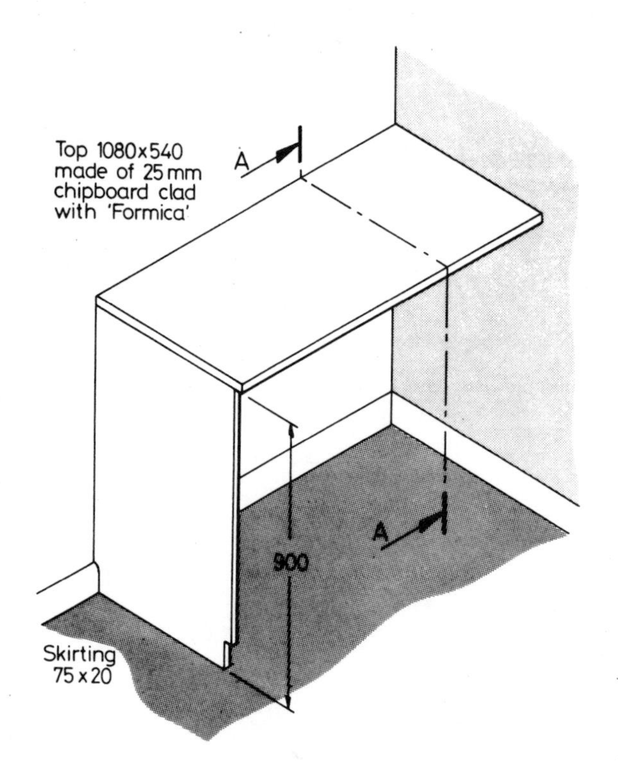

2. Right is a front view and plan of a small wind vane to be fitted to the roof of a building. On a sheet of A2 size paper draw a vertical line 175 mm from the right hand border. Make ALL the following drawings.

In the space to the right of the line you have drawn, make THREE FREEHAND sketches showing suitable different designs for a fitting at C to allow the vane to rotate freely.

In the space to the left of the line you have drawn, and using your freehand sketches as a guide, draw to a scale of 1:1:

(i) the given front view and plan to include details of the fitting you have designed;

(ii) the sectional view BB showing clearly the method of allowing the vane to rotate.

Add hidden detail and dimensions where necessary.

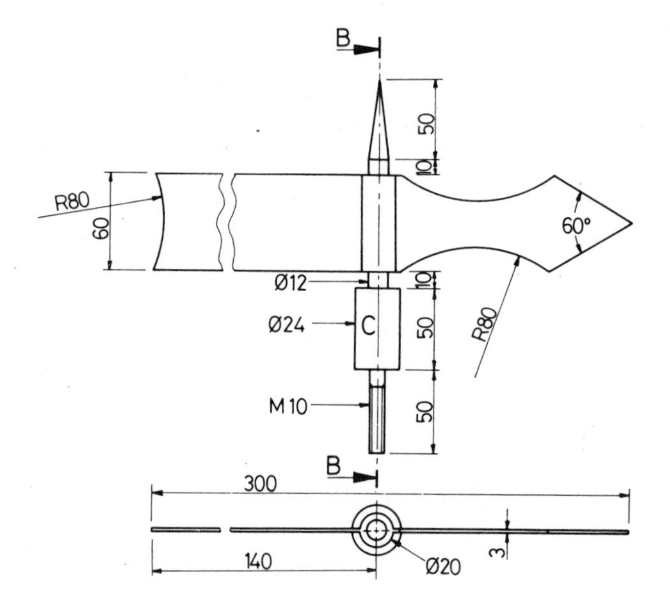

3. Below is a sketch of a corner in a bedroom. A shelf fitment is to be made and fixed to the side of the wardrobe in the position shown. The following items will need to be placed on the shelves when they have been fitted in place:

(i) a cup and saucer;
(ii) a clock;
(iii) a transistor radio;
(iv) a book or two.

On a sheet of A2 size paper draw a vertical line 200 mm from the right-hand border. Make ALL the following drawings.

In the space to the right of the line you have drawn:

(a) Draw FREEHAND sketches showing suitable designs for the required fitment. Your sketches should show details of any jointing methods.

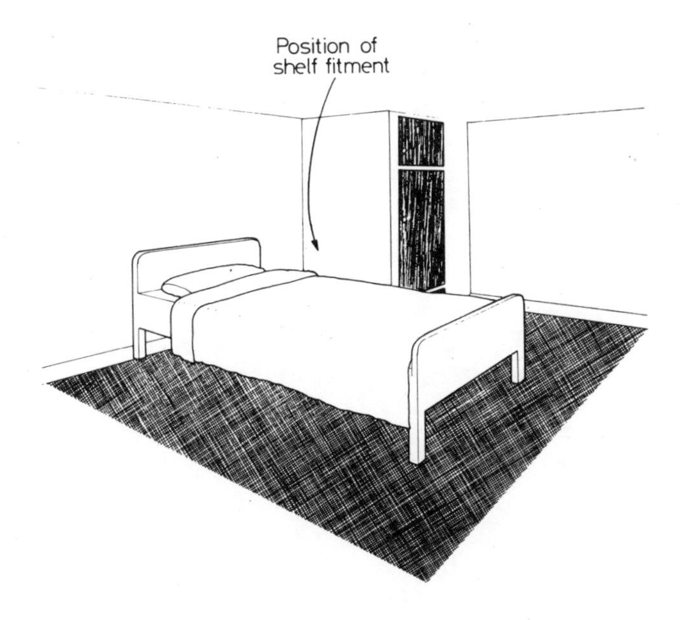

(b) Draw FREEHAND sketches to show suitable methods of securing the fitment to the side of the wardrobe.

In the space to the left of the line you have drawn:

(c) Using your sketches as a guide, draw accurately with instruments and working to a suitable scale of your own choice:
(i) a front view of the whole fitment;
(ii) an end view as seen from the right hand end;
(iii) a plan.

Add hidden detail and dimension where necessary.

4. Select a window catch – either at school or at home. Examine the catch carefully and then proceed as follows.

(a) Make neat freehand drawings of the assembled catch.
(b) If possible, unscrew the catch from its seating in the window and make freehand drawings of its various parts.
(c) From which materials has the catch been made?
(d) By what processes was it manufactured?
(e) Make freehand drawings showing any modifications which you consider would improve the design of the catch to improve its effectiveness.
(f) Make an orthographic projection of the catch, drawn with the aid of instruments. Your drawing should contain at least two views and include your suggested modifications. Include essential dimensions.

5. A photograph of part of a shop window display showing 'snowflakes' cut from aluminium foil is shown below. The shapes of the 'snowflakes' are based on regular hexagons.

Design similar snowflakes to any suitable size based on the shape of regular pentagons.

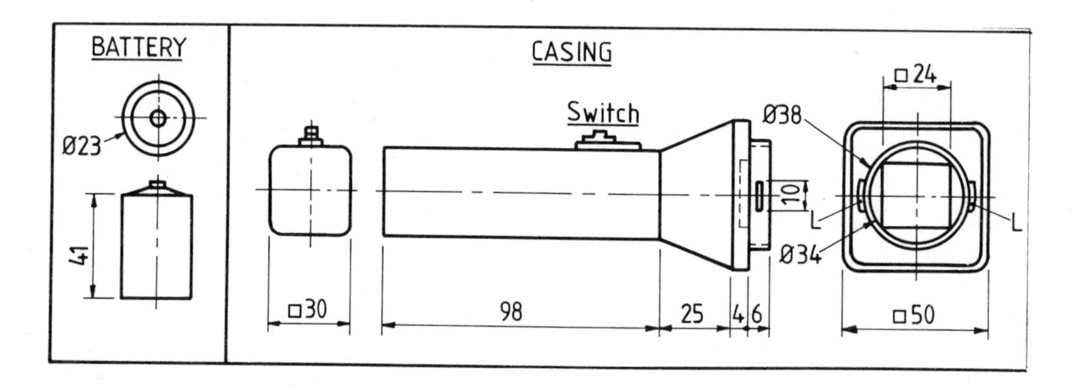

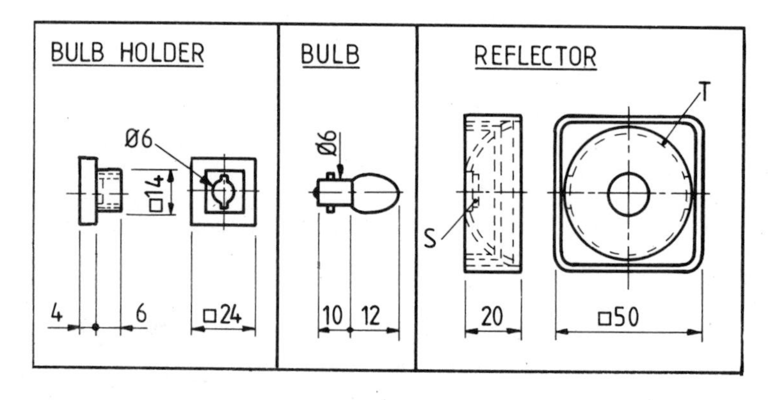

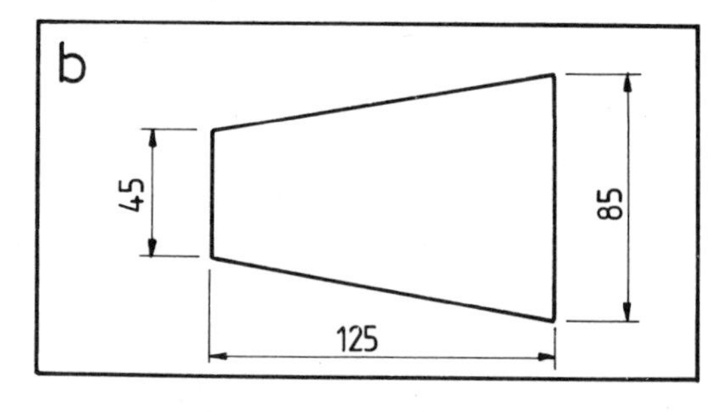

9. Orthographic views of the component parts of a torch are given above.

A photograph of a similar torch is also given. The photograph is intended to show the assembly of the parts of the torch; details and dimensions must be taken from the drawings.

The torch is to be sold with an instruction leaflet showing how it is taken apart so that either new batteries or a bulb can be fitted.

Note: The bulb fits in the bulb holder; the bulb holder fits in the 25 mm square recess in the casing and the reflector clips on the casing with the lugs L in the sockets S of the plastic ring T.

(a) Draw scale 1:1 using instruments, an exploded pictorial view of the torch with the reflector to the front in a form suitable for the purpose of the instruction sheet. Although two batteries are used in the torch, only one should be included in your answer.

Detail should be shown clearly and simply. Printed instructions may be added if you wish.

Colour and/or shading should be used to enhance your drawing.

(b) The company selling the torch, Capital Lighting Industries, require a new logo or trademark. The shape of the logo is to be in the form of a shaft of light in which the initial letters of the company are to be incorporated.

The outline of the beam of light is given (drawing b).

Design and draw a suitable logo to fit the shape.

Design sketches may be drawn on your answer sheet.

Marks will be given for the effective use of colour and/or shading.

(Specimen GCSE question, Welsh Joint Education Committee)

Projects for Design and Communication

Examples of sheets of drawings from design projects have been included throughout this book. They have been included to give pupils and students examples of the type of graphics involved in projects in Design and Communication courses in schools and colleges.

The submission of projects and/or coursework to examiners for marking is a very important part of all the examinations at GCSE level set under the subject title *CDT:Design and Communication.* As much as 40% of the total marks available for the examination are awarded for projects and/or coursework. This percentage varies according to the examination board which is setting the examination, but is never less than 30% and is much more likely to be 40%. You can judge for yourself from these percentages the importance of the project or coursework which you have to submit to examiners for your chances of obtaining a good grade in the examination.

You will be asked to choose a project title:

1. from a list published by the examination board;

   or

2. from discussion with your teachers or lecturers;

   or

3. from your own ideas,

   or

4. from a combination of 1, 2 and 3. For example, you may be required to submit a major project selected from a list published by the examination board, plus a 'mini' project suited to your own interests.

Whichever method of choice is allowed, it is advisable to choose your project titles with care, making sure you are suiting your own interests and choosing those with which you feel confident you can produce good, interesting design work. Remember that a major project may involve you in graphics work over a period of a term, or a term and a half, or even over a period of two terms. It is not in your own best interests to get involved in a project over such a long period of time unless you find the work interesting.

Sources of information and research

Information on which research into a design investigation can be based is available from a number of sources. Some suggestions are given below. When making use of such sources you must remember that the slavish copying of other people's design ideas is to be deplored. However, other people's designs can form a basis from which you can develop your own designs.

The major source of information will be books, from which you can extract information such as details of suitable materials, methods of construction, types of fittings available, statistical information, graphical techniques, historical details and the like. If the books you are seeking are not in your school or college library, go to your local public library. When looking for books in libraries you may well have to refer to the library indexes or catalogues. These are commonly based on the Dewey decimal system of numbering all possible subjects found in non-fiction books. A few examples of the system are given below, but the whole classification is very extensive; to learn more about it, ask at the library you are visiting.

Examples of Dewey decimal subject numbers

Aeroplanes	629.1	
Bridges	624	
Computing	510.78	540.81

Electronics	621.381
Graphics	741.6
Materials	620.1
Mechanics	620.
Plastics technology	668.4
Technical drawing	604.2
Technology research	607.2
Television	621.388

Other sources of information

- Articles in magazines and some newspapers often provide up-to-date information not easily found elsewhere.
- A study of the brochures and operating manuals supplied with many modern products, may well provide suggestions.
- The graphics seen regularly on television programmes are a rich source of modern graphical methods.
- A visit to any type of exhibition always provides a basis for fresh ideas.
- The study of photographs and paintings will give rise to many suggestions for proportions, composition and colour.
- Close observation of one's own environment – at school, at home, in the street, when travelling from place to place, when visiting friends – brings fresh ideas about designs.
- The designs on greetings cards may be of value to you.
- Symbols on electrical, electronics and mechanical equipment. Symbols on vehicle controls.
- Shapes in nature should be observed.
- Shape and form from architecture.
- Graphics derived from running computer programs.
- Patterns seen in wallpapers, in carpets, in dress materials, in curtain materials.

When thinking about the investigation into a design, discussions with your school friends or with teachers will always provide 'food for thought'. Sometimes the best method of solving a design problem is to form a small discussion group of three or four people to talk over the problem. Ideas developed from discussion can be sketched or notes taken with rough drawings which can be accurately drawn at a later date. It is a mistake to make such discussion groups too large – three or four is quite sufficient.

When investigating the problems arising from your attempt at answering a design brief, carry a notebook and pencil around with you. Jot down ideas as they come to mind. Another good method of 'storing' ideas is to speak them into a tape recorder and so have the store on tape. Many professional designers carry a micro hand tape recorder with them. Such tape recorders fit into a pocket. Ideas as they come to mind can be spoken into the recorder no matter where the designer is at that moment.

Design and communication projects

The flow chart given on page 179 is a summary of steps which need to be followed when you are developing a project under the title *Design and Communication*.

When submitting a project or your coursework for examination, pay attention to the following:

1. All the sheets of drawings and notes should be neatly presented within covers, in a folio or as a folder.
2. A good, well produced cover whose design reflects the contents of the folder/folio will assist in gaining good marks.
3. Use as many different drawing techniques as possible to give good evidence that you are familiar with a range of methods of drawing.
4. Mounting and presentation of graphics are important BUT remember the major consideration is DESIGN. You must show clear evidence that your project is design based.
5. Use an Index, sheet numbers or clear lettering on each sheet to make the progression of your design drawings clear. Follow the route – Situation; Design Brief; Investigation; Solutions; Realisation; Models; Evaluation.
6. You should consider the inclusion of a model or models as an essential part of your Design and Communication project.
7. Photographs of finished designs are an important feature of many design projects.
8. At least 50% of the marks given for your project will be awarded for the design element.

Suggestions for project topics

There is no limit to the number of topics which can be suggested for projects. The diagram on page 179 gives some topics. Further suggestions are given below. Some of these enlarge upon the topics given in the column *Suggestions* in the flow diagram.

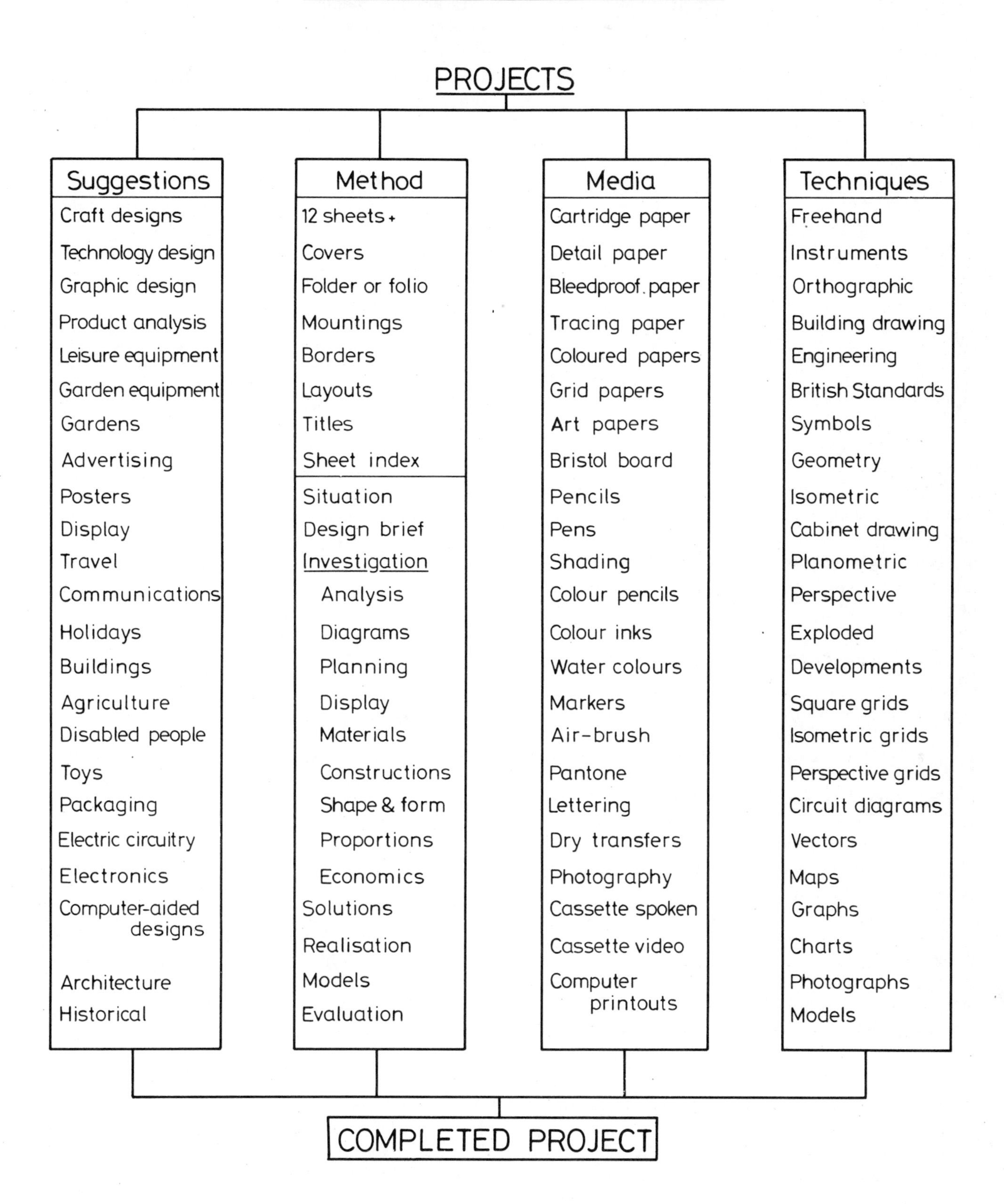

Flow chart showing how to design a graphics project

Craft designs
Designs to be made in wood, in metal, in plastics or in combinations of such materials. Designs to be made in leather, fabrics, ceramics, glass, concrete, clays and such materials. Designs for clothing, for embroidery, for knitwear.

Technology designs
Designs for methods of control by electrics, by electronics, pneumatics or hydraulics, designs for computer control. Designs of structures and mechanical systems. Aerodynamics. Logic control.

Graphic design
Posters, advertisements, designs for informing people of forthcoming changes in, for example, bus and train routes or new buildings to be constructed in a given area.

Product analysis
Analysis of modern products together with evaluations of their design and suggestions for improvements in their design. Examples are the drawing equipment you are using; lighting equipment; parts of vehicles; shape and form of vehicles; telephones; TV sets; computer equipment.

Leisure equipment
Designs for sport activities; for camping and/or caravanning; for transporting luggage and equipment when travelling; designs of sports and games fields and pitches; childrens' play areas; childrens' toys.

Garden equipment
Analysis of design of items such as spades, forks, rakes, shears, cold frames, greenhouses. Designs for such items.

Gardens
Designs for your own garden at home; garden areas in towns; gardens in parks and other open spaces; childrens' play areas within schools and/or parks. Garden ponds.

Travel
Projects connected with road, rail, air, bus, coach travel; mapping; travel guides and brochures.

Communications
How people communicate with each other – speech; signs; signalling; symbols; telephone; radio.

Holidays
'My holiday' would form the basis of a good design project, showing how leisure pursuits can be designed. Projects giving information as to where, for how long, how they travel to an area; leisure pursuits within a holiday area.

Buildings
My home; building patterns within a community; different forms of dwellings; town planning; different forms of dwellings; town planning; planning of, say, a car park or a town by-pass; additions to homes, including applications to local authorities.

Agriculture
Field patterns within a farming area; underlying geological structures; farm machinery; farm buildings; changes in farming patterns within an area; land drainage; footpaths and bridle paths within a given area.

Disabled people
Designs to assist disabled people in sitting, sleeping, eating, moving from place to place, reading, leisure, travel.

Conservation
Designs for energy conservation – gas, electricity, coal, wind, wave, solar, conservation of land, of animals, of food sources, of water supplies.

Pets and their management
Feeding patterns; cleaning and grooming; cages; kennels; run areas; different types of pets; the influence of pets on people and vice versa.

Architectural
Study of designs of modern large buildings; of buildings of historic importance in an area; of a cathedral; of one's local church; of the local town hall.

Historical
Time charts; family tree charts; importance and distribution of earthworks in an area.

Games
The design of board games; design of children's games; new computer based games.

Further suggestions

Design for a demonstration area within a school.

Design for an exhibition area in the foyer of a school.

The design of items of jewellery.

The design of additions to one's bicycle – e.g. panniers; map reading stand; bicycle stand.

Methods of manufacture – e.g. of iron and steel; of aluminium; of certain types and forms of plastics.

A study of how graphic design affects society generally.

A study of methods of knock-down construction.

Design-and-make models showing how one form of transmission can be changed to other forms.

Purely engineering projects – e.g. tools for bending sheet metals; tools for cutting sheet metals; tools for casting molten metals.

Index